T0270918

206 Probability on Real Lie Algebras

CAMBRIDGE TRACTS IN MATHEMATICS

GENERAL EDITORS

B. BOLLOBÁS, W. FULTON, A. KATOK, F. KIRWAN, P. SARNAK,

B. SIMON, B. TOTARO

A complete list of books in the series can be found at www.cambridge.org/mathematics.
Recent titles include the following:

Probability on Real Lie Algebras

UWE FRANZ
Université de Franche-Comté

NICOLAS PRIVAULT
Nanyang Technological University, Singapore

CAMBRIDGE
UNIVERSITY PRESS

CAMBRIDGE
UNIVERSITY PRESS

32 Avenue of the Americas, New York, NY 10013-2473, USA

Cambridge University Press is part of the University of Cambridge.

It furthers the University's mission by disseminating knowledge in the pursuit of
education, learning, and research at the highest international levels of excellence.

www.cambridge.org
Information on this title: www.cambridge.org/9781107128651

First published 2016

A catalogue record for this publication is available from the British Library

Library of Congress Cataloging in Publication Data
Franz, Uwe.
Probability on real Lie algebras / Uwe Franz, Université de Franche-Comté, Nicolas
Privault, Nanyang Technological University, Singapore.
pages cm. – (Cambridge tracts in mathematics)
Includes bibliographical references and index.
ISBN 978-1-107-12865-1 (hardback : alk. paper)
1. Lie algebras. 2. Probabilities. I. Privault, Nicolas. II. Title.
QA252.3.F72 2015
512'.482–dc23   2015028912

ISBN 978-1-107-12865-1 Hardback

To Aline Mio.
To Christine Jing and Sophie Wanlu.

Contents

Notation

$\mathbb{K} = \mathbb{R}$, *resp.* $\mathbb{K} = \mathbb{C}$, denote the fields of real, *resp.* complex, numbers.

$\mathcal{B}(\mathbb{R})$ denotes the *Borel σ-algebra, i.e.*, the σ-algebra generated by the open subsets of $\mathbb{R}$.

$\bar{z}$ denotes the complex conjugate of $z \in \mathbb{C}$.

$i = \sqrt{-1}$ denotes the complex square root of -1.

$\Im(z)$ denotes the imaginary part of $z \in \mathbb{C}$.

$\Re(z)$ denotes the real part of $z \in \mathbb{C}$.

$\operatorname{sgn} x \in \{-1, 0, 1\}$ denotes the sign of $x \in \mathbb{R}$.

$\delta_{n,m} = \mathbf{1}_{\{n=m\}} = \begin{cases} 1 & \text{if } n = m, \\ 0 & \text{if } n \neq m, \end{cases}$ is the Kronecker symbol.

ϵ_x denotes the Dirac measure at the point x.

ℓ^2 denotes the space of complex-valued square-summable sequences.

$\mathfrak{h}$ denotes a (complex) separable Hilbert space and $\mathfrak{h}_{\mathbb{C}}$ denotes its complexification when $\mathfrak{h}$ is real.

"$\circ$" denotes the symmetric tensor product in Hilbert spaces.

$\Gamma_s(\mathfrak{h})$ denotes the symmetric Fock space over the real (*resp.* complexified) Hilbert space $\mathfrak{h}$ (*resp.* $\mathfrak{h}_{\mathbb{C}}$).

$B(\mathfrak{h})$ denotes the algebra of bounded operators over a Hilbert space $\mathfrak{h}$.

$\operatorname{tr} \rho$ denotes the trace of the operator ρ.

$|X|$ denotes the absolute value of a normal operator X, with $|X| := (X^*X)^{1/2}$ when X is not normal.

$|\phi\rangle\langle\psi|$ with $\phi, \psi \in \mathfrak{h}$ denotes the rank one operator on the Hilbert space $\mathfrak{h}$, defined by $|\phi\rangle\langle\psi|(v) = \langle\psi, v\rangle\phi$ for $v \in \mathfrak{h}$.

$[\cdot, \cdot]$ denotes the commutator $[X, Y] = XY - YX$.

$\{\cdot, \cdot\}$ denotes the anti-commutator $\{X, Y\} = XY + YX$.

Ad, *resp.* ad X, denote the adjoint action on a Lie group, *resp.* Lie algebra.

$\mathcal{S}(\mathbb{R})$ denotes the Schwartz space of rapidly decreasing smooth functions.

$\mathcal{C}_0(\mathbb{R})$ denotes the set of continuous functions on $\mathbb{R}$, vanishing at infinity.

$\mathcal{C}_b^\infty(\mathbb{R})$ denotes the set of infinitely differentiable functions on $\mathbb{R}$ which are bounded together with all their derivatives.

$H^{p,\kappa}(\mathbb{R}^2)$ denotes the Sobolev space of orders $\kappa \in \mathbb{N}$ and $p \in [2, \infty]$.

$\Gamma(x) := \int_0^\infty t^{x-1}e^{-t}dt$ denotes the standard gamma function.

$J_m(x)$ denotes the Bessel function of the first kind of order $m \geq 0$.

Preface

This monograph develops a pedagogical approach to the role of noncommutativity in probability theory, starting in the first chapter at a level suitable for graduate and advanced undergraduate students. The contents also aim at being relevant to the physics student and to the algebraist interested in connections with probability and statistics.

Our presentation of noncommutativity in probability revolves around concrete examples of relations between algebraic structures and probability distribution, especially via recursive relations among moments and their generating functions. In this way, basic Lie algebras such as the Heisenberg–Weyl algebra $\mathfrak{hw}$, the oscillator algebra $\mathfrak{osc}$, the special linear algebra $sl(2, \mathbb{R})$, and other Lie algebras such as $so(2)$ and $so(3)$, can be connected with classical probability distributions, notably the Gaussian, Poisson, and gamma distributions, as well as some other infinitely divisible distributions.

Based on this framework, the Chapters 1–3 allow the reader to directly manipulate examples and as such they remain accessible to advanced undergraduates seeking an introduction to noncommutative probability. This setting also allows the reader to become familiar with more advanced topics, including the notion of couples of noncommutative random variables via the use of Wigner densities, in relation with quantum optics.

The following chapters are more advanced in nature, and are targeted to the graduate and research levels. They include the results of recent research on quantum Lévy processes and the noncommutative Malliavin [75] calculus. The Malliavin calculus is introduced in both the commutative and noncommutative settings and contributes to a better understanding of the smoothness properties of Wigner densities.

While this text is predominantly based on research literature, part of the material has been developed for teaching in the course "Special topics in statistics" at the Nanyang Technological University, Singapore, in the second semester of academic year 2013–2014. We thank the students and participants for useful questions and suggestions.

We thank Souleiman Omar Hoche and Michaël Ulrich for their comments, suggestions, and corrections of an earlier version of these notes. During the writing of this book, UF was supported by the ANR Project OSQPI (ANR-11-BS01-0008) and by the Alfried Krupp Wissenschaftskolleg in Greifswald. NP acknowledges the support of NTU MOE Tier 2 Grant No. M4020140.

Uwe Franz
Nicolas Privault

Introduction

Mathematics is the tool specially suited for dealing with abstract
concepts of any kind and there is no limit to its power in this field.
(P.A.M. Dirac, in The Principles of Quantum Mechanics.*)*

Quantum probability addresses the challenge of merging the apparently distinct domains of algebra and probability, in view of physical applications.
Those fields typically involve radically different types of thinking, which are
not often mastered simultaneously. Indeed, the framework of algebra is often
abstract and noncommutative while probability addresses the "fluctuating"
but classical notion of a random variable, and requires a good amount of
statistical intuition. On the other hand, those two fields combined yield natural
applications, for *e.g.*, quantum mechanics. On a more general level, the
noncommutativity of operations is a common real life phenomenon which
can be connected to classical probability via quantum mechanics. Algebraic
approaches to probability also have applications in theoretical computer
science, cf. *e.g.*, [39].

In the framework of this noncommutative (or algebraic) approach to probability, often referred to as quantum probability, real-valued random variables
on a classical probability space become special examples of noncommutative
(or quantum) random variables. For this, a real-valued random variable on a
probability space $(\Omega, \mathcal{F}, P)$ is viewed as (unbounded) self-adjoint multiplication operators acting on the Hilbert space $L^2(\Omega, P)$. This has led to the
suggestion by several authors to develop a theory of quantum probability
within the framework of operators and group representations in a Hilbert space,
cf. [87] and references therein.

In this monograph, our approach is to focus on the links between the
commutation relations within a given noncommutative algebra $\mathcal{A}$ on the one
hand, and the combinatorics of the moments of a given probability distribution

on the other hand. This approach is exemplified in Chapters 1–3. In this respect our point of view is consistent with the description of quantum probability by P.A. Meyer in [80] as a _set of prescriptions to extract probability from algebra_, based on various choices for the algebra $\mathcal{A}$.

For example, it is a well-known fact that the Gaussian distribution arises from the Heisenberg–Weyl algebra which is generated by three elements $\{P, Q, I\}$ linked by the commutation relation

$$[P, Q] = PQ - QP = 2iI.$$

It turns out similarly that other infinitely divisible distributions such as the gamma and continuous binomial distributions can be constructed via noncommutative random variables using representations of the special linear algebra $\mathfrak{sl}_2(\mathbb{R})$, or more simply on the affine algebra viewed as a sub-algebra of $\mathfrak{sl}_2(\mathbb{R})$. Other (joint) probability laws can be deduced in this setting; _e.g._, one can construct noncommutative couples of random variables with gamma and continuous binomial marginals. Similarly, the Poisson distribution can be obtained in relation with the oscillator algebra. In Chapters 4 and 6, those basic examples are revisited and extended in the more general framework of quantum random variables on real Lie algebras. We often work on real Lie algebras given by complex Lie algebras with an involution, because calculations are more convenient on complexifications. The real Lie algebras can then be recovered as real subspaces of anti-Hermitian elements.

Since the elements of a Lie algebra $\mathfrak{g}$ can be regarded as functions on its dual $\mathfrak{g}^*$ it might be more precise to view a random variable $j : \mathfrak{g} \to \mathcal{A}$ as taking values in $\mathfrak{g}^*$. In this sense, this book deals with "probability on duals of real Lie algebras", which would better reflect the implicit dualisation in the definition of quantum probability spaces and quantum random variables. For simplicity of exposition we nonetheless prefer to work with the less precise terminology "probability on real Lie algebras". We refer to [10] and the references therein for further discussion and motivation of "noncommutative (or quantum) mathematics".

The notion of _joint distribution_ for random vectors is of capital importance in classical probability theory. It also has an analog for couples of noncommutative random variables, through the definition of the (not necessarily positive) _Wigner_ [124] density functions. In Chapter 7 we present a construction of joint densities for noncommutative random variables, based on functional calculus on real Lie algebras using the general framework of [7] and [8], in particular on the affine algebra. In that sense our presentation is also connected to the framework of standard quantum mechanics and quantum optics, where Wigner densities have various applications in, *e.g.*, time-frequency analysis, see, *e.g.*, the references given in [29] and [7].

Overall, this monograph puts more emphasis on noncommutative "problems with fixed time" as compared with "problems in moving time"; see, *e.g.*, [31] and [32] for a related organisation of topics in classical probability and stochastic calculus. Nevertheless, we also include a discussion of noncommutative stochastic processes via quantum Lévy processes in Chapter 8. Lévy processes, or stochastic processes with independent and stationary increments, are used as models for random fluctuations, *e.g.*, in physics and finance. In quantum physics the so-called quantum noises or quantum Lévy processes occur, *e.g.*, in the description of quantum systems coupled to a heat bath [47] or in the theory of continuous measurement [53]. See also [122] for a model motivated by lasers, and [2, 106] for the theory of Lévy processes on involutive bialgebras. Those contributions extend, in a sense, the theory of factorisable representations of current groups and current algebras as well as the theory of classical Lévy processes with values in Euclidean space or, more generally, semigroups. For a historical survey on the theory of factorisable representions and its relation to quantum stochastic calculus, see [109, section 5]. In addition, many interesting classical stochastic processes can be shown to arise as components of quantum Lévy processes, cf. *e.g.*, [1, 18, 42, 105].

We also intend to connect noncommutative probability with the Malliavin calculus, which was originally designed by P. Malliavin, cf. [75], as a tool to provide sufficient conditions for the smoothness of partial differential equation solutions using probabilistic arguments, see Chapter 9 for a review of its construction. Over the years, the Malliavin calculus has developed into many directions, including anticipating stochastic calculus and extensions of stochastic calculus to fractional Brownian motion, cf. [84] and references therein.

The Girsanov theorem is an important tool in stochastic analysis and the Malliavin calculus, and we derive its noncommutative, or algebraic version in Chapter 10, starting with the case of noncommutative Gaussian processes. By differentiation, Girsanov-type identities can be used to derive integration by parts formulas for the Wigner densities associated to the noncommutative processes, by following Bismut's argument, cf. [22]. In Chapter 10 we will demonstrate on several examples how quasi-invariance formulas can be obtained in such a situation. This includes the Girsanov formula for Brownian motion, as well as a quasi-invariance result of the gamma processes [111, 112], which actually appeared first in the context of factorisable representations of current groups [114], and a quasi-invariance formula for the Meixner process.

In Chapter 11 we present the construction of noncommutative Malliavin calculus on the Heisenberg–Weyl algebra [43], [44], which generalises the Gaussian Malliavin calculus to Wigner densities, and allows one to prove the smoothness of joint Wigner distributions with Gaussian marginals using Sobolev spaces over $\mathbb{R}^2$. Here, noncommutative Gaussian processes can be built as the couple of the position and momentum Brownian motions on the Fock space. We also provide a treatment of other probability laws, including noncommutative couples of random variables with gamma and continuous binomial marginals based on the affine algebra. More generally, the long term goal in this field is to extend the hypoellipticity results of the Malliavin calculus to noncommutative quantum processes. In this chapter, we also point out the relationship between noncommutative and commutative differential calculi. In the white noise case, *i.e.*, if the underlying Hilbert space is the L^2-space of some measure space, the classical divergence operator defines an anticipating stochastic integral, known as the Hitsuda–Skorohod integral.

Several books on other extensions of the Malliavin calculus have been recently published, such as [86] which deals with infinitesimal (nonstandard) analysis and [56] which deals with Lévy processes. See [108] for a recent introduction to quantum stochastic calculus with connections to noncommutative geometry. See also [26] and [123] for recent introductions to quantum stochastics based on quantum stochastic calculus and quantum Markov semigroups.

The outline of the book is as follows (we refer the reader to [55] for an introduction to the basic concepts of quantum theory used in this book). In Chapter 1 we introduce the boson Fock space and we show how the first moments of the associated normal distribution can be computed using basic noncommutative calculus. Chapter 2 collects the background material on real Lie algebras and their representations. In Chapter 3 we consider fundamental examples of probability distributions (Gaussian, Poisson, gamma), and their connections with the Heisenberg–Weyl and oscillator algebras, as well as with the special linear algebra $\mathfrak{sl}_2(\mathbb{R})$, generated by the annihilation and creation operators on the boson Fock space. This will also be the occasion to introduce other representations based on polynomials. After those introductory sections, the construction of noncommutative random variables as operators acting on Lie algebras will be formalised in Chapter 4, based in particular on the notion of spectral measure. Quantum stochastic integrals are introduced in Chapter 5. In Chapter 6 we revisit the approaches of Chapters 3 and 4, relating Lie algebraic relations and probability distributions, in the unified framework of the splitting lemma, see chapter 1 of [38]. The problem of defining joint densities of couples of noncommutative random variables is treated in Chapter 7 under the angle of Weyl calculus, and Lévy processes on real Lie algebras are considered in Chapter 8. The classical, commutative Malliavin calculus is introduced in Chapter 9, and an introduction to quasi-invariance and the Girsanov theorem for noncommutative Lévy processes is given in Chapter 10. The noncommutative counterparts of the Malliavin calculus for Gaussian distributions, and then for gamma and other related probability densities are treated in Chapters 11 and 12, respectively, including the case of $so(3)$.

1

Boson Fock space

You don't know who he was? Half the particles in the universe obey
him!
(Reply by a physics professor when a student asked who Bose was.)

We start by introducing the elementary boson Fock space together with
its canonically associated creation and annihilation operators on a space of
square-summable sequences, and in the more general setting of Hilbert spaces.
The boson Fock space is a simple and fundamental quantum model which will
be used in preliminary calculations of Gaussian moments on the boson Fock
space, based on the commutation and duality relations satisfied by the creation
and annihilation operators. Those calculations will also serve as a motivation
for the general framework of the subsequent chapters.

1.1 Annihilation and creation operators

Consider the space of square-summable sequences

$$\ell^2 := \Gamma(\mathbb{C}) = \left\{ f \colon \mathbb{N} \to \mathbb{C} : \sum_{k=0}^{\infty} |f(k)|^2 < \infty \right\}$$

with the inner product

$$\langle f, g \rangle_{\ell^2} := \sum_{k=0}^{\infty} \overline{f(k)} g(k), \qquad f, g \in \ell^2,$$

and orthonormal basis $(e_n)_{n \in \mathbb{N}}$ given by the Kronecker symbols

$$e_n(k) := \delta_{k,n} = \begin{cases} 1 & k = n, \\ 0 & k \neq n, \end{cases}$$

$k, n \in \mathbb{N}$.

Definition 1.1.1 *Let $\sigma > 0$. The annihilation and creation operators are the linear operators a^- and a^+ implemented on ℓ^2 by letting*

$$a^+ e_n := \sigma \sqrt{n+1}\, e_{n+1}, \qquad a^- e_n := \sigma \sqrt{n}\, e_{n-1}, \qquad n \in \mathbb{N}.$$

Note that the above definition means that $a^- e_0 = 0$.

The sequence space ℓ^2 endowed with the annihilation and creation operators a^- and a^+ is called the *boson (or bosonic) Fock space*. In the physical interpretation of the boson Fock space, the vector e_n represents a physical n-particle state. The term "boson" refers to the Bose–Einstein statistics and in particular to the possibility for n particles to share the same state e_n, and Fock spaces are generally used to model the quantum states of identical particles in variable number.

As a consequence of Definition 1.1.1 the *number operator* a° defined as $a^\circ := a^+ a^-$ has eigenvalues given by

$$a^\circ e_n = a^+ a^- e_n = \sigma^2 \sqrt{n} a^+ e_{n-1} = n\sigma^2 e_n, \qquad n \in \mathbb{N}. \tag{1.1}$$

Noting the relation

$$a^- a^+ e_n = \sigma \sqrt{n+1} a^- e_{n+1} = \sigma^2 (n+1) e_n,$$

in addition to (1.1), we deduce the next proposition.

Proposition 1.1.2 *We have the commutation relation*

$$[a^+, a^-] e_n = \sigma^2 e_n, \qquad n \in \mathbb{N}.$$

Quantum physics provides a natural framework for the use of the non-commutative operators a^- and a^+, by connecting them with the statistical intuition of probability. Indeed, the notion of physical measurement is noncommutative in nature; think, *e.g.*, of measuring the depth of a pool *vs.* measuring water temperature: each measurement will perturb the next one in a certain way, thus naturally inducing noncommutativity. In addition, noncommutativity gives rise to the impossibility of making measurements with infinite precision, and the physical interpretation of quantum mechanics is essentially probabilistic as a given particle only has a *probability density* of being in a given state/location. In the sequel we take $\sigma = 1$.

Given $f = (f(n))_{n \in \mathbb{N}}$ and $g = (g(n))_{n \in \mathbb{N}}$ written as

$$f = \sum_{n=0}^{\infty} f(n) e_n \quad \text{and} \quad g = \sum_{n=0}^{\infty} g(n) e_n,$$

we have

$$a^+ f = \sum_{n=0}^{\infty} f(n) a^+ e_n = \sum_{n=0}^{\infty} f(n) \sqrt{n+1}\, e_{n+1} = \sum_{n=1}^{\infty} f(n-1) \sqrt{n}\, e_n$$

and

$$a^- f = \sum_{n=0}^{\infty} f(n) a^- e_n = \sum_{n=1}^{\infty} f(n) \sqrt{n} \, e_{n-1} = \sum_{n=0}^{\infty} f(n+1) \sqrt{n+1} \, e_n,$$

hence we have

$$(a^+ f)(n) = \sqrt{n} f(n-1), \quad \text{and} \quad (a^- f)(n) = \sqrt{n+1} f(n+1). \tag{1.2}$$

This shows the following duality relation between a^- and a^+.

Proposition 1.1.3 *For all $f, g \in \ell^2$ with finite support in $\mathbb{N}$ we have*

$$\langle a^- f, g \rangle_{\ell^2} = \langle f, a^+ g \rangle_{\ell^2}.$$

Proof: By (1.2) we have

$$
\begin{aligned}
\langle a^- f, g \rangle_{\ell^2} &= \sum_{n=0}^{\infty} \overline{(a^- f)(n)} g(n) \\
&= \sum_{n=0}^{\infty} \sqrt{n+1} \, \overline{f(n+1)} g(n) \\
&= \sum_{n=1}^{\infty} \sqrt{n} \, \overline{f(n)} g(n-1) \\
&= \sum_{n=1}^{\infty} \overline{f(n)} (a^+ g)(n) \\
&= \langle f, a^+ g \rangle_{\ell^2}.
\end{aligned}
$$

$\square$

We also define the *position* and *momentum* operators

$$Q := a^- + a^+ \quad \text{and} \quad P := i(a^+ - a^-),$$

which satisfy the commutation relation

$$[P, Q] = PQ - QP = -2I_d.$$

To summarise the results of this section, the Hilbert space $H = \ell^2$ with inner product $\langle \cdot, \cdot \rangle_{\ell^2}$ has been equipped with two operators a^- and a^+, called *annihilation and creation operators* and acting on the elements of H such that

a) a^- and a^+ are dual of each other in the sense that

$$\langle a^- u, v \rangle_{\ell^2} = \langle u, a^+ v \rangle_{\ell^2},$$

and this relation will also be written as $(a^+)^* = a^-$, with respect to the inner product $\langle \cdot, \cdot \rangle_{\ell^2}$.

b) the operators a^- and a^+ satisfy the commutation relation

$$[a^+, a^-] = a^+a^- - a^-a^+ = \sigma^2 I_d,$$

where Id is the identity operator.

1.2 Lie algebras on the boson Fock space

In this section we characterise the Lie algebras made of linear mappings

$$Y: \ell^2 \longmapsto \ell^2,$$

written on the orthonormal basis $(e_n)_{n \in \mathbb{N}}$ of the boson Fock space ℓ^2 as

$$Ye_n = \gamma_n e_{n+1} + \epsilon_n e_n + \eta_n e_{n-1}, \qquad n \in \mathbb{N}, \tag{1.3}$$

where $\gamma_n, \epsilon_n, \eta_n \in \mathbb{C}$, with $\eta_0 = 0$ and $\gamma_n \neq 0$, $n \in \mathbb{N}$. We assume that Y is Hermitian, *i.e.*, $Y^* = Y$, or equivalently

$$\bar{\gamma}_n = \eta_{n+1} \quad \text{and} \quad \epsilon_n \in \mathbb{R}, \qquad n \in \mathbb{N}.$$

For example, the *position* and *moment* operators

$$Q := a^- + a^+ \quad \text{and} \quad P := i(a^+ - a^-)$$

can be written as

$$Qe_n = a^- e_n + a^+ e_n = \sqrt{n}e_{n-1} + \sqrt{n+1}e_{n+1},$$

i.e., $\gamma_n = \sqrt{n+1}$, $\epsilon_n = 0$, and $\eta_n = \sqrt{n}$, while

$$Pe_n = i(a^+ e_n - a^- e_n) = i\sqrt{n+1}e_{n+1} - i\sqrt{n}e_{n-1},$$

i.e., $\gamma_n = i\sqrt{n+1}$, $\epsilon_n = 0$, and $\eta_n = -i\sqrt{n}$.

In the sequel we consider the sequence $(P_n)_{n \in \mathbb{N}}$ of polynomials given by

$$P_n(Y) := \sum_{k=0}^{n} \alpha_{k,n} Y^k, \qquad n \in \mathbb{N}.$$

Proposition 1.2.1 *The condition*

$$e_n = P_n(Y)e_0, \qquad n \in \mathbb{N}, \tag{1.4}$$

defines a unique sequence $(P_n)_{n \in \mathbb{N}}$ of polynomials that satisfy the three-term recurrence relation

$$xP_n(x) = \gamma_n P_{n+1}(x) + \epsilon_n P_n(x) + \eta_n P_{n-1}(x), \qquad n \in \mathbb{N}, \tag{1.5}$$

from which the sequence $(P_n)_{n \in \mathbb{N}}$ can be uniquely determined based on the initial condition $P_{-1} = 0$, $P_1 = 1$.

Proof: The relation (1.3) and the condition (1.4) show-that

$$YP_n(Y)e_0 = \gamma_n P_{n+1}(Y)e_0 + \epsilon_n P_n(Y)e_0 + \eta_n P_{n-1}(Y)e_0$$
$$= \gamma_n e_{n+1} + \epsilon_n e_n + \eta_n e_{n-1},$$

which implies the recurrence relation (1.5). $\qquad\qquad\qquad\square$

For example, the monomial Y^n satisfies

$$\langle e_n, Y^n e_0 \rangle_{\ell^2} = \gamma_0 \cdots \gamma_{n-1}, \qquad n \in \mathbb{N},$$

hence since $\gamma_n \neq 0$, $n \in \mathbb{N}$, we have in particular

$$1 = \langle e_n, e_n \rangle_{\ell^2}$$
$$= \langle e_n, P_n(Y)e_0 \rangle_{\ell^2}$$
$$= \sum_{k=0}^{n} \alpha_{k,n} \langle e_n, Y^k e_0 \rangle_{\ell^2}$$
$$= \alpha_{n,n} \langle e_n, Y^n e_0 \rangle_{\ell^2}$$
$$= \alpha_{n,n} \gamma_1 \cdots \gamma_n, \qquad n \in \mathbb{N}.$$

In the case where $Y = Q$ is the position operator, imposing the relation

$$e_n = P_n(Q)e_0, \qquad n \in \mathbb{N},$$

i.e., (1.4), shows that

$$QP_n(Q)e_0 = \sqrt{n+1}P_{n+1}(Q)e_0 + \sqrt{n}P_{n-1}(Q)e_0,$$

hence the three-term recurrence relation (1.5) reads

$$xP_n(x) = \sqrt{n+1}P_{n+1}(x) + \sqrt{n}P_{n-1}(x),$$

for $n \in \mathbb{N}$, with initial condition $P_{-1} = 0$, $P_1 = 1$, hence $(P_n)_{n \in \mathbb{N}}$ is the family of normalised Hermite polynomials, cf. Section 12.1.

Definition 1.2.2 *By a probability law of Y in the fundamental state e_0 we will mean a probability measure μ on $\mathbb{R}$ such that*

$$\int_{\mathbb{R}} x^n \mu(dx) = \langle e_0, Y^n e_0 \rangle_{\ell^2}, \qquad n \in \mathbb{N},$$

which is also called the spectral measure of Y evaluated in the state $Y \mapsto \langle e_0, Y e_0 \rangle_{\ell^2}$.

In this setting the *moment generating function* defined as

$$t \longmapsto \langle e_0, e^{tY} e_0 \rangle_{\ell^2}$$

will be used to determine the probability law μ of Y in the state e_0.

We note that in this case the polynomials $P_n(x)$ are orthogonal with respect to $\mu(dx)$, since

$$\int_{-\infty}^{\infty} P_n(x) P_m(x) \mu(dx) = \langle e_0, P_n(Y) P_m(Y) e_0 \rangle_{\ell^2}$$
$$= \langle P_n(Y) e_0, P_m(Y) e_0 \rangle_{\ell^2}$$
$$= \langle e_n, e_m \rangle_{\ell^2}$$
$$= \delta_{n,m}, \qquad n, m \in \mathbb{N}.$$

1.3 Fock space over a Hilbert space

More generally, the boson Fock space also admits a construction upon any real separable Hilbert space $\mathfrak{h}$ with complexification $\mathfrak{h}_{\mathbb{C}}$, and in this more general framework it will simply be called the Fock space.

The basic structure and operators of the Fock space over $\mathfrak{h}$ are similar to those of the simple boson Fock space, however it allows for more degrees of freedom. The boson Fock space ℓ^2 defined earlier corresponds to the symmetric Fock space over the one-dimensional real Hilbert space $\mathfrak{h} = \mathbb{R}$. We will use the conjugation operator

$$^{-}: \mathfrak{h}_{\mathbb{C}} \to \mathfrak{h}_{\mathbb{C}}$$

on the complexification

$$\mathfrak{h}_{\mathbb{C}} := \mathfrak{h} \oplus i\mathfrak{h} = \{h_1 + ih_2 : h_1, h_2 \in \mathfrak{h}\},$$

of $\mathfrak{h}$, defined by letting

$$\overline{h_1 + ih_2} := h_1 - ih_2, \qquad h_1, h_2 \in \mathfrak{h}.$$

This conjugate operation satisfies

$$\langle \overline{h}, \overline{k} \rangle_{\mathfrak{h}_{\mathbb{C}}} = \overline{\langle h, k \rangle}_{\mathfrak{h}_{\mathbb{C}}} = \langle k, h \rangle_{\mathfrak{h}_{\mathbb{C}}}, \qquad h, k \in \mathfrak{h}_{\mathbb{C}}.$$

The elements of $\mathfrak{h}$ are characterised by the property $\overline{h} = h$, and we will call them real. The next definition uses the notion of the symmetric tensor product "$\circ$" in Hilbert spaces.

Definition 1.3.1 *The* symmetric Fock space *over $\mathfrak{h}_{\mathbb{C}}$ is defined by the direct sum*

$$\Gamma_s(\mathfrak{h}) = \bigoplus_{n \in \mathbb{N}} \mathfrak{h}_{\mathbb{C}}^{\circ n}.$$

We denote by $\mathbf{\Omega} := 1 + 0 + \cdots$ the vacuum vector in $\Gamma_s(\mathfrak{h})$. The symmetric Fock space is isomorphic to the complexification of the Wiener space $L^2(\Omega)$ associated to $\mathfrak{h}$ in Section 9.2.

The exponential vectors

$$\mathcal{E}(f) := \sum_{n=0}^{\infty} \frac{f^{\otimes n}}{\sqrt{n!}}, \qquad f \in \mathfrak{h}_{\mathbb{C}},$$

are total in $\Gamma_s(\mathfrak{h})$, and their scalar product in $\Gamma_s(\mathfrak{h})$ is given by

$$\langle \mathcal{E}(k_1), \mathcal{E}(k_2) \rangle_{\mathfrak{h}_{\mathbb{C}}} = e^{\langle k_1, k_2 \rangle_{\mathfrak{h}_{\mathbb{C}}}}.$$

1.3.1 Creation and annihilation operators on $\Gamma_s(\mathfrak{h})$

The annihilation, creation, position, and momentum operators $a^-(h)$, $a^+(h)$, $Q(h)$, $P(h)$, $h \in \mathfrak{h}$, can be defined as unbounded and closed operators on the Fock space over $\mathfrak{h}$, see, *e.g.*, [17, 79, 87]. The creation and annihilation operators $a^+(h)$ and $a^-(h)$ are mutually adjoint, and the position and momentum operators

$$Q(h) = a^-(\overline{h}) + a^+(h) \quad \text{and} \quad P(h) = i\big(a^-(\overline{h}) - a^+(h)\big)$$

are self-adjoint if $h \in \mathfrak{h}$ is real. The commutation relations of creation, annihilation, position, and momentum are

$$\begin{cases} [a(h), a^+(k)] = \langle h, k \rangle_{\mathfrak{h}_{\mathbb{C}}}, \\[2mm] [a(h), a(k)] = [a^+(h), a^+(k)] = 0, \\[2mm] [Q(h), Q(k)] = [P(h), P(k)] = 0, \\[2mm] [P(h), Q(k)] = 2i\langle \overline{h}, k \rangle_{\mathfrak{h}_{\mathbb{C}}}. \end{cases}$$

The operators $a^-(h), a^+(h), Q(h), P(h)$ are unbounded, but their domains contain the exponential vectors $\mathcal{E}(f), f \in \mathfrak{h}_{\mathbb{C}}$. We will need to compose them with bounded operators on $\Gamma_s(\mathfrak{h})$, and in order to do so we will adopt the following convention. Let

$$\mathcal{L}\big(\mathcal{E}(\mathfrak{h}_{\mathbb{C}}), \Gamma_s(\mathfrak{h})\big)$$
$$= \Big\{ B \in \mathrm{Lin}\big(\mathrm{span}(\mathcal{E}(\mathfrak{h}_{\mathbb{C}})), \Gamma_s(\mathfrak{h})\big) : \exists B^* \in \mathrm{Lin}\big(\mathrm{span}(\mathcal{E}(\mathfrak{h}_{\mathbb{C}})), \Gamma_s(\mathfrak{h})\big)$$
$$\text{such that } \big\langle \mathcal{E}(f), B\mathcal{E}(g)\big\rangle_{\mathfrak{h}_{\mathbb{C}}} = \big\langle B^*\mathcal{E}(f), \mathcal{E}(g)\big\rangle_{\mathfrak{h}_{\mathbb{C}}} \text{ for all } f, g \in \mathfrak{h}_{\mathbb{C}} \Big\},$$

denote the space of linear operators that are defined on the exponential vectors and that have an "adjoint" that is also defined on the exponential vectors. Obviously the operators $a^-(h), a^+(h), Q(h), P(h), U(h_1, h_2)$ belong to $\mathcal{L}\big(\mathcal{E}(\mathfrak{h}_{\mathbb{C}}), \Gamma_s(\mathfrak{h})\big)$. We will say that an expression of the form

$$\sum_{j=1}^{n} X_j B_j Y_j,$$

with $X_1, \ldots, X_n, Y_1, \ldots, Y_n \in \mathcal{L}\big(\mathcal{E}(\mathfrak{h}_{\mathbb{C}}), \Gamma_s(\mathfrak{h})\big)$ and $B_1, \ldots, B_n \in \mathcal{B}\big(\Gamma_s(\mathfrak{h})\big)$ defines a bounded operator on $\Gamma_s(\mathfrak{h})$, if there exists a bounded operator $M \in \mathcal{B}\big(\Gamma_s(\mathfrak{h})\big)$ such that

$$\big\langle \mathcal{E}(f), M\mathcal{E}(g)\big\rangle_{\mathfrak{h}_{\mathbb{C}}} = \sum_{j=1}^{n} \big\langle X_j^* \mathcal{E}(f), B_j Y_j \mathcal{E}(g)\big\rangle_{\mathfrak{h}_{\mathbb{C}}}$$

holds for all $f, g \in \mathfrak{h}_{\mathbb{C}}$. If it exists, this operator is unique because the exponential vectors are total in $\Gamma_s(\mathfrak{h})$, and we will then write

$$M = \sum_{j=1}^{n} X_j B_j Y_j.$$

1.3.2 Weyl operators

The Weyl operators $U(h_1, h_2)$ are defined by

$$U(h_1, h_2) = \exp\big(iP(h_1) + iQ(h_2)\big) = \exp\big(i\big(a^-(\overline{h_2} - i\overline{h_1}) + a^+(h_2 - ih_1)\big)\big),$$

and they satisfy

$$U(h_1, h_2)U(k_1, k_2) = \exp i\big(\langle \overline{h_2}, k_1 \rangle_{\mathfrak{h}_{\mathbb{C}}} - \langle \overline{h_1}, k_2 \rangle_{\mathfrak{h}_{\mathbb{C}}}\big)U(h_1 + h_2, k_1 + k_2).$$

Furthermore, we have $U(h_1, h_2)^* = U(-\overline{h}_1, -\overline{h}_2)$ and $U(h_1, h_2)^{-1} = U(-h_1, -h_2)$. We see that $U(h_1, h_2)$ is unitary, if h_1 and h_2 are real. These operators act on the vacuum state $\Omega = \mathcal{E}(0)$ as

$$U(h_1, h_2)\mathbf{\Omega} = \exp\left(-\frac{\langle \overline{h}_1, h_1 \rangle_{\mathfrak{h}_\mathbb{C}} + \langle \overline{h}_2, h_2 \rangle_{\mathfrak{h}_\mathbb{C}}}{2}\right) \mathcal{E}(h_1 + ih_2)$$

and on the exponential vectors $\mathcal{E}(f)$ as

$$U(h_1, h_2)\mathcal{E}(f)$$
$$= \exp\left(-\langle \overline{f}, h_1 + ih_2 \rangle_{\mathfrak{h}_\mathbb{C}} - \frac{\langle \overline{h}_1, h_1 \rangle_{\mathfrak{h}_\mathbb{C}} + \langle \overline{h}_2, h_2 \rangle_{\mathfrak{h}_\mathbb{C}}}{2}\right) \mathcal{E}(f + h_1 + ih_2).$$

Exercises

Exercise 1.1 *Moments of the normal distribution.*

In this exercise we consider an example in which the noncommutativity property of a^- and a^+ naturally gives rise to a fundamental example of probability distribution, *i.e.*, the normal distribution.

In addition to that we will assume the existence of a unit vector $\mathbf{1} \in \mathfrak{h}$ (fundamental or empty state) such that $a^-\mathbf{1} = 0$ and $\langle \mathbf{1}, \mathbf{1} \rangle_{\mathfrak{h}} = 1$. In particular, this yields the rule

$$\langle a^+ u, \mathbf{1} \rangle_{\mathfrak{h}} = \langle u, a^-\mathbf{1} \rangle_{\mathfrak{h}} = 0.$$

Based on this rule, check by an elementary computation that the first four moments of the centered $\mathcal{N}(0, \sigma^2)$ can be recovered from $\langle Q^n \mathbf{1}, \mathbf{1} \rangle_{\mathfrak{h}}$ with $n = 1, 2, 3, 4$.

In the following chapters this problem will be addressed in a systematic way by considering other algebras and probability distributions as well as the problem of *joint distributions* such as the distribution of the couple (P, Q).

2

Real Lie algebras

Algebra is the offer made by the devil to the mathematician. The devil says: "I will give you this powerful machine, it will answer any question you like. All you need to do is give me your soul: give up geometry and you will have this marvelous machine".

(M. Atiyah, Collected works.)

In this chapter we collect the definition and properties of the real Lie algebras that will be needed in the sequel. We consider in particular the Heisenberg–Weyl Lie algebra $\mathfrak{hw}$, the oscillator Lie algebra $\mathfrak{osc}$, and the Lie algebras $\mathfrak{sl}_2(\mathbb{R})$, $so(2)$, and $so(3)$ as particular cases. Those examples and their relationships with classical probability distributions will be revisited in more details in the subsequent chapters.

2.1 Real Lie algebras

Definition 2.1.1 *A Lie algebra $\mathfrak{g}$ over a field $\mathbb{K}$ is a $\mathbb{K}$-vector space with a linear map $[\cdot,\cdot]\colon \mathfrak{g} \times \mathfrak{g} \longrightarrow \mathfrak{g}$ called* Lie *bracket that satisfies the following two properties.*

1. Anti-symmetry: for all $X, Y \in \mathfrak{g}$, we have

$$[X, Y] = -[Y, X].$$

2. Jacobi identity: for all $X, Y, Z \in \mathfrak{g}$, we have

$$\big[X, [Y, Z]\big] + \big[Y, [Z, X]\big] + \big[Z, [X, Y]\big] = 0.$$

For $\mathbb{K} = \mathbb{R}$, we call $\mathfrak{g}$ a real Lie algebra, *for $\mathbb{K} = \mathbb{C}$ a* complex Lie algebra.

Definition 2.1.2 *Let $\mathfrak{g}$ be a complex Lie algebra. An* involution on $\mathfrak{g}$ is a *conjugate linear map $*\colon \mathfrak{g} \longrightarrow \mathfrak{g}$ such that* *i)* $(X^*)^* = X$ *for all* $X \in \mathfrak{g}$,

ii) $[X, Y]^* = -[X^*, Y^*]$ *for all* $X, Y \in \mathfrak{g}$.

In the sequel we will only consider *real Lie algebras*, i.e., Lie algebras over either the field $\mathbb{K} = \mathbb{R}$ of real numbers, or involutive Lie algebras over the field $\mathbb{K} = \mathbb{C}$ of complex numbers.

Remark 2.1.3 *Let* $\mathfrak{g}$ *be a real Lie algebra. Then the complex vector space*

$$\mathfrak{g}_{\mathbb{C}} := \mathbb{C} \otimes_{\mathbb{R}} \mathfrak{g} = \mathfrak{g} \oplus i\mathfrak{g}$$

is a complex Lie algebra with the Lie bracket

$$[X + iY, X' + iY'] := [X, X'] - [Y, Y'] + i([X, Y'] + [Y, X']),$$

for $X, X', Y, Y' \in \mathfrak{g}$. *In addition,*

1. the conjugate linear map

$$*: \mathfrak{g}_{\mathbb{C}} \longrightarrow \mathfrak{g}_{\mathbb{C}}$$
$$Z = X + iY \longmapsto Z^* = -X + iY$$

defines an involution on $\mathfrak{g}_{\mathbb{C}}$, *i.e., it satisfies*

$$(Z^*)^* = Z \qquad and \qquad [Z_1, Z_2]^* = [Z_2^*, Z_1^*]$$

for all $Z, Z_1, Z_2 \in \mathfrak{g}_{\mathbb{C}}$

2. the functor[1] $\mathfrak{g} \longmapsto (\mathfrak{g}_{\mathbb{C}}, *)$ *is an isomorphism between the category of real Lie algebras and the category of involutive complex Lie algebras. The inverse functor associates to an involutive complex Lie algebra* $(\mathfrak{g}, *)$ *the real Lie algebra*

$$\mathfrak{g}_{\mathbb{R}} = \{X \in \mathfrak{g} : X^* = -X\},$$

where the Lie bracket on $\mathfrak{g}_{\mathbb{R}}$ *is the restriction of the Lie bracket of* $\mathfrak{g}$. *Note that* $[\cdot, \cdot]$ *leaves* $\mathfrak{g}_{\mathbb{R}}$ *invariant, since, if* $X^* = -X$, $Y^* = -Y$, *then*

$$[X, Y]^* = -[X^*, Y^*] = -[(-X), (-Y)] = -[X, Y].$$

2.1.1 Adjoint action

In addition to the Lie algebra $\mathfrak{g}$ we will consider the *Lie group* generated by all exponentials of the form $g_t := e^{tY/2}$, $Y \in \mathfrak{g}$.

---

[1] This functor is used for the equivalence of *real Lie algebras* and *complex Lie algebras with an involution* by associating a complex Lie algebra with an involution to every real Lie algebra, and vice versa. Categories are outside the scope of this book.

The adjoint action of $g_t := e^{tY/2}$ on X is defined by

$$X(t) = \mathrm{Ad}_{g_t}(X) := e^{tY/2} X e^{-tY/2}, \qquad t \in \mathbb{R}.$$

2.2 Heisenberg–Weyl Lie algebra $\mathfrak{hw}$

The Heisenberg–Weyl Lie algebra $\mathfrak{hw}$ is the three-dimensional Lie algebra with basis $\{P, Q, E\}$ satisfying the *involution*

$$P^* = P, \qquad Q^* = Q, \qquad E^* = E,$$

and the *commutation relations*

$$[P, Q] = -2iE, \quad [P, E] = [Q, E] = 0.$$

2.2.1 Boson Fock space representation

As in Chapter 1, the Heisenberg–Weyl Lie algebra can be implemented using a^- and a^+ through the *position and moment* operators

$$Q = a^- + a^+ \quad \text{and} \quad P = i(a^+ - a^-),$$

which satisfy

$$[P, Q] = PQ - QP = -2iE,$$

with $E = \sigma^2 I$, and both have Gaussian laws.

2.2.2 Matrix representation

Under the relations $Q = a^- + a^+$ and $P = i(a^+ - a^-)$, the Heisenberg–Weyl Lie algebra $\mathfrak{hw}$ has the matrix representation

$$a^- = \begin{bmatrix} 0 & \sigma & 0 \\ 0 & 0 & 0 \\ 0 & 0 & 0 \end{bmatrix}, \quad a^+ = \begin{bmatrix} 0 & 0 & 0 \\ 0 & 0 & \sigma \\ 0 & 0 & 0 \end{bmatrix}, \quad E = \begin{bmatrix} 0 & 0 & \sigma^2 \\ 0 & 0 & 0 \\ 0 & 0 & 0 \end{bmatrix}.$$

Here the exponential $\exp(\alpha a^- + \beta a^+ + \gamma E)$ can be computed as

$$\exp\left(\begin{bmatrix} 0 & \alpha\sigma & \gamma\sigma^2 \\ 0 & 0 & \beta\sigma \\ 0 & 0 & 0 \end{bmatrix} \right) = \begin{bmatrix} 1 & \alpha\sigma & \gamma\sigma^2 + \alpha\beta\sigma^2/2 \\ 0 & 1 & \beta\sigma \\ 0 & 0 & 1 \end{bmatrix},$$

which is however not sufficient in order to recover the Gaussian moment generating function of Q hinted at in Exercise 1.1.

2.2.3 Representation on $\mathfrak{h} = L^2(\mathbb{R}, dt)$

A representation of $\mathfrak{hw}$ on $\mathfrak{h} = L^2(\mathbb{R}, dt)$ can also be constructed by letting

$$P\phi(t) := \frac{2}{i}\phi'(t) \quad \text{and} \quad Q\phi(t) := t\phi(t), \qquad t \in \mathbb{R}, \quad \phi \in \mathcal{S}(\mathbb{R}).$$

2.3 Oscillator Lie algebra osc

In addition to P, Q, and E we consider a fourth symmetric basis element N to the Heisenberg–Weyl Lie algebra $\mathfrak{hw}$, and we impose the relations

$$[N, P] = -iQ, \quad [N, Q] = iP, \quad [N, E] = 0,$$

and the involution $N^* = N$. This yields the *oscillator* Lie algebra

$$\mathfrak{osc} = \text{span}\{N, P, Q, E\}.$$

2.3.1 Matrix representation

The oscillator algebra $\mathfrak{hw}$ has the matrix representation

$$a^- = \begin{bmatrix} 0 & \sigma & 0 \\ 0 & 0 & 0 \\ 0 & 0 & 0 \end{bmatrix}, \quad a^+ = \begin{bmatrix} 0 & 0 & 0 \\ 0 & 0 & \sigma \\ 0 & 0 & 0 \end{bmatrix},$$

$$E = \begin{bmatrix} 0 & 0 & \sigma^2 \\ 0 & 0 & 0 \\ 0 & 0 & 0 \end{bmatrix}, \quad N = \begin{bmatrix} 0 & 0 & 0 \\ 0 & 1 & 0 \\ 0 & 0 & 0 \end{bmatrix}.$$

2.3.2 Boson Fock space representation

The oscillator Lie algebra osc can be written as the four dimensional Lie algebra with basis

$$\{N, a^+, a^-, E\},$$

where N and E are given on the boson Fock space ℓ^2 by

$$Ne_n = \frac{1}{\lambda}ne_n \quad \text{and} \quad Ee_n = \lambda e_n, \qquad n \in \mathbb{N},$$

where $N = \lambda^{-1}a^\circ$, $\lambda > 0$, and a° is the *number* operator. Recall that here the creation and annihilation operators

$$a^- e_{n-1} = \sqrt{n}e_{n-1}, \quad \text{and} \quad a^+ e_n = \sqrt{n+1}e_{n+1},$$

satisfy

$$a^+ = \frac{Q + iP}{2} \quad \text{and} \quad a^- = \frac{Q - iP}{2}.$$

The Lie bracket $[\cdot, \cdot]$ satisfies

$$[N, a^{\pm}] = \pm a^{\pm}, \quad [a^-, a^+] = E, \quad [E, N] = [E, a^{\pm}] = 0.$$

2.3.3 The harmonic oscillator

Due to the relation

$$N = \frac{1}{\lambda} a^+ a^- = \frac{P^2 + Q^2 - 2\sigma^2 I}{4\lambda},$$

the operator N is also known as the Hamiltonian of the *harmonic oscillator*. This is by analogy with classical mechanics where the Hamiltonian of the harmonic oscillator is given by

$$H = \frac{m}{2} |\dot{x}|^2 + \frac{k}{2} x^2 = \frac{1}{2m} p^2 + \frac{k}{2} x^2 = T + U,$$

with x the position of the particle (= elongation of the spring from its rest position), $\dot{x}$ its velocity, $p = m\dot{x}$ its momentum, and the two terms T and U are respectively the kinetic energy

$$T = \frac{1}{2} m |\dot{x}|^2 = \frac{1}{2m} p^2$$

and the energy stored in the spring

$$U = \frac{k}{2} x^2,$$

with m the mass of the particle and k Hooke's constant, a characteristic of the spring.

2.4 Lie algebra $\mathfrak{sl}_2(\mathbb{R})$

Consider the three-dimensional real Lie algebra

$$\mathfrak{sl}_2(\mathbb{R}) = \text{span} \{B^+, B^-, M\}$$

with basis B^+, B^-, M, Lie bracket

$$[M, B^{\pm}] = \pm 2B^{\pm}, \qquad [B^-, B^+] = M,$$

and the involution $(B^+)^* = B^-$, $M^* = M$. Letting

$$X = B^+ + B^- + M,$$

we will check that X has a gamma distribution with parameter $\beta > 0$, provided

$$Me_0 = \beta e_0 \qquad \text{and} \qquad B^- e_0 = 0.$$

2.4.1 Boson Fock space representation

For any $c > 0$ we can define a representation of $\mathfrak{sl}_2(\mathbb{R})$ on ℓ^2 by

$$\begin{cases} \rho_c(B^+)e_k = \sqrt{(k+c)(k+1)}e_{k+1}, \\ \rho_c(M)e_k = (2k+c)e_k, \\ \rho_c(B^-)e_k = \sqrt{k(k+c-1)}e_{k-1}, \end{cases}$$

where $e_0, e_1, \ldots$ is an orthonormal basis of ℓ^2.

Letting

$$M := \frac{1}{2} + a^\circ, \quad B^- := \frac{1}{2}(a^-)^2, \quad B^+ := \frac{1}{2}(a^+)^2,$$

generates the representation

$$[B^-, B^+] = M, \qquad [M, B^-] = -2B^-, \qquad [M, B^+] = 2B^+.$$

of $\mathfrak{sl}_2(\mathbb{R})$. On the other hand, by defining

$$a^+ := \frac{1}{2}(a^+)^2 + \frac{1}{2}a^\circ \quad \text{and} \quad a^- := \frac{1}{2}(a^-)^2 + \frac{1}{2}a^\circ,$$

we have

$$a^- + a^+ = B^- + B^+ + M - \frac{1}{2}, \quad \text{and} \quad i(a^- - a^+) = i(B^- - B^+),$$

and $a^- + a^+ + 1/2$ has a gamma law, while $i(a^- - a^+)$ has a continuous binomial law with parameter $1/2$, under the conditions

$$Me_0 = \beta e_0 \qquad \text{and} \qquad B^- e_0 = 0.$$

2.4.2 Matrix representation

The Lie algebra $\mathfrak{sl}_2(\mathbb{R})$ can be represented by 2×2 matrices with trace zero, *i.e.*,

$$B^- = \begin{bmatrix} 0 & i \\ 0 & 0 \end{bmatrix}, \quad B^+ = \begin{bmatrix} 0 & 0 \\ i & 0 \end{bmatrix}, \quad M = \begin{bmatrix} -1 & 0 \\ 0 & 1 \end{bmatrix},$$

however this matrix representation is not compatible with the involution of the Lie algebra. On the other hand, taking

$$B^- = \begin{bmatrix} 0 & 0 \\ 1 & 0 \end{bmatrix}, \quad B^+ = \begin{bmatrix} 0 & 1 \\ 0 & 0 \end{bmatrix}, \quad M = \begin{bmatrix} -1 & 0 \\ 0 & 1 \end{bmatrix},$$

satisfies the correct involution, but with the different commutation relation $[M, B^\pm] = \mp 2B^\pm$.

2.4.3 Adjoint action

Lemma 2.4.1 *Letting $Y = B^- - B^+$, the adjoint action of $g_t := e^{tY/2}$ on X_β is given by*

$$e^{tY/2} X_\beta e^{-tY/2} = e^{t(\mathrm{ad}Y)/2} X_\beta = \big(\cosh(t) + \beta \sinh(t)\big) X_{\gamma(\beta,t)},$$

where

$$\gamma(\beta, t) = \frac{\beta \cosh(t) + \sinh(t)}{\cosh(t) + \beta \sinh(t)}, \qquad t \in \mathbb{R}_+.$$

See Section 4.4 of [46] for a proof of Lemma 2.4.1.

2.4.4 Representation of $\mathfrak{sl}_2(\mathbb{R})$ on $L^2_{\mathbb{C}}(\mathbb{R}_+, \gamma_\beta(\tau) d\tau)$

Denoting by

$$\gamma_\beta(\tau) = \mathbf{1}_{\{\tau \geq 0\}} \frac{\tau^{\beta-1}}{\Gamma(\beta)} e^{-\tau}, \qquad \tau \in \mathbb{R},$$

the gamma probability density function on $\mathbb{R}$ with shape parameter $\beta > 0$, a representation $\{M, B^-, B^+\}$ of $\mathfrak{sl}_2(\mathbb{R})$ can be constructed by letting

$$M := \beta + 2\tilde{a}^\circ, \quad B^- := \tilde{a}^- - \tilde{a}^\circ, \quad B^+ := \tilde{a}^+ - \tilde{a}^\circ,$$

where $\tilde{a}^- = \tau \dfrac{\partial}{\partial \tau}$, i.e.,

$$\tilde{a}^- f(\tau) = \tau f'(\tau), \qquad f \in C^\infty_c(\mathbb{R}),$$

as in [93], [95], [97]. The adjoint $\tilde{a}^+$ of $\tilde{a}^-$ with respect to the gamma density $\gamma_\beta(\tau)$ satisfies

$$\int_0^\infty g(\tau) \tilde{a}^- f(\tau) \gamma_\beta(\tau) d\tau = \int_0^\infty f(\tau) \tilde{a}^+ g(\tau) \gamma_\beta(\tau) d\tau, \quad f, g \in C^\infty_c(\mathbb{R}),$$

$$\tag{2.1}$$

and is given by $\tilde{a}^+ = (\tau - \beta) - \tilde{a}^-$, i.e.,

$$\tilde{a}^+ f(\tau) = (\tau - \beta)f(\tau) - \tau\frac{\partial f}{\partial \tau}(\tau) = (\tau - \beta)f(\tau) - \tilde{a}^- f(\tau).$$

The operator $\tilde{a}^\circ$ defined as

$$\tilde{a}^\circ = \tilde{a}^+ \frac{\partial}{\partial \tau} = -(\beta - \tau)\frac{\partial}{\partial \tau} - \tau\frac{\partial^2}{\partial \tau^2}$$

has the Laguerre polynomials L_n^β with parameter β as eigenfunctions:

$$\tilde{a}^\circ L_n^\beta(\tau) = nL_n^\beta(\tau), \quad n \in \mathbb{N},$$

and the multiplication operator $\tilde{a}^- + \tilde{a}^+ = \tau - \beta$ has a compensated gamma law in the vacuum state $\mathbf{1}_{\mathbb{R}_+}$ in $L_{\mathbb{C}}^2(\mathbb{R}_+, \gamma_\beta(\tau)d\tau)$. Letting

$$\begin{cases} \tilde{Q} = B^- + B^+ = \tilde{a}^- + \tilde{a}^+ - 2\tilde{a}^\circ, \\ \tilde{P} = i(B^- - B^+) = i(\tilde{a}^- - \tilde{a}^+), \\ M = \tau - \tilde{Q} = \tau - B^- - B^+ = \tau - \tilde{a}^- - \tilde{a}^+ + 2\tilde{a}^\circ, \end{cases}$$

i.e.,

$$\begin{cases} \tilde{Q} = \tau - \beta + 2(\beta - \tau)\frac{\partial}{\partial \tau} + 2\tau\frac{\partial^2}{\partial \tau^2}, & (2.2a) \\ \tilde{P} = 2i\tau\frac{\partial}{\partial \tau} - i(\tau - \beta), \\ M = \beta - 2(\beta - \tau)\frac{\partial}{\partial \tau} - 2\tau\frac{\partial^2}{\partial \tau^2}, & (2.2b) \end{cases}$$

we have

$$[\tilde{P}, \tilde{Q}] = 2iM, \qquad [\tilde{P}, M] = 2i\tilde{Q}, \qquad [\tilde{Q}, M] = -2i\tilde{P},$$

and $\tilde{Q} + M$ is the multiplication operator

$$\tilde{Q} + M = \tau,$$

hence $\tilde{Q} + M$ has the gamma law with parameter β in the vacuum state $\mathbf{\Omega} = \mathbf{1}_{\mathbb{R}_+}$ in $L_{\mathbb{C}}^2(\mathbb{R}_+, \gamma_\beta(\tau)d\tau)$.

We will show in Chapter 6 that when $|\alpha| < 1$, the law (or spectral measure) of $\alpha M + \tilde{Q}$ is absolutely continuous with respect to the Lebesgue measure on $\mathbb{R}$. In particular, for $\alpha = 0$, $\tilde{Q}$ and $\tilde{P}$ have continuous binomial distributions and $M + \tilde{Q}$ and $M - \tilde{Q}$ are gamma distributed when $\alpha = \pm 1$. On the other hand, $\tilde{Q} + \alpha M$ has a geometric distribution, when $|\alpha| > 1$, cf. [1], and Exercise 6.3.

2.4.5 Construction on the one-dimensional Gaussian space - $\beta = 1/2$

When $\beta = 1/2$, writing $\tau = x^2/2$, the operators $\tilde{a}^-$, $\tilde{a}^+$, $\tilde{a}^\circ$ are identified to the operators

$$\tilde{a}^\circ_\tau = \frac{1}{2}\alpha^+_x \alpha^-_x, \quad \tilde{a}^-_\tau = \frac{1}{2}Q\alpha^-_x, \quad \tilde{a}^+_\tau = \frac{1}{2}\alpha^+_x Q,$$

acting on the variable x, where

$$Q = \alpha^-_x + \alpha^+_x \quad \text{and} \quad P = i(\alpha^-_x - \alpha^+_x),$$

and

$$\alpha^-_x = \frac{\partial}{\partial x} \quad \text{and} \quad \alpha^+_x = x - \frac{\partial}{\partial x},$$

i.e., Q is multiplication by x and $P = -ix + 2i\dfrac{\partial}{\partial x}$, with $[P, Q] = 2iI$, and we have

$$\begin{cases} \tilde{a}^\circ_\tau f(\tau) = \dfrac{1}{2}\alpha^+_x \alpha^-_x f\left(\dfrac{x^2}{2}\right), \\[2mm] \tilde{a}^-_\tau f(\tau) = \dfrac{1}{2}Q\alpha^-_x f\left(\dfrac{x^2}{2}\right), \\[2mm] \tilde{a}^+_\tau f(\tau) = \dfrac{1}{2}\alpha^+_x Q f\left(\dfrac{x^2}{2}\right). \end{cases}$$

The above relations have been exploited in various contexts, see, *e.g.*, [64], [66], [91]. In [91], these relations have been used to construct a Malliavin calculus on Poisson space directly from the Gaussian case. In [66] they are used to prove logarithmic Sobolev inequalities for the exponential measure.

Taking $\beta = 1/2$, a representation $\{M, B^-, B^+\}$ of $\mathfrak{sl}_2(\mathbb{R})$ can be constructed as

$$\begin{cases} M = \dfrac{1}{2} + 2\tilde{a}^\circ_\tau = \dfrac{\alpha^-_x \alpha^+_x + \alpha^+_x \alpha^-_x}{2} = \dfrac{P^2 + Q^2}{4}, \\[3mm] B^- = \tilde{a}^-_\tau - \tilde{a}^\circ_\tau = \dfrac{1}{2}(\alpha^-_x)^2, \\[3mm] B^+ = \tilde{a}^+_\tau - \tilde{a}^\circ_\tau = \dfrac{1}{2}(\alpha^+_x)^2. \end{cases}$$

In fact, letting

$$\hat{Q} := B^- + B^+ = \frac{1}{2}((\alpha^-_x)^2 + (\alpha^+_x)^2) = \frac{P^2 - Q^2}{4}$$

and

$$\hat{P} := i(B^- - B^+) = \frac{i}{2}((\alpha_x^-)^2 - (\alpha_x^+)^2) = \frac{PQ + QP}{4},$$

we have the commutation relations

$$[M, \hat{P}] = -2i\hat{Q}, \qquad [M, \hat{Q}] = 2i\hat{P}, \qquad [\hat{P}, \hat{Q}] = 2iM,$$

and

$$\hat{Q} + \alpha M = \frac{\alpha + 1}{2}\frac{P^2}{2} + \frac{\alpha - 1}{2}\frac{Q^2}{2},$$

and

$$M + \alpha\hat{Q} = \left(\frac{\alpha + 1}{2}\right)\frac{P^2}{2} + \left(\frac{1 - \alpha}{2}\right)\frac{Q^2}{2}.$$

2.4.6 Construction on the two-dimensional Gaussian space - $\beta = 1$

When $\beta = 1$ and $\gamma_1(\tau)$ is the exponential probability density we let

$$\alpha_x^- := \frac{\partial}{\partial x}, \quad \alpha_y^- := \frac{\partial}{\partial y}, \quad \alpha_x^+ := x - \frac{\partial}{\partial x}, \quad \alpha_y^+ := y - \frac{\partial}{\partial y}$$

denote the partial annihilation and creation operators on the two–variable boson Fock space

$$\Gamma(\mathbb{C}e_1 \oplus \mathbb{C}e_2) \simeq L_{\mathbb{C}}^2\left(\mathbb{R}^2; \frac{1}{2\pi}e^{-(x^2 + y^2)/2}dxdy\right).$$

The next lemma is valid when $\beta = 1$, in which case the exponential random variable $\tau = (x^2 + y^2)/2$ can be represented as

$$\tau = \frac{1}{2}((\alpha_x^+ + \alpha_x^-)^2 + (\alpha_y^+ + \alpha_y^-)^2).$$

Lemma 2.4.2 *The operators* $\tilde{a}^-$, $\tilde{a}^+$, $\tilde{a}^\circ$ *are identified to operators on* $\Gamma(\mathbb{C}e_1 \oplus \mathbb{C}e_2)$, *acting on the variable* $\tau = \frac{1}{2}(x^2 + y^2)$ *by the relations*

$$\begin{cases} \tilde{a}^\circ = \frac{1}{2}(\alpha_x^+ \alpha_x^- + \alpha_y^+ \alpha_y^-), & (2.3a) \\[2mm] \tilde{a}^+ = -\frac{1}{2}((\alpha_x^+)^2 + (\alpha_y^+)^2) - \tilde{a}^\circ, & (2.3b) \\[2mm] \tilde{a}^- = -\frac{1}{2}((\alpha_x^-)^2 + (\alpha_y^-)^2) - \tilde{a}^\circ, & (2.3c) \end{cases}$$

and we have

$$\tilde{P} = \frac{i}{2}(\tilde{a}^- - \tilde{a}^+) = \frac{i}{2}((\alpha_x^+)^2 + (\alpha_y^+)^2 - (\alpha_x^-)^2 - (\alpha_y^-)^2).$$

Proof: From (2.3b) and (2.3c) we have

$$\left((\alpha_x^-)^2 + (\alpha_y^+)^2\right) f\left(\frac{x^2+y^2}{2}\right)$$

$$= \left(\left(x - \frac{\partial}{\partial x}\right)\left(x - \frac{\partial}{\partial x}\right) + \left(y - \frac{\partial}{\partial y}\right)\left(y - \frac{\partial}{\partial y}\right)\right) f\left(\frac{x^2+y^2}{2}\right)$$

$$= \left(x^2 - x\frac{\partial}{\partial x} - 1 - x\frac{\partial}{\partial x} + \partial_x^2 + y^2 - y\frac{\partial}{\partial y} - 1 - y\frac{\partial}{\partial y} + \partial_y^2\right) f\left(\frac{x^2+y^2}{2}\right)$$

$$= \left(-2(1 + x^2 + y^2)f(\tau) - (x^2+y^2)f'(\tau) + 2f'(\tau) + x^2 + y^2\right) f''(\tau)$$

$$= -2((1-\tau)f(\tau) + \tau f'(\tau) - (1-\tau)f'(\tau) - \tau f''(\tau))$$

$$= -2(\tilde{a}^+ + \tilde{a}^\circ)f(\tau),$$

and

$$\left((\alpha_x^-)^2 + (\alpha_y^-)^2\right) f\left(\frac{x^2+y^2}{2}\right) = \left(\frac{\partial}{\partial x}\frac{\partial}{\partial x} + \frac{\partial}{\partial y}\frac{\partial}{\partial y}\right) f\left(\frac{x^2+y^2}{2}\right)$$

$$= \left(2f'(\tau) + (x^2+y^2)\right)f''(\tau)$$

$$= -2(-\tau f'(\tau) - (1-\tau)f'(\tau) - \tau f''(\tau))$$

$$= -2(\tilde{a}^- + \tilde{a}^\circ)f(\tau).$$

$\square$

2.5 Affine Lie algebra

The affine algebra can be viewed as the sub-algebra of $\mathfrak{sl}_2(\mathbb{R})$ generated by

$$X_1 = \begin{bmatrix} 1 & 0 \\ 0 & 0 \end{bmatrix}, \qquad X_2 = \begin{bmatrix} 0 & 1 \\ 0 & 0 \end{bmatrix},$$

with the commutation relation

$$[X_1, X_2] = X_2,$$

and the affine group can be constructed as the group of 2×2 matrices of the form

$$g = e^{x_1 X_1 + x_2 X_2} = \begin{bmatrix} a & b \\ 0 & 1 \end{bmatrix} = \begin{bmatrix} e^{x_1} & x_2 e^{x_1/2}\,\mathrm{sinch}\,(x_1/2) \\ 0 & 1 \end{bmatrix},$$

$a > 0, b \in \mathbb{R}$, where

$$\mathrm{sinch}\,x = \frac{\sinh x}{x}, \qquad x \in \mathbb{R}.$$

The affine group also admits a classical representation on $L^2(\mathbb{R})$ given by

$$(U(g)\phi)(t) = a^{-1/2}\phi\left(\frac{t-b}{a}\right), \quad \phi \in L^2(\mathbb{R}),$$

where

$$g = \begin{bmatrix} a & b \\ 0 & 1 \end{bmatrix}, \quad a > 0, \quad b \in \mathbb{R},$$

and the modified representation on $\mathfrak{h} = L^2_{\mathbb{C}}(\mathbb{R}, \gamma_\beta(|\tau|)d\tau)$ defined by

$$(\hat{U}(g)\phi)(\tau) = \phi(a\tau)e^{ib\tau}e^{-(a-1)|\tau|/2}a^{\beta/2}, \quad \phi \in L^2_{\mathbb{C}}(\mathbb{R}, \gamma_\beta(|\tau|)d\tau), \quad (2.4)$$

obtained by Fourier transformation and a change of measure. We have

$$\hat{U}(X_1)\phi(\tau) = \frac{d}{dt}\bigg|_{t=0}\hat{U}(e^{itX_1})\phi(\tau) = -i\frac{P}{2}\phi(\tau),$$

with $P = i(\beta - |\tau|) + 2i\tau\frac{\partial}{\partial\tau}$, and

$$\hat{U}(X_2)\phi(\tau) = \frac{d}{dt}\bigg|_{t=0}\hat{U}(e^{itX_2})\phi(\tau) = i\tau\phi(\tau) = i(Q+M)\phi(\tau),$$

$\tau \in \mathbb{R}$, where P, Q, and M are defined in (2.2a)-(2.2b). In other words we have

$$\hat{U}(X_1) = -\frac{i}{2}P \quad \text{and} \quad \hat{U}(X_2) = i(Q+M),$$

hence we have

$$\hat{U}(e^{x_1X_1+x_2X_2}) = \exp\left(-ix_1\frac{P}{2} + ix_2(Q+M)\right),$$

under the identification

$$X_1 = \frac{1}{2}(B^- + B^+) \quad \text{and} \quad X_2 = i\left(B^- + B^+ + 2M - \frac{1}{2}\right).$$

2.6 Special orthogonal Lie algebras

In this section we focus on special orthogonal Lie algebras $so(2)$ and $so(3)$.

2.6.1 Lie algebra $so(2)$

The Lie algebra $so(2)$ of $SO(2)$ is *commutative* and generated by

$$\xi_1 = \begin{bmatrix} 0 & -1 \\ 1 & 0 \end{bmatrix}.$$

By direct exponentiation we have

$$g_t = \exp\left(\begin{bmatrix} 0 & -\theta \\ \theta & 0 \end{bmatrix}\right) = \begin{bmatrix} \cos\theta & -\sin\theta \\ \sin\theta & \cos\theta \end{bmatrix}, \qquad \theta \in \mathbb{R}_+.$$

2.6.2 Lie algebra *so*(3)

The Lie algebra $so(3)$ of $SO(3)$ is noncommutative and has a basis consisting of the three anti-Hermitian elements ξ_1, ξ_2, ξ_3, with the relations

$$[\xi_1, \xi_2] = \xi_3, \qquad [\xi_2, \xi_3] = \xi_1, \qquad [\xi_3, \xi_1] = \xi_2.$$

Let $x = \begin{bmatrix} x_1 \\ x_2 \\ x_3 \end{bmatrix} \in \mathbb{R}$, then

$$\xi(x) = (x_1\xi_1 + x_2\xi_2 + x_3\xi_3)$$

defines a general element of $so(3)$, it is anti-Hermitian, *i.e.*, $\xi(x)^* = -\xi(x)$. We can take *e.g.*,

$$\xi_1 = \begin{bmatrix} 0 & 1 & 0 \\ -1 & 0 & 0 \\ 0 & 0 & 0 \end{bmatrix}, \quad \xi_2 = \begin{bmatrix} 0 & 0 & -1 \\ 0 & 0 & 0 \\ 1 & 0 & 0 \end{bmatrix}, \quad \xi_3 = \begin{bmatrix} 0 & 0 & 0 \\ 0 & 0 & 1 \\ 0 & -1 & 0 \end{bmatrix}.$$

We note that by Rodrigues' rotation formula, every $g \in SO(3)$ can be parameterised as

$$g = e^{x\xi_1 + y\xi_2 + z\xi_3}$$

$$= e^{a(z,-y,x)}$$

$$= I_d + \sin(\phi)\, a(u_1, u_2, u_3) + (1 - \cos\phi)\, a(u_1, u_2, u_3)^2,$$

for some $x, y, z \in \mathbb{R}$, where

$$a(u_1, u_2, u_3) = \begin{bmatrix} 0 & -u_3 & -u_2 \\ u_3 & 0 & -u_1 \\ u_2 & u_1 & 0 \end{bmatrix},$$

and $\phi = \sqrt{x^2 + y^2 + z^2}$ is the angle of rotation about the axis

$$(u_1, u_2, u_3) := \frac{1}{\sqrt{x^2 + y^2 + z^2}} (z, -y, x) = (\cos\alpha, \sin\alpha\cos\theta, \sin\alpha\sin\theta) \in S^2.$$

2.6.3 Finite-dimensional representations of $so(3)$

We consider a family of finite-dimensional representations of $so(3)$ in terms of the basis ξ_0, ξ_+, ξ_- of $so(3)$ defined by

$$\xi_0 = 2i\xi_3, \qquad \xi_+ = i(\xi_1 + i\xi_2), \qquad \xi_- = i(\xi_1 - i\xi_2).$$

In this basis the commutation relations of $so(3)$ take the form

$$[\xi_0, \xi_\pm] = \pm 2\xi_\pm \qquad \text{and} \qquad [\xi_+, \xi_-] = \xi_0,$$

and $\xi_0^* = \xi_0$, $\xi_+^* = \xi_-$, $\xi_-^* = \xi_+$. This is close to a representation of $\mathfrak{sl}_2(\mathbb{R})$, although not with the correct involution. Letting $n \in \mathbb{N}$ be a positive integer and given $e_{-n}, e_{-n-2}, \ldots, e_{n-2}, e_n$ an orthonormal basis of an $n + 1$-dimensional Hilbert space, we define a representation of $so(3)$ by

$$
\begin{cases}
\xi_0 e_k = k e_k, & \text{(2.5a)} \\[2mm]
\xi_+ e_k = \begin{cases} 0 & \text{if } k = n, \\[2mm] \dfrac{1}{2}\sqrt{(n-k)(n+k+2)}e_{k+2} & \text{else.} \end{cases} & \text{(2.5b)} \\[6mm]
\xi_- e_k = \begin{cases} 0 & \text{if } k = -n, \\[2mm] \dfrac{1}{2}\sqrt{(n+k)(n-k+2)}e_{k-2} & \text{else.} \end{cases} & \text{(2.5c)}
\end{cases}
$$

In order to get back a representation in terms of the basis ξ_1, ξ_2, ξ_3, we have

$$\xi_1 = -\frac{i}{2}(\xi_- + \xi_+), \quad \xi_2 = \frac{1}{2}(\xi_- - \xi_+), \quad \xi_3 = -\frac{i}{2}\xi_0.$$

2.6.4 Two-dimensional representation of $so(3)$

For $n = 1$, we get the two-dimensional representation

$$\xi_0 = \begin{bmatrix} 1 & 0 \\ 0 & -1 \end{bmatrix}, \qquad \xi_+ = \begin{bmatrix} 0 & 1 \\ 0 & 0 \end{bmatrix}, \qquad \xi_- = \begin{bmatrix} 0 & 0 \\ 1 & 0 \end{bmatrix},$$

with respect to the basis $\{e_1, e_{-1}\}$, or

$$\xi_1 = -\frac{i}{2}\begin{bmatrix} 0 & 1 \\ 1 & 0 \end{bmatrix}, \quad \xi_2 = -\frac{1}{2}\begin{bmatrix} 0 & 1 \\ -1 & 0 \end{bmatrix}, \quad \xi_3 = -\frac{i}{2}\begin{bmatrix} 1 & 0 \\ 0 & -1 \end{bmatrix}.$$

In this representation we get

$$\xi(x) = -\frac{i}{2}\begin{bmatrix} x_3 & x_1 - ix_2 \\ x_1 + ix_2 & -x_3 \end{bmatrix}.$$

2.6.5 Adjoint action

We note that

$$C = \xi_1^2 + \xi_2^2 + \xi_3^3$$

commutes with the basis elements ξ_1, ξ_2, ξ_3, *e.g.*,

$$
\begin{aligned}
[\xi_1, C] &= [\xi_1, \xi_1^2] + [\xi_1, \xi_2^2] + [x_1, \xi_3^3] \\
&= 0 + \xi_3 \xi_2 + \xi_2 \xi_3 - \xi_2 \xi_3 - \xi_3 \xi_2 \\
&= 0,
\end{aligned}
$$

where we used the Leibniz formula for the commutator, *i.e.*, the fact that we always have

$$[a, bc] = [a, b]c + b[a, c].$$

The element C is called the *Casimir operator*.

Let us now study the commutator of two general elements $\xi(x), \xi(y)$ of $so(3)$, with $x, y \in \mathbb{R}^3$. We have

$$
\begin{aligned}
[\xi(x), \xi(y)] &= [x_1 \xi_1 + x_2 \xi_2 + x_3 \xi_3, y_1 \xi_1 + y_2 \xi_2 + y_3 \xi_3] \\
&= (x_2 y_3 - x_3 y_2)\xi_2 + (x_3 y_1 - x_1 y_3)\xi_2 + (x_1 y_2 - x_2 y_1)\xi_3 \\
&= \xi(x \times y),
\end{aligned}
$$

where $x \times y$ denotes the cross product or vector product of two vectors x and y in three-dimensional space,

$$
x \times y = \begin{bmatrix} x_2 y_3 - x_3 y_2 \\ x_3 y_1 - x_1 y_3 \\ x_1 y_2 - x_2 y_1 \end{bmatrix}.
$$

This shows that the element $\exp\left(\xi(x)\right)$ of the Lie group $SO(3)$ acts on $so(3)$ as a rotation. More precisely, we have the following result.

Lemma 2.6.1 *Let $x, y \in \mathbb{R}^3$, then we have*

$$\mathrm{Ad}\Big(\exp\big(\xi(x)\big)\Big)\big(\xi(y)\big) = \xi\big(R_x(y)\big)$$

where R_x denotes a rotation around the axis given by x, by an angle $\|x\|$.

Proof: Recall that the adjoint action of a Lie group of matrices on its Lie algebra is defined by

$$\mathrm{Ad}\big(\exp(X)\big)(Y) = \exp(X) Y \exp(-X).$$

It is related to the adjoint action of the Lie algebra on itself,

$$\mathrm{ad}(X)Y = [X, Y]$$

by

$$\mathrm{Ad}\big(\exp(X)\big)(Y) = \exp\big(\mathrm{ad}(X)\big)(Y).$$

We already checked that

$$\mathrm{ad}\big(\xi(x)\big)\big(\xi(y)\big) = [\xi(x), \xi(y)] = \xi(x \times y).$$

We now have to compute the action of the exponential of $\mathrm{ad}\big(\xi(x)\big)$, we will choose a convenient basis for this purpose. Let

$$e_1 = \frac{x}{||x||},$$

choose for e_2 any unit vector orthogonal to e_1, and set

$$e_3 = e_1 \times e_2.$$

Then we have

$$x \times e_j = \begin{cases} 0 & \text{if } j = 1, \\ e_3 & \text{if } j = 2, \\ -e_2 & \text{if } j = 3. \end{cases}$$

We check that the action of $\mathrm{Ad}\big(\xi(x)\big)$ on this basis is given by

$$\mathrm{Ad}\Big(\exp\big(\xi(x)\big)\Big)\big(\xi(e_1)\big) = \sum_{n=0}^{\infty} \frac{1}{n!}\big(\mathrm{ad}(\xi(x))\big)^n\big(\xi(e_1)\big)$$

$$= \xi(e_1) + \xi\underbrace{(x \times e_1)}_{=0} + \frac{1}{2}\xi\big(\underbrace{x \times (x \times e_1)}_{=0}\big) + \cdots$$

$$= \xi\big(R_x(e_1)\big),$$

$$\mathrm{Ad}\Big(\exp\big(\xi(x)\big)\Big)\big(\xi(e_2)\big) = \xi(e_2) + \xi\underbrace{(x \times e_2)}_{=||x||e_3} + \frac{1}{2}\xi\big(\underbrace{x \times (x \times e_2)}_{=-||x||^2 e_2}\big) + \cdots$$

$$= \xi\big(\cos(||x||)e_2 + \sin(||x||)e_3\big)$$

$$= \xi\big(R_x(e_2)\big),$$

$$\mathrm{Ad}\Big(\exp\big(\xi(x)\big)\Big)\big(\xi(e_3)\big) = \xi(e_3) + \xi\underbrace{(x \times e_3)}_{=-||x||e_2} + \frac{1}{2}\xi\big(\underbrace{x \times (x \times e_3)}_{=-||x||^2 e_3}\big) + \cdots$$

$$= \xi\big(\cos(||x||)e_3 - \sin(||x||)e_2\big)$$

$$= \xi\big(R_x(e_3)\big). \qquad \square$$

Notes

Relation (2.3a) has been used in [92] to study the relationship between the stochastic calculus of variations on the Wiener and Poisson spaces, cf. also [64].

Exercises

Exercise 2.1 Consider the Weyl type representation, defined as follows for a subgroup of $\mathfrak{sl}_2(\mathbb{R})$. Given $z = u + iv \in \mathbb{C}$, $u < 1/2$, define the operator W_z as

$$W_z f(x) = \frac{1}{\sqrt{1 - 2u}} f\left(\frac{x}{1 - 2u}\right) \exp\left(-\frac{ux}{1 - 2u} + iv(1 - x)\right).$$

1. Show that the operator W_z is unitary on L^2 with $W_0 = I_d$ and that for any

$$\lambda = \kappa + i\zeta, \quad \lambda' = \kappa' + i\zeta' \in \mathbb{C}$$

we have

$$\begin{cases} W_\lambda W_{\lambda'} = W_{\kappa + \kappa' - 2\kappa\kappa' + \Im(\zeta + \zeta'/(1 - 2\kappa))}, \\[2mm] \dfrac{dW_{t\lambda}}{dt}\bigg|_{t=0} = \lambda \tilde{a}^+ - \bar{\lambda} \tilde{a}_{\underline{=}}^- - \lambda \tilde{P}, \quad \lambda = \kappa + i\zeta \in \ell^2(\mathbb{N}; \mathbb{C}), \\[2mm] W_{is} W_u = \exp\left(2\dfrac{ius\tau}{1 - 2u}\right) W_u W_{is}, \quad u < 1/2, \ s \in \mathbb{R}. \end{cases}$$

   Conclude that W_λ can be extended to $L^2(\mathbb{R}; \mathbb{C})$, provided $|\kappa| < 1/2$, and

$$\begin{bmatrix} 1/a & b \\ 0 & a \end{bmatrix} \longmapsto W_{(1 - a^2)/2 + ib/a}, \quad a \in \mathbb{R} \setminus \{0\}, \quad b \in \mathbb{R},$$

   is a representation of the subgroup of $SL(2, \mathbb{R})$ made of upper-triangular matrices.

2. Show that the representation $(W_\lambda)_\lambda$ contains the commutation relations between $\tilde{a}^+$ and $\tilde{a}^-$, *i.e.*, we have

$$\tilde{P}\tilde{Q} = -\frac{d}{dt}\frac{d}{ds} W_t W_{is}\big|_{t=s=0} \quad \text{and} \quad \tilde{Q}\tilde{P} = -\frac{d}{dt}\frac{d}{ds} W_{is} W_t\big|_{t=s=0}.$$

3

Basic probability distributions on Lie algebras

The theory of probabilities is at bottom nothing but common sense
reduced to calculus.
(P.S. de Laplace, in Théorie Analytique des Probabilités.*)*

In this chapter we show how basic examples of continuous and discrete
probability distributions can be constructed from real Lie algebras, based on
the annihilation and creation operators a^-, a^+, completed by the number
operator $a^\circ = a^+ a^-$. In particular, we study in detail the relationship between
the Gaussian and Poisson distributions and the Heisenberg–Weyl and oscillator
Lie algebras $\mathfrak{hw}$ and $\mathfrak{osc}$, which generalises the introduction given in Chapter 1.
We work on the Fock space over a real separable Hilbert space $\mathfrak{h}$, and we also
examine a situation where the gamma and continuous binomial distributions
appear naturally on $\mathfrak{sl}_2(\mathbb{R})$, in relation with integration by parts with respect to
the gamma distribution.

3.1 Gaussian distribution on $\mathfrak{hw}$

Since the Heisenberg–Weyl Lie algebra $\mathfrak{hw}$ is based on the creation and
annihilation operators a^- and a^+ introduced on the boson Fock space ℓ^2 in
Chapter 1, we start with an extension of those operators to an arbitrary complex
Hilbert space.

Namely, we consider

a) a complex Hilbert space $\mathfrak{h}$ equipped with a *sesquilinear* inner product $\langle \cdot, \cdot \rangle$,
such that

$$\langle z\,u, v \rangle = \bar{z}\langle u, v \rangle, \qquad z \in \mathbb{C},$$

and $\langle u, v \rangle = \overline{\langle v, u \rangle}$,

b) two operators a^- and a^+, called *annihilation and creation operators* acting on the elements of $\mathfrak{h}$, such that

i) a^- and a^+ are dual of each other in the sense that

$$\langle a^- u, v \rangle = \langle u, a^+ v \rangle, \qquad u, v \in \mathfrak{h} \tag{3.1}$$

which will also be written $(a^+)^* = a^-$, for the scalar product $\langle \cdot, \cdot \rangle$, and

ii) the operators a^- and a^+ satisfy the commutation relation

$$[a^-, a^+] = a^- a^+ - a^+ a^- = E, \tag{3.2}$$

where E commutes with a^- and a^+,

c) a unit vector $e_0 \in \mathfrak{h}$ (fundamental or empty state) such that $a^- e_0 = 0$ and $\langle e_0, e_0 \rangle = 1$.

We will show that under the conditions

$$a^- e_0 = 0 \quad \text{and} \quad E e_0 = \sigma^2 e_0,$$

(e.g., when $E = \sigma^2 I_{\mathfrak{h}}$ where $I_{\mathfrak{h}}$ is the identity of $\mathfrak{h}$), the operator $Q = a^- + a^+$ has a Gaussian law in the sense that it yields the moment generating function

$$t \longmapsto \langle e_0, e^{tQ} e_0 \rangle = e^{t^2 \sigma^2 / 2}, \qquad t \in \mathbb{R}_+,$$

which extends in particular the example of Exercise 1.1 to moments of all orders. Similarly, we could show that $P = i(a^- - a^+)$ also has a Gaussian law in the state e_0, see Exercise 3.1.

Next, we will consider several representations for the aforementioned noncommutative framework.

3.1.1 Gaussian Hilbert space representation

A way to implement the Heisenberg–Weyl algebra and the above operators a^- and a^+ is to take

$$\mathfrak{h} := L^2_{\mathbb{C}} \left(\mathbb{R}; \frac{1}{\sqrt{2\pi\sigma^2}} e^{-x^2/(2\sigma^2)} dx \right)$$

$$= \left\{ f: \mathbb{R} \to \mathbb{C} : \int_{-\infty}^{\infty} |f(x)|^2 e^{-x^2/(2\sigma^2)} dx < \infty \right\}$$

under the inner product

$$\langle u, v \rangle_{\mathfrak{h}} := \frac{1}{\sqrt{2\pi\sigma^2}} \int_{-\infty}^{\infty} \bar{u}(x) v(x) e^{-x^2/(2\sigma^2)} dx,$$

by letting

$$a^- := \sigma^2 \frac{\partial}{\partial x} \quad \text{and} \quad a^+ := x - \sigma^2 \frac{\partial}{\partial x}$$

and by defining e_0 to be the constant function equal to one, *i.e.*, $e_0(x) = 1$, $x \in \mathbb{R}$, which satisfies the conditions $\langle e_0, e_0 \rangle_{\mathfrak{h}} = 1$ and $a^- e_0 = 0$. A standard integration by parts shows that

$$
\begin{aligned}
\langle a^- u, v \rangle_{\mathfrak{h}} &= \frac{\sigma^2}{\sqrt{2\pi\sigma^2}} \int_{-\infty}^{\infty} \bar{u}'(x) v(x) e^{-x^2/(2\sigma^2)} dx \\
&= \frac{1}{\sqrt{2\pi\sigma^2}} \int_{-\infty}^{\infty} \bar{u}(x)(xv(x) - \sigma^2 v'(x)) e^{-x^2/(2\sigma^2)} dx \\
&= \langle u, a^+ v \rangle_{\mathfrak{h}},
\end{aligned}
$$

i.e., (3.1) is satisfied, and

$$
\begin{aligned}
[a^-, a^+] u(x) &= a^- a^+ u(x) - a^+ a^- u(x) \\
&= a^-(xu(x) - \sigma^2 u'(x)) - \sigma^2 a^+ u'(x) \\
&= \sigma^2 \frac{\partial}{\partial x}(xu(x) - \sigma^2 u'(x)) - \sigma^2 xu'(x) + \sigma^4 u''(x) \\
&= \sigma^2 u(x),
\end{aligned}
$$

hence (3.2) is satisfied.

In this representation, we easily check that the position and momentum operators Q and P are written as

$$Q = a^- + a^+ = xI_{\mathfrak{h}} \quad \text{and} \quad P = i(a^+ - a^-) = i\left(xI_{\mathfrak{h}} - 2\sigma^2 \frac{\partial}{\partial x}\right),$$

and that

$$\langle e_0, Q^n e_0 \rangle_{\mathfrak{h}} = \frac{1}{\sqrt{2\pi\sigma^2}} \int_{-\infty}^{\infty} x^n e^{-x^2/(2\sigma^2)} dx$$

is indeed the centered Gaussian moment of order $n \in \mathbb{N}$, which recovers in particular the first four Gaussian moments computed in Exercise 1.1. In addition, the moment generating function of Q in the state e_0, defined by

$$\langle e_0, e^{tQ} e_0 \rangle_{\mathfrak{h}} = \sum_{n=0}^{\infty} \frac{t^n}{n!} \langle e_0, Q^n e_0 \rangle_{\mathfrak{h}},$$

satisfies

$$\langle e_0, e^{tQ} e_0 \rangle_{\mathfrak{h}} = \frac{1}{\sqrt{2\pi\sigma^2}} \int_{-\infty}^{\infty} e^{tx} e^{-x^2/(2\sigma^2)} dx = \exp\left(\frac{1}{2}\sigma^2 t^2\right).$$

3.1.2 Hermite representation

In this section we implement the representation of the Heisenberg–Weyl algebra constructed on the boson Fock space in Section 1.2 using the Hermite polynomials $H_n(x; \sigma^2)$ with parameter $\sigma^2 > 0$, which define the orthononomal sequence

$$e_n(x) := \frac{1}{\sigma^n \sqrt{n!}} H_n(x; \sigma^2)$$

in $\mathfrak{h} := L^2_{\mathbb{C}}\left(\mathbb{R}; \frac{1}{\sqrt{2\pi\sigma^2}} e^{-x^2/(2\sigma^2)} dx\right)$, *i.e.* we have

$$\langle e_n, e_m \rangle_{\mathfrak{h}} = \delta_{n,m}, \qquad n, m \in \mathbb{N}.$$

In addition, the Hermite polynomials are known to satisfy the relations

$$a^- H_n(x; \sigma^2) = \sigma^2 \frac{\partial H_n}{\partial x}(x; \sigma^2) = n\sigma^2 H_{n-1}(x; \sigma^2),$$

and

$$a^+ H_n(x; \sigma^2) = \left(x - \sigma^2 \frac{\partial}{\partial x}\right) H_n(x; \sigma^2) = H_{n+1}(x; \sigma^2),$$

i.e.,

$$a^- e_n = \sigma \sqrt{n}\, e_{n-1}, \qquad a^+ e_n = \sigma \sqrt{n+1}\, e_{n+1}, \qquad n \in \mathbb{N},$$

with

$$a^- e_0 = 0.$$

We also note that the relation

$$e_n = \frac{1}{\sigma^n \sqrt{n!}} (a^+)^n e_0$$

reads

$$H_n(x) = (a^+)^n H_0(x) = (a^+)^n e_0 = \sigma^n \sqrt{n!}\, e_n,$$

i.e.,

$$e_n = \frac{1}{\sigma^n \sqrt{n!}} H_n(Q) e_0, \qquad n \in \mathbb{N},$$

which is Condition (1.4) of Proposition 1.2.1.

3.2 Poisson distribution on osc

The generic Hermitian element of the *oscillator* Lie algebra

$$\mathfrak{osc} = \text{span}\,\{N, P, Q, E\}$$

can be written in the form

$$X_{\alpha,\zeta,\beta} = \alpha N + \zeta a^+ + \overline{\zeta} a^- + \beta E,$$

with $\alpha, \beta \in \mathbb{R}$ and $\zeta \in \mathbb{C}$. We will show that

$$X := X_{1,1,1} = N + a^+ + a^- + E$$

has a *Poisson* distribution with parameter $\lambda > 0$ under the conditions

$$a^- e_0 = 0 \quad \text{and} \quad E e_0 = \lambda e_0,$$

i.e., we take $\sigma = \sqrt{\lambda}$. We start by checking this fact on the first moments. We have

$$
\begin{aligned}
\langle X^n e_0, e_0 \rangle &= \langle X^{n-1} e_0, X e_0 \rangle \\
&= \langle X^{n-1} e_0, a^+ e_0 \rangle + \lambda \langle X^{n-1} e_0, e_0 \rangle \\
&= \langle a^- X^{n-1} e_0, e_0 \rangle + \lambda \langle X^{n-1} e_0, e_0 \rangle.
\end{aligned}
$$

On the other hand, we note the commutation relation

$$[a^-, X] = [a^-, N] + [a^-, a^+] = a^- + \lambda I_{\mathfrak{h}},$$

which implies

$$
\begin{aligned}
\langle X^{n+1} e_0, e_0 \rangle &= \langle X^n e_0, X e_0 \rangle \\
&= \langle X^n e_0, a^+ e_0 \rangle + \lambda \langle X^n e_0, e_0 \rangle \\
&= \langle a^- X^n e_0, e_0 \rangle + \lambda \langle X^n e_0, e_0 \rangle \\
&= \langle X a^- X^{n-1} e_0, e_0 \rangle + \langle a^- X^{n-1} e_0, e_0 \rangle + \lambda \langle X^{n-1} e_0, e_0 \rangle + \lambda \langle X^n e_0, e_0 \rangle,
\end{aligned}
$$

and recovers by induction the Poisson moments given by the Touchard polynomials $T_n(\lambda)$ as

$$\mathbf{E}_\lambda[Z^n] = T_n(\lambda) = \sum_{k=0}^{n} \lambda^k S(n, k), \qquad n \in \mathbb{N},$$

cf. Relation (A.8) in the Appendix A.2.

The representation of $\mathfrak{osc}$ on the boson Fock space ℓ^2 is given by

$$
\begin{cases}
Ne_n = ne_n, \\[2mm]
a^+ e_n = \sqrt{n+1}\lambda e_{n+1}, \\[2mm]
a^- e_{n-1} = \sqrt{n}\lambda e_{n-1},
\end{cases}
$$

where $N := \frac{a^\circ}{\lambda} = \frac{a^+ a^-}{\lambda}$ is the *number* operator.

3.2.1 Poisson Hilbert space representation

We choose $\mathfrak{h} = \ell^2(\mathbb{N}, p_\lambda)$ where

$$
p_\lambda(k) = e^{-\lambda}\frac{\lambda^k}{k!}, \qquad k \in \mathbb{N},
$$

is the Poisson distribution, with the inner product

$$
\langle f, g \rangle := \sum_{k=0}^{\infty} f(k)g(k)p_\lambda(k) := e^{-\lambda}\sum_{k=0}^{\infty} f(k)g(k)\frac{\lambda^k}{k!}.
$$

In the sequel we will use the finite difference operator Δ defined as

$$
\Delta f(k) = f(k+1) - f(k), \qquad k \in \mathbb{N}, \tag{3.3}
$$

cf. Section 9.3 for its generalisation to spaces of configurations under a Poisson random measure.

Let $\lambda > 0$. In the next proposition we show that the operators a^- and a^+ defined by

$$
a^- f(k) := \lambda \Delta f(k) = \lambda(f(k+1) - f(k)), \tag{3.4}
$$

and

$$
a^+ f(k) := kf(k-1) - \lambda f(k), \tag{3.5}
$$

satisfy the Conditions (3.1)-(3.2) above with $E = \lambda I_{\mathfrak{h}}$.

Proposition 3.2.1 *The operators a^- and a^+ defined in (3.4)-(3.5) satisfy the commutation relation*

$$
[a^-, a^+] = \lambda I_{\mathfrak{h}}
$$

and the involution $(a^-)^* = a^+$.

Proof: The commutation relation follows from (A.3b) and (A.3c). Next, by the Abel transformation of sums we have

$$\langle a^- f, g \rangle = \lambda e^{-\lambda} \sum_{k=0}^{\infty} (f(k+1) - f(k)) g(k) \frac{\lambda^k}{k!}$$

$$= \lambda e^{-\lambda} \sum_{k=0}^{\infty} f(k+1) g(k) \frac{\lambda^k}{k!} - \lambda e^{-\lambda} \sum_{k=0}^{\infty} f(k) g(k) \frac{\lambda^k}{k!}$$

$$= e^{-\lambda} \sum_{k=1}^{\infty} f(k) g(k-1) \frac{\lambda^k}{(k-1)!} - \lambda e^{-\lambda} \sum_{k=0}^{\infty} f(k) g(k) \frac{\lambda^k}{k!}$$

$$= e^{-\lambda} \sum_{k=1}^{\infty} f(k) (k g(k-1) - \lambda g(k)) \frac{\lambda^k}{k!}$$

$$= \langle f, a^+ g \rangle.$$ $\square$

We also note that the number operator $N := a^+ a^- / \lambda$ satisfies

$$Nf(k) = \frac{1}{\lambda} a^+ a^- f(k)$$

$$= a^+ f(k+1) - a^+ f(k)$$

$$= kf(k) - \lambda f(k+1) - (kf(k-1) - \lambda f(k)$$

$$= kf(k) - \lambda f(k+1) - kf(k-1) + \lambda f(k)$$

$$= -\lambda f(k+1) + (k + \lambda + 1) f(k) - kf(k-1), \qquad k \in \mathbb{N}.$$

This shows that

$$(N + a^+ + a^- + E) f(k) = -\lambda f(k+1) + (k + \lambda) f(k) - kf(k-1)$$
$$+ \lambda (f(k+1) - f(k)) + kf(k-1) - \lambda f(k) + \lambda f(k)$$
$$= kf(k), \tag{3.6}$$

hence

$$N + a^+ + a^- + E$$

has a *Poisson* distribution with parameter $\lambda > 0$ in the vacuum state **1**.

3.2.2 Poisson-Charlier representation on the boson Fock space

In this section we use the Lie algebra representation based on the boson Fock space in Section 1.2, together with the Charlier polynomials $C_n(k; \lambda)$ defined in Section A.1.3.2 in appendix. First, we note that the functions

$$e_n(k) := \frac{1}{\lambda^{n/2}\sqrt{n!}}C_n(k;\lambda), \qquad n \in \mathbb{N},$$

form an orthonormal sequence in $\mathfrak{h} = \ell^2(\mathbb{N}, p_\lambda)$. Next, we note that the annihilation and creation operators a^- and a^+ defined in (3.4)-(3.5) satisfy

$$a^- C_n(k;\lambda) = \lambda(C_n(k+1,\lambda) - C_n(k,\lambda)) = n\lambda C_{n-1}(k,\lambda),$$

and

$$a^+ C_n(k,\lambda) = kC_n(k-1,\lambda) - \lambda C_n(k,\lambda) = C_{n+1}(k,\lambda),$$

hence we have

$$a^- e_n := \sqrt{\lambda n}e_{n-1}, \quad \text{and} \quad a^+ e_n := \sqrt{\lambda(n+1)}e_{n+1},$$

as in the ℓ^2 representation of the boson Fock space, and this yields

$$a^+ a^- C_n(k,\lambda) = n\lambda a^+ C_{n-1}(k,\lambda) = n\lambda C_n(k,\lambda).$$

In addition, the commutation relation of Proposition 3.2.1 can be recovered as follows:

$$
\begin{aligned}
[a^-, a^+]f(k) &= a^- a^+ f(k) - a^+ a^- f(k) \\
&= a^-(kf(k-1) - \lambda f(k)) - \lambda a^+(f(k+1) - f(k)) \\
&= \lambda((k+1)f(k) - \lambda f(k+1)) - \lambda(kf(k-1) - \lambda f(k)) \\
&\quad - \lambda k(f(k) - f(k-1)) + \lambda^2(f(k+1) - f(k)) \\
&= \lambda f(k),
\end{aligned}
$$

showing that $[a^-, a^+] = \lambda I_{\mathfrak{h}}$.

Similarly, the duality

$$\langle a^- f, g \rangle = \langle f, a^+ g \rangle$$

of Proposition 3.2.1 for the inner product $\langle f, g \rangle_{\mathfrak{h}}$ can be recovered by the similar Abel transformation of sums

$$\langle a^- f, C_n(\cdot, \lambda) \rangle = \lambda \sum_{k=0}^{\infty} (f(k+1) - f(k)) C_n(k, \lambda) \frac{\lambda^k}{k!}$$

$$= -\lambda f(0) C_n(0, \lambda)) + \lambda \sum_{k=1}^{\infty} f(k) (k C_n(k-1, \lambda) - \lambda C_n(k, \lambda)) \frac{\lambda^{k-1}}{k!}$$

$$= f(0) C_{n+1}(0, \lambda)) + \sum_{k=1}^{\infty} f(k) C_{n+1}(k, \lambda) \frac{\lambda^k}{k!}$$

$$= \sum_{k=0}^{\infty} f(k) C_{n+1}(k, \lambda) \frac{\lambda^k}{k!}$$

$$= \langle f, a^+ C_n(\cdot, \lambda) \rangle,$$

with $C_n(0, \lambda) = (-\lambda)^n$. We also check that (3.6) can equivalently be recovered as

$$(N + a^+ + a^- + E) C_n(k, \lambda)$$
$$= k(C_n(k, \lambda) - C_n(k-1, \lambda)) - \lambda(C_n(k+1, \lambda) - C_n(k, \lambda))$$
$$+ \lambda(C_n(k+1, \lambda) - C_n(k, \lambda))$$
$$+ k C_n(k-1, \lambda) - \lambda C_n(k, \lambda) + \lambda C_n(k, \lambda)$$
$$= k C_n(k, \lambda),$$

hence $N + a^+ + a^- + \lambda E$ has a *Poisson* distribution with parameter $\lambda > 0$.

3.2.3 Adjoint action

The next lemma will be used for the Girsanov theorem in Chapter 10.

Lemma 3.2.2 *Letting $Y = i(w a^+ + \overline{w} a^-)$, the adjoint action of $g_t := e^{tY}$ on*

$$X_{\alpha, \zeta, \beta} = \alpha N + \zeta a^+ + \overline{\zeta} a^- + \beta E$$

is given by

$$e^{tY} X_{\alpha, \zeta, \beta} e^{-tY} = \alpha N + (\zeta - i\alpha w t) a^+ + (\overline{\zeta} + i\alpha \overline{w} t) a^-$$
$$+ \left(\beta + 2t\Im(w\overline{\zeta}) + \alpha |w|^2 t^2 \right) E, \tag{3.7}$$

$t \in \mathbb{R}_+$, *where $\Im(z)$ denotes the imaginary part of z.*

Proof: The adjoint action

$$X(t) := e^{tY} X_{\alpha, \zeta, \beta} e^{-tY}$$

of $g_t := e^{tY}$ on $X_{\alpha,\zeta,\beta}$ solves the differential equation

$$\dot{X}(t) = \frac{d}{dt}\mathrm{Ad}_{g_t}(X) = [Y, X(t)].$$

Looking for a solution of the form

$$X(t) = a(t)N + z(t)a^+ + \bar{z}(t)a^- + b(t)E,$$

we get the system

$$\begin{cases} \dot{a}(t) = 0, \\[2mm] \dot{z}(t) = -i\alpha w, \\[2mm] \dot{b}(t) = i(\bar{w}z - w\bar{z}), \end{cases}$$

of ordinary differential equations with initial conditions

$$a(0) = \alpha, \quad z(0) = \zeta, \quad b(0) = \beta,$$

whose solution yields (3.7). □

3.3 Gamma distribution on $\mathfrak{sl}_2(\mathbb{R})$

In this section we revisit the representation of $\mathfrak{sl}_2(\mathbb{R})$ on $\mathfrak{h} = L^2_{\mathbb{C}}(\mathbb{R}, \gamma_\beta(x)dx)$ with the inner product

$$\langle f, g \rangle := \int_0^\infty f(x)g(x)\gamma_\beta(x)dx,$$

introduced in Section 2.4, in connection with the gamma probability density function

$$\gamma_\beta(x) = \frac{x^{\beta-1}}{\Gamma(\beta)}e^{-x}\mathbf{1}_{\{x\geq 0\}}$$

on $\mathbb{R}$, with shape parameter $\beta > 0$. We have $\tilde{a}^- = x\dfrac{\partial}{\partial x}$, i.e.,

$$\tilde{a}^-f(x) = xf'(x), \qquad f \in \mathcal{C}_b^\infty(\mathbb{R}).$$

The adjoint $\tilde{a}^+$ of $\tilde{a}^-$ with respect to the gamma density $\gamma_\beta(x)$ on $\mathbb{R}$ satisfies

$$\langle \tilde{a}^-f, g \rangle_{\mathfrak{h}} = \int_0^\infty g(x)\tilde{a}^-f(x)\gamma_\beta(x)dx$$

$$= \int_0^\infty xg(x)f'(x)\gamma_\beta(x)dx$$

$$= \int_0^\infty f(x)(xg(x) - \beta g(x) - xg'(x))\gamma_\beta(x)dx$$

$$= \langle f, \tilde{a}^+ g \rangle_\mathfrak{h}, \qquad f, g \in C_b^\infty(\mathbb{R}),$$

hence we have

$$\tilde{a}^+ = x - \beta - \tilde{a}^-,$$

i.e.,

$$\tilde{a}^+ f(x) = (x - \beta)f(x) - x\frac{\partial}{\partial x}f(x) = (x - \beta)f(x) - \tilde{a}^- f(x).$$

In other words, the multiplication operator $\tilde{a}^- + \tilde{a}^+ = \tau - \beta$ has a compensated gamma distribution in the vacuum state e_0 in $L_\mathbb{C}^2(\mathbb{R}_+, \gamma_\beta(\tau)d\tau)$.

The operator $\tilde{a}^\circ$ defined as

$$\tilde{a}^\circ = \tilde{a}^+ \frac{\partial}{\partial x} = -(\beta - x)\frac{\partial}{\partial x} - x\frac{\partial^2}{\partial x^2}$$

has the Laguerre polynomials L_n^β with parameter β as eigenfunctions:

$$\tilde{a}^\circ L_n^\beta(x) = nL_n^\beta(x), \quad n \in \mathbb{N}. \tag{3.8}$$

Recall that the basis $\{M, B^-, B^+\}$ of $\mathfrak{sl}_2(\mathbb{R})$, which satisfies

$$[B^-, B^+] = M, \quad [M, B^-] = -2B^-, \quad [M, B^+] = 2B^+,$$

can be constructed as

$$M = \beta + 2\tilde{a}^\circ, \quad B^- = \tilde{a}^- - \tilde{a}^\circ, \quad B^+ = \tilde{a}^+ - \tilde{a}^\circ.$$

For example, for the commutation relation $[M, B^-] = -2B^-$ we note that

$$[M, B^-] = 2[\tilde{a}^\circ, \tilde{a}^-]$$

$$= -2[(\beta - x)\partial + x\partial^2, x\partial]$$

$$= -2(\beta - x)\partial(x\partial) - 2x\partial^2(x\partial) + 2x\partial((\beta - x)\partial + x\partial^2)$$

$$= -2(\beta - x)\partial - 2(\beta - x)x\partial^2 - 2x\partial(\partial + x\partial^2)$$

$$\quad - 2x\partial + 2x(\beta - x)\partial^2 + 2x\partial^2 + 2x^2\partial^3$$

$$= -2(\beta - x)\partial - 2(\beta - x)x\partial^2 - 2x\partial^2 - 2x^2\partial^3 - 2x\partial^2$$

$$\quad - 2x\partial + 2x(\beta - x)\partial^2 + 2x\partial^2 + 2x^2\partial^3$$

$$= -2\beta\partial - 2x\partial^2$$

$$= -2(x\partial + (\beta - x)\partial + x\partial^2)$$

$$= -2B^-.$$

We check that

$$B^- + B^+ = \tilde{a}^- + \tilde{a}^+ - 2\tilde{a}^\circ = x - \beta + 2(\beta - x)\frac{\partial}{\partial x} + 2x\frac{\partial^2}{\partial x^2}$$

and

$$i(B^- - B^+) = i(\tilde{a}^- - \tilde{a}^+) = 2ix\frac{\partial}{\partial x} - i(x - \beta),$$

hence

$$B^- + B^+ + M = \beta + \tilde{a}^- + \tilde{a}^+ = x, \qquad (3.9)$$

identifies with the multiplication by x, therefore it has a gamma distribution with parameter β.

3.3.1 Probability distributions

As a consequence of (3.9) we find that $Q + M$ identifies to the multiplication operator

$$Q + M = \tau,$$

hence $Q + M$ has the gamma distribution with parameter β in the vacuum state e_0 in $L^2_{\mathbb{C}}(\mathbb{R}_+, \gamma_\beta(\tau)d\tau)$. In this way we can also recover the moment generating function

$$\langle e_0, e^{t(B^- + B^+ + M)}e_0\rangle = \int_0^\infty e^{tx}\gamma_\beta(x)dx = \frac{1}{(1 - t)^\beta}, \qquad t < 1,$$

which is the moment generating function of the gamma distribution with parameter $\beta > 0$.

More generally, the distribution (or spectral measure) of $\alpha M + Q$ has been completely determined in [1], depending on the value of $\alpha \in \mathbb{R}$:

– When $\alpha = \pm 1$, $M + Q$ and $M - Q$ have gamma distributions.
– For $|\alpha| < 1$, $Q + \alpha M$ has an absolutely continuous distribution and in particular for $\alpha = 0$, Q and P have continuous binomial distributions.
– When $|\alpha| > 1$, $Q + \alpha M$ has a Pascal distribution, cf. Case (iii) on page 41.

In order to define an inner product on span $\{v_n : n \in \mathbb{N}\}$ such that $M^* = M$ and $(B^-)^* = B^+$, the v_n have to be mutually orthogonal, and their norms have to satisfy the recurrence relation

$$||v_{n+1}||^2 = \langle B^+v_n, v_{n+1}\rangle = \langle v_n, B^-v_{n+1}\rangle = (n + 1)(n + \lambda)||v_n||^2. \quad (3.10)$$

It follows that there exists an inner product on span $\{v_n : n \in \mathbb{N}\}$ such that the lowest weight representation with

$$Me_0 = \lambda e_0, \qquad B^- e_0 = 0,$$

is a $*$-representation, if and only if the coefficients $(n + 1)(n + \lambda)$ in Equation (3.10) are non-negative for all $n \in \mathbb{N}$, *i.e.*, if and only if $\lambda \geq 0$.

For $\lambda = 0$ we get the trivial one-dimensional representation

$$B^+_{(0)} e_0 = B^-_{(0)} e_0 = M_{(0)} e_0 = 0$$

since $||v_1||^2 = 0$, and for $\lambda > 0$ we get

$$
\begin{cases}
B^+_{(\lambda)} e_n = \sqrt{(n+1)(n+\lambda)}\, e_{n+1}, & \text{(3.11a)} \\[2mm]
M_{(\lambda)} e_n = (2n + \lambda) e_n, & \\[2mm]
B^-_{(\lambda)} e_n = \sqrt{n(n+\lambda-1)}\, e_{n-1}, & \text{(3.11b)}
\end{cases}
$$

where $(e_n)_{n \in \mathbb{N}}$ is an orthonormal basis of ℓ^2. Letting

$$Y_{(\lambda)} := B^+_{(\lambda)} + B^-_{(\lambda)} + \lambda M_{(\lambda)}, \qquad \lambda \in \mathbb{R},$$

defines an essentially self-adjoint operator, and $Y_{(\lambda)}$ is a compound Poisson random variable with characteristic exponent

$$\Psi(u) = \langle e_0, \left(e^{iuY_{(\lambda)}} - 1 \right) e_0 \rangle.$$

Our objective in the sequel is to determine the Lévy measure of $Y_{(\lambda)}$, *i.e.*, to determine the measure μ on $\mathbb{R}$ for which we have

$$\Psi(u) = \int_{-\infty}^{\infty} \left(e^{iux} - 1 \right) \mu(dx).$$

This is the spectral measure of $Y_{(\lambda)}$ evaluated in the state $Y \mapsto \langle e_0, Y e_0 \rangle$.

3.3.2 Laguerre polynomial representation

Recall that the polynomial representation defined in Section 1.2 relies on the condition

$$e_n = p_n(Y_{(\lambda)}) e_0, \qquad n \in \mathbb{N},$$

which yields a sequence of orthogonal polynomials with respect to μ, since

$$\int_{-\infty}^{\infty} p_n(x)p_m(x)\mu(dx) = \langle e_0, p_n(Y_{(\lambda)})p_m(Y_{(\lambda)})e_0\rangle$$

$$= \langle p_n(Y_{(\lambda)})e_0, p_m(Y_{(\lambda)})e_0\rangle$$

$$= \delta_{nm},$$

for $n, m \in \mathbb{N}$. Looking at (3.11a)-(3.11b) and the definition of $Y_{(\lambda)}$, we can easily identify the three-term recurrence relation satisfied by the p_n as

$$Y_{(\lambda)}e_n = \sqrt{(n+1)(n+\lambda)}e_{n+1} + \beta(2n+\lambda)e_n + \sqrt{n(n+\lambda-1)}e_{n-1},$$

$n \in \mathbb{N}$. Therefore, Proposition 1.2.1 shows that the rescaled polynomials

$$P_n := \prod_{k=1}^{n} \sqrt{\frac{k}{k+\lambda}} p_n, \qquad n \in \mathbb{N},$$

satisfy the recurrence relation

$$(n+1)P_{n+1} + (2\beta n + \beta\lambda - x)P_n + (n+\lambda-1)P_{n-1} = 0, \qquad (3.12)$$

with initial condition $P_{-1} = 0, P_1 = 1$.
We can distinguish three cases according to the value of β, cf. [1].

i) $|\beta| = 1$: In this case we have, up to rescaling, Laguerre polynomials, *i.e.*,

$$P_n(x) = (-\beta)^n L_n^{(\lambda-1)}(\beta x)$$

where the Laguerre polynomials $L_n^{(\alpha)}$ are defined as in [63, Equation (1.11.1)], with in particular

$$L_n^{(0)}(x) = \sum_{k=0}^{n} \binom{n}{k}(-1)^k \frac{x^k}{k!}, \qquad x \in \mathbb{R}_+.$$

The measure μ can be obtained by normalising the measure of orthogonality of the Laguerre polynomials, it is equal to

$$\mu(dx) = \frac{|x|^{\lambda-1}}{\Gamma(\lambda)} e^{-\beta x} \mathbf{1}_{\beta\mathbb{R}_+} dx.$$

If $\beta = +1$, then this measure is, up to a normalisation parameter, the usual gamma distribution (with parameter λ) of probability theory.

ii) $|\beta| < 1$: In this case we find the Meixner-Pollaczek polynomials after rescaling,

$$P_n(x) = P_n^{(\lambda/2)}\left(\frac{x}{2\sqrt{1-\beta^2}}; \pi - \arccos\beta\right).$$

For the definition of these polynomials see, *e.g.*, [63, Equation (1.7.1)]. For the measure μ we get

$$\mu(dx) = C \exp\left(\frac{(\pi - 2\arccos\beta)x}{2\sqrt{1-\beta^2}}\right)\left|\Gamma\left(\frac{\lambda}{2} + \frac{ix}{2\sqrt{1-\beta^2}}\right)\right|^2 dx,$$

where C has to be chosen such that μ is a probability measure.

iii) $|\beta| > 1$: In this case we get the Meixner polynomials after rescaling,

$$P_n(x) = (-c\,\mathrm{sgn}\beta)^n \prod_{k=1}^{n} \frac{k+\lambda-1}{k} M_n\left(\frac{x}{1/c-c}\,\mathrm{sgn}\beta - \frac{\lambda}{2}; \lambda; c^2\right)$$

where

$$c = |\beta| - \sqrt{\beta^2 - 1}.$$

The definition of these polynomials can be found, *e.g.*, in [63, Equation (1.9.1)]. The density μ is again the measure of orthogonality of the polynomials P_n (normalised to a probability measure). We therefore find the probability distribution

$$\mu = C \sum_{n=0}^{\infty} c^{2n} \frac{(\lambda)_n}{n!} \delta_{x_n}$$

where

$$x_n = \left(n + \frac{\lambda}{2}\right)\left(\frac{1}{c} - c\right)\mathrm{sgn}\beta, \qquad \text{for } n \in \mathbb{N}$$

and

$$\frac{1}{C} = \sum_{n=0}^{\infty} c^{2n} \frac{(\lambda)_n}{n!} = (1 - c^2)^{-\lambda}.$$

Here, $(\lambda)_n$ denotes the Pochhammer symbol

$$(\lambda)_n := \lambda(\lambda+1)\cdots(\lambda+n-1), \qquad n \in \mathbb{N}.$$

The representation (3.11a)-(3.11b) of $\mathfrak{osc}$ on ℓ^2 can be built by defining an orthonormal basis $(e_n)_{n\in\mathbb{N}}$ of $L^2(\mathbb{R}_+, \gamma_\lambda(\tau)d\tau)$ using the Laguerre polynomials, as

$$e_n(x) = (-1)^n \sqrt{\frac{n!\Gamma(\lambda)}{(n+\lambda-1)!}} L_n^{\lambda-1}(x), \qquad n \in \mathbb{N}.$$

The relation

$$B^- L_n^{\lambda-1}(x) = x \frac{\partial}{\partial x} L_n^{\lambda-1}(x) - n L_n^{\lambda-1}(x) = -(n+\lambda-1) L_{n-1}^{\lambda-1}(x),$$

$n \geq 1$, shows that

$$B^- e_n = (-1)^n \sqrt{\frac{n! \Gamma(\lambda)}{(n+\lambda-1)!}} B^- L_n^{\lambda-1}(x)$$

$$= -(n+\lambda-1)(-1)^n \sqrt{\frac{n! \Gamma(\lambda)}{(n+\lambda-1)!}} L_{n-1}^{\lambda-1}(x)$$

$$= -\sqrt{n(n+\lambda-1)}(-1)^n \sqrt{\frac{(n-1)! \Gamma(\lambda)}{(n+\lambda-2)!}} L_{n-1}^{\lambda-1}(x)$$

$$= \sqrt{n(n+\lambda-1)} e_{n-1}(x),$$

$n \geq 1$, and similarly by the recurrence relation

$$(x-\lambda-2n) L_n^{\lambda-1}(x) + (n+\lambda-1) L_{n-1}^{\lambda-1}(x) + (n+1) L_{n+1}^{\beta-1}(x) = 0,$$

(see (3.12)) we have

$$B^+ L_n^\beta(x) = (\tilde{a}^+ - \tilde{a}^\circ) L_n^{\beta-1}(x)$$

$$= (x-\beta) L_n^{\beta-1}(x) - x \frac{\partial}{\partial x} L_n^{\beta-1}(x) - n L_n^{\beta-1}(x)$$

$$= (x-\beta) L_n^{\beta-1}(x) - n L_n^{\beta-1}(x) + (n+\beta-1) L_{n-1}^{\beta-1}(x) - n L_n^{\beta-1}(x)$$

$$= (x-\beta-2n) L_n^{\beta-1}(x) + (n+\beta-1) L_{n-1}^{\beta-1}(x)$$

$$= -(n+1) L_{n+1}^{\beta-1}(x)$$

$$= (n+1)(-1)^n \sqrt{\frac{(n+\beta)!}{(n+1)! \Gamma(\beta)}} e_{n+1}(x)$$

$$= (-1)^n \sqrt{(n+\beta)(n+1)} \sqrt{\frac{(n+\beta-1)!}{n! \Gamma(\beta)}} e_{n+1}(x),$$

hence

$$B^+ e_n(x) = \sqrt{(n+\beta)(n+1)} e_{n+1}(x).$$

3.3.3 The case $\beta = 1$

When $\beta = 1$ the operators $\tilde{a}^\circ$, $\tilde{a}^-$ and $\tilde{a}^+$ satisfy

$$\begin{cases} \tilde{a}^- = -x\dfrac{\partial}{\partial x}, \\[2mm] \tilde{a}^+ = x - 1 - x\dfrac{\partial}{\partial x}, \\[2mm] \tilde{a}^\circ = (x-1)\dfrac{\partial}{\partial x} - x\dfrac{\partial^2}{\partial x^2}, \end{cases} \quad i.e. \quad \begin{cases} \tilde{a}^- L_n(x) = -x\displaystyle\sum_{k=0}^{n-1} L_k(x), \\[2mm] \tilde{a}^+ L_n(x) = nL_n(x) - (n+1)L_{n+1}(x), \\[2mm] \tilde{a}^\circ L_n(x) = nL_n(x), \end{cases}$$

with

$$\tilde{a}^+ + \tilde{a}^- = 1 - x,$$

and the commutation relations

$$[\tilde{a}^+, \tilde{a}^-] = -x, \qquad [\tilde{a}^\circ, \tilde{a}^+] = \tilde{a}^\circ + \tilde{a}^+, \qquad [\tilde{a}^-, \tilde{a}^\circ] = \tilde{a}^\circ + \tilde{a}^-.$$

We have noted earlier that $i(\tilde{a}^- - \tilde{a}^+)$ has a continuous binomial distribution (or spectral measure) in the vacuum state **1**, with hyperbolic cosine density $(2\cosh \pi\xi/2)^{-1}$, in relation to a representation of the subgroup of $\mathfrak{sl}_2(\mathbb{R})$ made of upper-triangular matrices.

Next, we also notice that although this type of distribution can be studied for every value of $\beta > 0$ in the above framework, the construction can also be specialised based on Lemma 2.4.2 for half-integer values of β using the annihilation and creation operators α_x^-, α_y^-, α_x^+, α_y^+ on the two-dimensional boson Fock space $\Gamma(\mathbb{C}e_1 \oplus \mathbb{C}e_2)$.

Defining the operator L as $L = -\tilde{Q} - 2\tilde{a}^\circ$ with

$$\tilde{Q} = 1 - x, \qquad \tilde{P} = -i(2x\partial_x + 1 - x),$$

we find that

$$[\tilde{a}^\circ, \tilde{P}] = iL, \qquad [\tilde{a}^\circ, L] = i\tilde{P}, \qquad [L, \tilde{P}] = 2iM,$$

and

$$[\tilde{P}, \tilde{Q}] = 2ix, \qquad [\tilde{a}^\circ, \tilde{Q}] = -i\tilde{P}, \tag{3.13}$$

hence

$$\left\{ \frac{i}{2}L, \ -\frac{i}{2}\tilde{P}, \ \frac{i}{2}M \right\}$$

generates the unitary representation

$$\left[-\frac{\tilde{P}}{2},\frac{M}{2}\right] = i\frac{L}{2}, \quad \left[\frac{L}{2},\frac{M}{2}\right] = i\frac{\tilde{P}}{2}, \quad \left[\frac{L}{2},-\frac{\tilde{P}}{2}\right] = -i\frac{M}{2},$$

also called the Segal–Shale–Weil representation of $\mathfrak{sl}_2(\mathbb{R})$. Indeed, the above relations can be proved by ordinary differential calculus as

$$(1 - x + x\partial_x)(-x\partial_x) - (-x\partial_x)(1 - x + x\partial_x)$$
$$= -(1-x)x\partial_x - x\partial_x - x^2\partial_x^2 + (1-x)x\partial_x - x + x^2\partial_x^2 + x\partial_x$$
$$= -x,$$

and

$$(-(1-x)\partial_x - x\partial_x^2)(1 - x + x\partial_x) - (1 - x + x\partial_x)(-(1-x)\partial_x - x\partial_x^2)$$
$$= -(1-x)^2\partial_x + 1 - x - (1-x)x\partial_x^2 - (1-x)\partial_x - x(-2\partial_x + (1-x)\partial_x^2)$$
$$\quad - x(2\partial_x^2 + x\partial_x^3) - (-(1-x)^2\partial_x - x(1-x)\partial_x^2 - x(1-x)\partial_x^2 + x\partial_x - x^2\partial_x^3 - x\partial_x^2)$$
$$= -(1-x)\partial_x - x\partial_x^2 + (1-x) + x\partial_x,$$

and

$$- x\partial_x(-(1-x)\partial_x - x\partial_x^2) - (-(1-x)\partial_x - x\partial_x^2)(-x\partial_x)$$
$$= x(1-x)\partial_x^2 - x\partial_x + x\partial_x^2 + x^2\partial_x^3 - ((1-x)\partial_x + (1-x)x\partial_x^2 + 2x\partial_x^2 + x^2\partial_x^3)$$
$$= -(1-x)\partial_x - x\partial_x^2 - x\partial_x.$$

3.3.4 Adjoint action

The next lemma will be used for the Girsanov theorem in Chapter 10.

Lemma 3.3.1 *Letting* $Y = B^- - B^+$, *the adjoint action of* $g_t := e^{tY}$ *on* X_β *is given by*

$$e^{tY/2}X_\beta e^{-tY/2} = e^{t(\mathrm{ad}Y)/2}X_\beta = \big(\cosh(t) + \beta\sinh(t)\big)X_{\gamma(\beta,t)},$$

where

$$\gamma(\beta,t) = \frac{\beta\cosh(t) + \sinh(t)}{\cosh(t) + \beta\sinh(t)}.$$

See Section 4.4 of [45] for a proof of Lemma 3.3.1.

Exercises

Exercise 3.1 Define the operators b^- and b^+ by

$$b^- = -ia^-, \qquad b^+ = ia^+.$$

1. Show that b^- and b^+ satisfy the same commutation relation

$$[b^-, b^+] = [-ia^-, ia^+] = [a^-, a^+] = \sigma^2 I_{\mathfrak{h}}$$

   as a^- and a^+, with the condition $b^- e_0 = 0$.
2. Show that we have the duality relation $\langle b^- u, v \rangle_{\mathfrak{h}} = \langle u, b^+ v \rangle_{\mathfrak{h}}$, $u, v \in \mathfrak{h}$.
3. Show that $P = i(a^+ - a^-)$ also has a Gaussian distribution in the fundamental state e_0.

Exercise 3.2   Moments of the Poisson distribution.

The goal of this exercise is to recover the first moments of the Poisson distribution from the commutation relations of the oscillator algebra and the relation $E e_0 = \lambda e_0$.

   In particular, show that $\langle X e_0, e_0 \rangle = \lambda$, $\langle X^2 e_0, e_0 \rangle = \lambda + \lambda^2$, and $\langle X^3 e_0, e_0 \rangle = \lambda + 3\lambda^2 + \lambda^3$.

Exercise 3.3   Classical gamma moments.

Consider a (classical) random variable X having the gamma distribution with shape parameter $\alpha > 0$, probability density function

$$\varphi_X(x) := \frac{x^{\alpha-1}}{\Gamma(\alpha)} e^{-\alpha x}, \qquad x > 0,$$

and moment generating function

$$\mathbf{E}[e^{tX}] = (1-t)^{-\alpha}, \qquad t < 1.$$

Show that the moment of order $n \in \mathbb{N}$ of X is given by

$$\mathbf{E}[X^n] = \underbrace{\alpha(\alpha+1)\cdots(\alpha+n-1)}_{n \text{ times}}. \qquad (3.14)$$

Hint: you may use the relation

$$\mathbf{E}[X^n] = \frac{\partial^n}{\partial t^n} \mathbf{E}[e^{tX}]_{|t=0}.$$

Exercise 3.4   Gamma moments on $\mathfrak{sl}_2(\mathbb{R})$.

Consider the Lie algebra $\mathfrak{sl}_2(\mathbb{R})$ with basis B^-, B^+, M, and the commutation relations

$$[B^-, B^+] = M, \qquad [M, B^-] = -2B^-, \qquad [M, B^+] = 2B^+,$$

under the involution

$$(B^-)^* = B^+, \qquad M^* = M. \qquad (3.15)$$

Next, consider a (Hilbert) space $\mathfrak{h}$ with inner product $\langle \cdot, \cdot \rangle$ and a representation of B^-, B^+, M on $\mathfrak{h}$ such that

$$B^- e_0 = 0, \qquad M e_0 = \alpha e_0,$$

for a certain unit vector $e_0 \in \mathfrak{h}$ such that $\langle e_0, e_0 \rangle = 1$. Recall that the involution (3.15) reads

$$\langle B^- u, v \rangle = \langle u, B^+ v \rangle \quad \text{and} \quad \langle M u, v \rangle = \langle u, M v \rangle, \qquad u, v \in \mathfrak{h}.$$

The goal of this question is to show that the first three moments of $B^- + B^+ + M$ in the state e_0 coincide with the moments (3.14) of a gamma distribution with shape parameter $\alpha > 0$ in the state e_0, *i.e.*,

1. for $n = 1$, show that

$$\langle e_0, (B^- + B^+ + M)e_0 \rangle = \mathbf{E}[X],$$

2. for $n = 2$, show that

$$\langle e_0, (B^- + B^+ + M)^2 e_0 \rangle = \mathbf{E}\left[X^2\right],$$

3. for $n = 3$, show that

$$\langle e_0, (B^- + B^+ + M)^3 e_0 \rangle = \mathbf{E}\left[X^3\right],$$

i.e., show that we have

$$\langle e_0, (B^- + B^+ + M)^n e_0 \rangle = \mathbf{E}[X^n], \qquad n = 0, 1, 2, 3,$$

where $\mathbf{E}[X^n]$ is given by the relation (3.14) of Question 1.

4

Noncommutative random variables

In these days the angel of topology and the devil of abstract algebra
fight for the soul of each individual mathematical domain.
(H. Weyl, "Invariants", Duke Mathematical Journal, 1939.)

Starting with this chapter we move from particular examples to the more general framework of noncommutative random variables, with an introduction to the basic concept of noncommutative probability space. In comparison with the previous chapters which were mostly concerned with distinguished families of distributions, we will see here how to construct arbitrary distributions in a noncommutative setting.

4.1 Classical probability spaces

Following the description given by K.R. Parthasarathy in Reference [87], the notion of a real-valued observable has three faces: (i) a spectral measure on the line, (ii) a self-adjoint operator in a Hilbert space, and (iii) a unitary representation of the real line as an additive group. The equivalence of these three descriptions is a consequence of von Neumann's spectral theorem for a, not necessarily bounded, self-adjoint operator and Stone's theorem on the infinitesimal generator of a one-parameter unitary group in a Hilbert space.

Before switching to this noncommutative picture we recall the framework of classical probability.

Definition 4.1.1 *A "classical" probability space is a triple* $(\Omega, \mathcal{F}, \mathbb{P})$ *where*

- Ω *is a set, the* sample space, *the set of all possible outcomes.*
- $\mathcal{F} \subseteq \mathcal{P}(\Omega)$ *is the* σ*-algebra of events.*

- $\mathbb{P}\colon \mathcal{F} \longrightarrow [0,1]$ *is a probability measure that assigns to each event its* probability.

This description of randomness is based on the idea that randomness is due to a lack of information: if we knew which $\omega \in \Omega$ is realised, then the randomness disappears.

Recall that to any real-valued random variable $X\colon (\Omega, \mathcal{F}, \mathbb{P}) \longrightarrow (\mathbb{R}, \mathcal{B}(\mathbb{R}))$, we can associate a probability measure $\mathbb{P}_X$ on $\mathbb{R}$ by

$$\mathbb{P}_X(B) := \mathbb{P}\big(X^{-1}(B)\big)$$

for $B \in \mathcal{B}(\mathbb{R})$ or

$$\int_{\mathbb{R}} f d\mathbb{P}_X = \int_{\Omega} f \circ X d\mathbb{P}$$

for $f\colon \mathbb{R} \longrightarrow \mathbb{R}$, a bounded measurable function. The probability measure $\mathbb{P}_X$ is called the *distribution* of X with respect to $\mathbb{P}$.

This construction is not limited to single random variables, as we can also define the *joint distribution* of an n-tuple $= (X_1, \ldots, X_n)$ of real random variables by

$$\mathbb{P}_{(X_1, \ldots, X_n)}(B) := \mathbb{P}(X^{-1}(B))$$

for $B \in \mathcal{B}(\mathbb{R}^n)$. We shall see that the distribution of a single noncommutative random variable can be defined similarly as in the "classical" (or commutative) case, but that the joint distribution of noncommuting random variables requires a more careful discussion.

Still quoting Reference [87], "real valued random variables on a classical probability space $(\Omega, \mathcal{F}, \mathbb{P})$ when viewed as selfadjoint multiplication operators in the Hilbert space $L^2(\Omega)$ are special examples in the quantum description. This suggests the possibility of developing a theory of quantum probability within the framework of operators and group representations in a Hilbert space."

4.2 Noncommutative probability spaces

Next is the most fundamental definition in quantum (or noncommutative) probability.

Definition 4.2.1 *A* quantum probability space *is a pair* $(\mathcal{A}, \Phi)$ *consisting of a unital associative *-algebra* $\mathcal{A}$ *and a positive normalised functional (called a* state*)* $\Phi\colon \mathcal{A} \longrightarrow \mathbb{C}$.

By "*unital associative *-algebra*" we mean that $\mathcal{A}$ is a vector space over the field of complex numbers $\mathbb{C}$, equipped with an associative bilinear multiplication

$$m\colon \mathcal{A} \times \mathcal{A} \longrightarrow \mathcal{A},$$
$$(a, b) \longmapsto m(a, b) = ab,$$

with an element $I_{\mathcal{A}}$ (called the *unit* of $\mathcal{A}$) such that $aI_{\mathcal{A}} = I_{\mathcal{A}}a = a$ for all $a \in \mathcal{A}$, and a map

$$*\colon \mathcal{A} \longrightarrow \mathcal{A},$$
$$a \longmapsto a^*,$$

(called an *involution*) such that

$$\begin{cases} (a^*)^* = a & (* \text{ is } \textit{involutive}), \\ (\lambda a + \mu b)^* = \bar{\lambda} a^* + \bar{\mu} b^* & (* \text{ is } \textit{conjugate linear}), \\ (ab)^* = b^* a^* & (* \text{ is } \textit{anti-multiplicative}), \end{cases}$$

for all $a, b \in \mathcal{A}$, $\lambda, \mu \in \mathbb{C}$. From now on, by "*algebra*" we will mean a unital associative *-algebra. By a *positive normalised functional* or *state* on an algebra we mean a map

$$\Phi\colon \mathcal{A} \longrightarrow \mathbb{C},$$
$$a \longmapsto \Phi(a),$$

such that

$$\begin{cases} \Phi(\lambda a + \mu b) = \lambda \Phi(a) + \mu \Phi(b) & (\Phi \text{ is } \textit{linear}), \\ \Phi(a^* a) \geq 0 & (\Phi \text{ is } \textit{positive}), \\ \Phi(I_{\mathcal{A}}) = 1 & (\Phi \text{ is } \textit{normalised}), \end{cases}$$

for all $a, b \in \mathcal{A}$, $\lambda, \mu \in \mathbb{C}$.

First, we note that the "classical" probability spaces described in Section 4.1 can be viewed as special cases of quantum probability spaces.

Example 4.2.2 (Classical $\subseteq$ Quantum) To a classical probability space $(\Omega, \mathcal{F}, \mathbb{P})$ we can associate a quantum probability space $(\mathcal{A}, \Phi)$ by taking

- $\mathcal{A} := L^\infty(\Omega, \mathcal{F}, \mathbb{P})$, the algebra of bounded measurable functions $f\colon \Omega \longrightarrow \mathbb{C}$, called the algebra of *random variables*. The involution is given by pointwise complex conjugation, $f^* = \bar{f}$, where $\bar{f}(\omega) = \overline{f(\omega)}$ for $\omega \in \Omega$.
- $\Phi\colon \mathcal{A} \ni f \longmapsto E(f) = \int_\Omega f d\mathbb{P}$, which assigns to each random variable its expected value.

4.2.1 Noncommutative examples

Other, genuinely noncommutative quantum probability spaces are motivated by quantum mechanics.

Example 4.2.3 (Quantum mechanics)   Let $\mathfrak{h}$ be a Hilbert space, with a unit vector ψ. Then the quantum probability space associated to $(\mathfrak{h}, \psi)$ is given by

- $\mathcal{A} = B(\mathfrak{h})$, the algebra bounded linear operators $X : \mathfrak{h} \longrightarrow \mathfrak{h}$. Self-adjoint (or normal) operators are called *quantum random variables*.
- $\Phi : B(\mathfrak{h}) \ni X \longmapsto \Phi(X) = \langle \psi, X\psi \rangle$.

Suppose now that $\mathfrak{h}$ is a finite dimensional complex Hilbert space, *i.e.*, $\mathfrak{h} = \mathbb{C}^n$ with the inner product

$$\langle x, y \rangle := \sum_{k=1}^{n} \overline{x}_k y_k$$

and the norm

$$\|x\| = \sqrt{\langle x, x \rangle}, \qquad x, y \in \mathbb{C}^n.$$

A linear operator $X \in B(\mathbb{C}^n)$ is simply a linear map

$$X : \mathbb{C}^n \longrightarrow \mathbb{C}^n,$$

or equivalently a matrix $X = (x_{jk})_{1 \le j,k \le n} \in M_n(\mathbb{C})$ that acts on a vector $v = (v_k)_{1 \le k \le n} \in \mathbb{C}^n$ by matrix multiplication,

$$Xv = \left(\sum_{k=1}^{n} x_{jk} v_k \right)_{j=1,\ldots,n},$$

i.e.,

$$\begin{bmatrix} x_{11} & x_{12} & \cdots & x_{1n} \\ x_{21} & x_{22} & \cdots & x_{2n} \\ \vdots & \vdots & \ddots & \vdots \\ x_{n1} & x_{n2} & \cdots & x_{nn} \end{bmatrix} \begin{bmatrix} v_1 \\ v_2 \\ \vdots \\ v_n \end{bmatrix} = \begin{bmatrix} x_{11}v_1 + x_{12}v_2 + \cdots + x_{1n}v_n \\ x_{21}v_1 + x_{22}v_2 + \cdots + x_{2n}v_n \\ \vdots \\ x_{n1}v_1 + x_{n2}v_2 + \cdots + x_{nn}v_n \end{bmatrix}.$$

The involution on $\mathcal{A} = B(\mathbb{C}^n) = M_n(\mathbb{C})$ is defined by complex conjugation and transposition, *i.e.*, $X^* = \overline{X}^T$, or equivalently by

$$\begin{bmatrix} x_{11} & x_{12} & \cdots & x_{1n} \\ x_{21} & x_{22} & \cdots & x_{2n} \\ \vdots & \vdots & \ddots & \vdots \\ x_{n1} & x_{n2} & \cdots & x_{nn} \end{bmatrix}^* = \begin{bmatrix} \overline{x}_{11} & \overline{x}_{21} & \cdots & \overline{x}_{n1} \\ \overline{x}_{12} & \overline{x}_{22} & \cdots & \overline{x}_{n2} \\ \vdots & \vdots & \ddots & \vdots \\ \overline{x}_{1n} & \overline{x}_{2n} & \cdots & \overline{x}_{nn} \end{bmatrix},$$

where $\bar{x}_{ij} = \overline{x_{ij}}$, $1 \le i,j \le n$. For any unit vector $\psi \in \mathfrak{h}$ we can define a state $\Phi: M_n(\mathbb{C}) \longrightarrow \mathbb{C}$ by

$$\Phi(X) = \langle \psi, X\psi \rangle, \qquad X \in M_n(\mathbb{C}).$$

The following example shows how to construct any Bernoulli distribution on the algebra $\mathcal{A} = M_2(\mathbb{C})$ of 2×2 complex matrices.

Example 4.2.4 $(M_2(\mathbb{C}))$ Let us consider $\mathcal{A} = M_2(\mathbb{C})$ with the state

$$\Phi(B) = \left\langle \begin{bmatrix} 1 \\ 0 \end{bmatrix}, B \begin{bmatrix} 1 \\ 0 \end{bmatrix} \right\rangle$$

for $B \in M_2(\mathbb{C})$, and the quantum random variable

$$X = \begin{bmatrix} a & b \\ \bar{b} & c \end{bmatrix}$$

with $a, c \in \mathbb{R}$ and $b \in \mathbb{C}$. Then the first three moments can be computed as follows:

$$\Phi(X) = \left\langle \begin{bmatrix} 1 \\ 0 \end{bmatrix}, \begin{bmatrix} a & b \\ \bar{b} & c \end{bmatrix} \begin{bmatrix} 1 \\ 0 \end{bmatrix} \right\rangle = a,$$

$$\Phi(X^2) = \left\langle \begin{bmatrix} 1 \\ 0 \end{bmatrix}, \begin{bmatrix} a & b \\ \bar{b} & c \end{bmatrix}^2 \begin{bmatrix} 1 \\ 0 \end{bmatrix} \right\rangle$$

$$= \left\langle \begin{bmatrix} 1 \\ 0 \end{bmatrix}, \begin{bmatrix} a^2 + |b|^2 & b(a+c) \\ \bar{b}(a+c) & c^2 + |b|^2 \end{bmatrix} \begin{bmatrix} 1 \\ 0 \end{bmatrix} \right\rangle$$

$$= a^2 + |b|^2,$$

$$\Phi(X^3) = a^3 + (2a+c)|b|^2, \tag{4.1}$$

$$\vdots$$

A natural question is then:

Can we find a general formula for the moments $\Phi(X^k)$ of X?

The answer is given by

$$\Phi(X^n) = \int x^n \mu_X(dx), \qquad n \in \mathbb{N}, \tag{4.2}$$

where μ_X is the probability measure on $\mathbb{R}$ defined by

$$\mu_X(dx) := a_1 \delta_{\lambda_1}(dx) + a_2 \delta_{\lambda_2}(dx), \tag{4.3}$$

where

$$a_1 = \frac{a - c + \sqrt{(a - c)^2 + |b|^2}}{2\sqrt{(a - c)^2 + |b|^2}}, \quad a_2 = \frac{c - a + \sqrt{(a - c)^2 + |b|^2}}{2\sqrt{(a - c)^2 + |b|^2}}. \quad (4.4)$$

and

$$\lambda_1 = \frac{a + c + \sqrt{(a - c)^2 + |b|^2}}{2}, \quad \lambda_2 = \frac{a + c - \sqrt{(a - c)^2 + |b|^2}}{2}. \quad (4.5)$$

Proof. The characteristic polynomial $P_X(z)$ of X is given by

$$P_X(z) = \det(zI - X)$$

$$= \det \begin{bmatrix} z - a & -b \\ -\bar{b} & z - c \end{bmatrix}$$

$$= (z - a)(z - c) - |b|^2$$

$$= z^2 - (a + c) + ca - |b|^2, \quad z \in \mathbb{C}.$$

We note that the zeroes λ_1, λ_2 of the characteristic polynomial P_X of X are given by (4.5), and they are real. Hence for any $z \in \mathbb{C}$ with $\Im z \neq 0$, we have $\det(zI - X) \neq 0$ and we can compute the inverse

$$R_X(z) = (zI - X)^{-1}$$

of $zI - X$, also called the *resolvent* of X, as

$$(zI - X)^{-1} = \frac{1}{z^2 - (a + c) + ca - |b|^2} \begin{bmatrix} z - c & -b \\ -\bar{b} & z - a \end{bmatrix}.$$

The expectation

$$\Phi(R_X(z)) = \left\langle \begin{bmatrix} 1 \\ 0 \end{bmatrix}, R_X(z) \begin{bmatrix} 1 \\ 0 \end{bmatrix} \right\rangle = \frac{z - c}{z^2 - (a + c) + ca - |b|^2}$$

of the resolvent in the state Φ can be written by partial fraction decomposition as follows:

$$\Phi(R_X(z)) = \frac{z - c}{(z - \lambda_1)(z - \lambda_2)} = \frac{a_1}{z - \lambda_1} + \frac{a_2}{z - \lambda_2}$$

with

$$a_1 = \lim_{z \to \lambda_1} (z - \lambda_1)\Phi(R_X(z))$$

$$= \frac{\lambda_1 - c}{\lambda_1 - \lambda_2}$$

$$= \frac{a - c + \sqrt{(a - c)^2 + |b|^2}}{2\sqrt{(a - c)^2 + |b|^2}},$$

and

$$a_2 = \lim_{z \to \lambda_2} (z - \lambda_2) \Phi\big(R_X(z)\big) = \frac{c - a + \sqrt{(a-c)^2 + |b|^2}}{2\sqrt{(a-c)^2 + |b|^2}},$$

as in (4.4).

Note that we have $0 \le a_1, a_2$ and $a_1 + a_2 = 1$, so that $\mu_X(dx)$ defined in (4.3) is indeed a probability measure on $\mathbb{R}$.

We have shown that the expectation of $R_X(z)$ satisfies

$$\Phi\big(R_X(z)\big) = \frac{z - c}{z^2 - (a+c) + ca - |b|^2}$$

$$= \frac{a_1}{z - \lambda_1} + \frac{a_2}{z - \lambda_2} = \int_{-\infty}^{\infty} \frac{1}{z - x} \mu_X(dx)$$

for $z \in \mathbb{C} \backslash \mathbb{R}$. From the geometric series

$$R_X(z) = (zI - X)^{-1} = \frac{1}{z}\left(I - \frac{X}{z}\right) = \sum_{n=0}^{\infty} \frac{X^n}{z^{n+1}},$$

and

$$\int \frac{1}{z - x} = \sum_{n=0}^{\infty} \frac{x^n}{z^{n+1}},$$

which converge uniformly for z sufficiently large, we get

$$\sum_{n=0}^{\infty} \Phi(X^n) z^{-n-1} = \sum_{n=0}^{\infty} \int x^n \mu_X(dx)$$

for z sufficiently large, and finally we check that (4.2) holds for all $n \in \mathbb{N}$. □

Remark 4.2.5 *The function $G_\mu : \mathbb{C} \backslash \mathbb{R} \longrightarrow \mathbb{C}$ defined by*

$$G_\mu(z) = \int_I \frac{1}{z - x} \mu(dx),$$

where I is an interval of the real line $\mathbb{R}$, is called the Cauchy–Stieltjes *transform of μ, cf. the appendix Section A.4.*

Definition 4.2.6 *Let $(\mathcal{A}, \Phi)$ be a quantum probability space and $X \in \mathcal{A}$ be a self-adjoint quantum random variable. Then we call a probability measure μ on $\mathbb{R}$ the* law *(or* distribution*) of X with respect to the state Φ if*

$$\Phi(X^k) = \int_{-\infty}^{\infty} x^k \mu(dx)$$

for all $k \in \mathbb{N}$.

Note that the law of a quantum variable X is defined with respect to the state Φ. If μ is the law of X in the state Φ, we shall write

$$\mathcal{L}_\Phi(X) = \mu.$$

In general, the law of a quantum random variable might not be unique (this is related to the *moment uniqueness (or determinacy) problem*, see Reference [4]).

In the previous example we have determined the law of $X = \begin{bmatrix} a & b \\ \bar{b} & c \end{bmatrix}$ with respect to the state Φ given by the vector $\begin{bmatrix} 1 \\ 0 \end{bmatrix}$, we found

$$\mathcal{L}_\Phi\left(\begin{bmatrix} a & b \\ \bar{b} & c \end{bmatrix}\right) = a_1\delta_{\lambda_1} + a_2\delta_{\lambda_2},$$

where a_1 and a_2 are given by (4.4).

We will now study the more general case of the distribution of Hermitian matrices. For this we shall use the spectral theorem and the functional calculus presented in Section 4.3.

Theorem 4.2.7 (Spectral theorem)   *A Hermitian linear map $X \in B(\mathbb{C}^n)$ can be written in the form*

$$X = \sum_{\lambda \in \sigma(X)} \lambda E_\lambda$$

where $\sigma(X)$ denotes the spectrum of X (= set of eigenvalues) and E_λ the orthogonal projection onto the eigenspace of X associated to the eigenvalue λ,

$$V(X, \lambda) = \{v \in \mathbb{C}^n : Xv = \lambda v\}.$$

4.3 Noncommutative random variables

Random variables over a quantum probability space $(\mathcal{A}, \Phi)$ can be defined in several ways. Hermitian elements $X \in \mathcal{A}$ can be considered noncommutative real-valued random variables, since the distribution of X in the state Φ defines a probability measure on $\mathbb{R}$. This is true in particular when $(\mathcal{A}, \Phi)$ is based on a classical probability space.

In general, we will need a more flexible notion that generalises the aforementioned setting. Recall that a random variable on a classical probability space $(\Omega, \mathcal{F}, \mathbb{P})$ with values in a measurable space $(M, \mathcal{M})$ is a measurable map

$$X : (\Omega, \mathcal{F}, \mathbb{P}) \longrightarrow (M, \mathcal{M}).$$

Such a map X induces a *-algebra homomorphism

$$j_X : L^\infty(M) \longrightarrow L^\infty(\Omega)$$

by the composition

$$j_X(f) = f \circ X, \quad f \in L^\infty(M).$$

In classical probability the composition $f \circ X$ is usually denoted by $f(X)$ by letting the function f act on the "variable" X. In quantum probability, on the other hand, we opt for the opposite (or dual) point of view, by letting the random variable X act on the function algebra $L^\infty(M)$. This leads to the following definition.

Definition 4.3.1 *A quantum (or noncommutative) random variable* on an *algebra $\mathcal{B}$ over a quantum probability space $(\mathcal{A}, \Phi)$ is a unital *-algebra homomorphism:*

$$j : \mathcal{B} \longrightarrow (\mathcal{A}, \Phi).$$

Note that by an "algebra" we actually refer to a "unital *-algebra". By unital *-algebra homomorphism we mean that j preserves the algebraic structure of $\mathcal{B}$, *i.e.*, j is

i) *linear*: we have

$$j(\lambda a + \mu b) = \lambda j(a) + \mu j(b), \qquad a, b \in \mathcal{B},$$

$\lambda, \mu \in \mathbb{C}$;

ii) *multiplicative*: we have

$$j(ab) = j(a)j(b), \qquad a, b \in \mathcal{B};$$

iii) *unit-preserving*: we have $j(I_\mathcal{B}) = I_\mathcal{A}$;
iv) *involutive*: we have $j(b^*) = j(b)^*$ for $b \in \mathcal{B}$.

Note that this definition extends the construction of quantum random variable given earlier when $(\mathcal{A}, \Phi)$ is based on a classical probability space. If $X \in \mathcal{A}$ is Hermitian, then we can define a quantum random variable j_X on the algebra $\mathbb{C}[x]$ of polynomials in a Hermitian variable x by setting

$$j_X(P(x)) = P(X), \qquad P \in \mathbb{C}[x].$$

Definition 4.3.2 *The state* $\Phi_j := \Phi \circ j$ *induced on* $\mathcal{B}$ *by*

$$j : \mathcal{B} \longrightarrow (\mathcal{A}, \Phi)$$

is called the distribution *(or law) of j with respect to* Φ.

When $\mathcal{B}$ is replaced by a real Lie algebra $\mathfrak{g}$, we have to modify the definition of a random variable.

Definition 4.3.3 *A* quantum (or noncommutative) random variable *on a real Lie algebra* $\mathfrak{g}$ *over a quantum probability space* $(\mathcal{A}, \Phi)$ *is a Hermitian Lie algebra homomorphism*

$$j : \mathfrak{g} \longrightarrow (\mathcal{A}, \Phi).$$

By a "Hermitian Lie algebra homomorphism" we mean that j has the following properties:

i) *Linearity*: we have

$$j(\lambda X + \mu Y) = \lambda j(X) + \mu j(Y), \qquad X, Y \in \mathfrak{g},$$

$\lambda, \mu \in \mathbb{R}$;

ii) *Lie algebra homomorphism*: we have

$$j([X, Y]) = j(X)j(Y) - j(Y)j(X), \qquad X, Y \in \mathfrak{g};$$

iii) *Hermitianity*: we have $j(X)^* = -j(X), X \in \mathfrak{g}$.

4.3.1 Where are noncommutative random variables valued?

We have now constructed various random variables and probability distributions from real Lie algebras. When restricted to a single Hermitian element, or to the commutative algebra it generates, noncommutative random variables are distributed over the real line, so we could think of these restrictions as real-valued random variables. But where do the random variables themselves really take their values?

In Definition 4.3.1 we have seen that the notion of quantum random variable

$$j : \mathcal{B} \longrightarrow (\mathcal{A}, \Phi)$$

extends the notion of classical X-valued random variable constructed on a classical probability space underlying $(\mathcal{A}, \Phi)$ and taking values in a space X, with $\mathcal{B}$ an algebra of functions on X. Since most of our Lie algebras are noncommutative, they cannot be genuine function algebras. However, the elements of a Lie algebra $\mathfrak{g}$ can be regarded as functions on its dual $\mathfrak{g}^*$, so that a random variable of the form

$$j : \mathfrak{g} \longrightarrow (\mathcal{A}, \Phi)$$

can be viewed as taking values in $\mathfrak{g}^*$. In that sense, the terminology "probability on duals of real Lie algebras" better reflects the dualisation which is implicit in the definition of quantum probability spaces and quantum random variables. For simplicity and convenience we nonetheless work with the less precise terminology of "probability on real Lie algebras".

4.4 Functional calculus for Hermitian matrices

Let $n \in \mathbb{N}$ and let $A \in M_n(\mathbb{C})$ a Hermitian matrix, *i.e.*, we have $A^* = A$, and let $f : \mathbb{R} \longrightarrow \mathbb{R}$ be any function. Then there are several equivalent methods to define $f(A)$, cf. *e.g.*, Exercise 4.1.

Example 4.4.1 Let $P \in B(\mathbb{C}^n)$ be an orthogonal projection, *i.e.*, P satisfies

$$P^2 = P = P^*.$$

If P is a non-trivial orthogonal projection, *i.e.*, $P \neq 0$ and $P \neq I$, then P has two eigenvalues $\lambda_1 = 0$ and $\lambda_2 = 1$, with the eigenspaces

$$\begin{cases} V(P, 0) = \ker(P) = \{v \in \mathbb{C}^n : Pv = 0\}, \\ V(P, 1) = \mathrm{range}(P) = \{v \in \mathbb{C}^n : \exists\, w \in \mathbb{C}^n \text{ such that } v = Pw\}. \end{cases}$$

The operator $f(P)$ depends only on the values of f at $\lambda_1 = 0$ and $\lambda_2 = 1$, and we have

$$f(P) = f(0)(I - P) + f(1)P,$$

since P is the orthogonal projection onto $V(P, 1) = \mathrm{range}(P)$ and $I - P$ is the orthogonal projection onto $V(P, 0) = \ker(P)$.

Let us now describe the law of a Hermitian matrix with respect to an arbitrary state.

Theorem 4.4.2 *Let Φ be a state on $M_n(\mathbb{C})$ and let $X \in M_n(\mathbb{C})$ be a Hermitian matrix with spectral decomposition*

$$X = \sum_{\lambda \in \sigma(X)} \lambda E_\lambda$$

Then the law of X with respect to Φ is given by

$$\mathcal{L}_\Phi(X) = \sum_{\lambda \in \sigma(X)} \Phi(E_\lambda)\delta_\lambda,$$

where $\sigma(X)$ is the spectrum of X.

Proof: For any function $f: \mathbb{R} \longrightarrow \mathbb{R}$ we have

$$f(X) = \sum_{\lambda \in \sigma(X)} f(\lambda)E_\lambda,$$

and therefore, by linearity of $\Phi: M_n(\mathbb{C}) \longrightarrow \mathbb{C}$,

$$\Phi\big(f(X)\big) = \sum_{\lambda \in \sigma(X)} f(\lambda)\Phi(E_\lambda) = \int f(x)\mu(dx)$$

with $\mu = \sum_{\lambda \in \sigma(X)} \Phi(E_\lambda)\delta_\lambda$. Since this is true in particular for the functions $f(x) = x^k$ with $k \in \mathbb{N}$, we can conclude that the law $\mathcal{L}_\Phi(X)$ of X in the state Φ is given by

$$\mathcal{L}_\Phi(X) = \sum_{\lambda \in \sigma(X)} \Phi(E_\lambda)\delta_\lambda.$$

$\square$

Example 4.4.3

a) If P is a non-trivial orthogonal projection, then $\sigma(X) = \{0, 1\}$ and we find

$$\mathcal{L}_\Phi(P) = \Phi(P)\delta_1 + \Phi(I - P)\delta_0.$$

Since in this sense orthogonal projections can only take the values 0 and 1, they can be considered as the quantum probabilistic analogue of *events*, *i.e.*, random experiments that have only two possible outcomes – "yes" and "no" (or "true" and "false").

b) Consider now the case where Φ is a vector state, *i.e.*,

$$\Phi(B) = \langle \psi, B\psi \rangle, \qquad B \in M_n(\mathbb{C}),$$

for some unit vector $\psi \in \mathbb{C}^n$. Let

$$\sum_{\lambda \in \sigma(X)} \lambda E_\lambda$$

be a quantum random variable in $(M_n(\mathbb{C}), \Phi)$, then the weights $\Phi(E_\lambda)$ in the law of X with respect to Φ,

$$\mathcal{L}_\Phi(X) = \sum_{\lambda \in \sigma(X)} \Phi(E_\lambda)\delta_\lambda,$$

are given by

$$\Phi(E_\lambda) = \langle \psi, E_\lambda \psi \rangle = ||E_\lambda \psi||^2,$$

i.e., the probability with which X takes a value λ with respect to the state associated to ψ is exactly the square of the length of the projection of ψ onto the eigenspace $V(X, \lambda)$,

$$\mathcal{L}_\Phi(X) = \sum_{\lambda \in \sigma(X)} ||E_\lambda \psi||^2 \delta_\lambda.$$

4.5 The Lie algebra *so*(3)

In this section we consider the real Lie algebra $so(3)$ with basis consisting of the three anti-Hermitian elements ξ_1, ξ_2, ξ_3 defined as

$$\xi_1 = \begin{bmatrix} 0 & 1 & 0 \\ -1 & 0 & 0 \\ 0 & 0 & 0 \end{bmatrix}, \quad \xi_2 = \begin{bmatrix} 0 & 0 & -1 \\ 0 & 0 & 0 \\ 1 & 0 & 0 \end{bmatrix}, \quad \xi_3 = \begin{bmatrix} 0 & 0 & 0 \\ 0 & 0 & 1 \\ 0 & -1 & 0 \end{bmatrix},$$

with the commutation relations

$$[\xi_1, \xi_2] = \xi_3, \qquad [\xi_2, \xi_3] = \xi_1, \qquad [\xi_3, \xi_1] = \xi_2.$$

Given $x = \begin{bmatrix} x_1 \\ x_2 \\ x_3 \end{bmatrix} \in \mathbb{R}$, we let $\xi(x) = (x_1\xi_1 + x_2\xi_2 + x_3\xi_3)$ define an anti-Hermitian general element of $so(3)$, *i.e.*, $\xi(x)^* = -\xi(x)$, cf. Section 2.6.

4.5.1 Two-dimensional representation of *so*(3)

In the two-dimensional representation,

$$\xi_0 = \begin{bmatrix} 1 & 0 \\ 0 & -1 \end{bmatrix}, \quad \xi_+ = \begin{bmatrix} 0 & 1 \\ 0 & 0 \end{bmatrix}, \quad \xi_- = \begin{bmatrix} 0 & 0 \\ 1 & 0 \end{bmatrix},$$

of $so(3)$ on $\mathfrak{h} = \mathbb{C}^2$ with respect to the basis $\{e_1, e_{-1}\}$, *i.e.*,

$$\xi_1 = -\frac{i}{2}\begin{bmatrix} 0 & 1 \\ 1 & 0 \end{bmatrix}, \quad \xi_2 = -\frac{1}{2}\begin{bmatrix} 0 & 1 \\ -1 & 0 \end{bmatrix}, \quad \xi_3 = -\frac{i}{2}\begin{bmatrix} 1 & 0 \\ 0 & -1 \end{bmatrix},$$

we get

$$\xi(x) = -\frac{i}{2}\begin{bmatrix} x_3 & x_1 - ix_2 \\ x_1 + ix_2 & -x_3 \end{bmatrix} \text{ and } J(x) = \frac{1}{2}\begin{bmatrix} x_3 & x_1 - ix_2 \\ x_1 + ix_2 & -x_3 \end{bmatrix}.$$

We note that two vectors ψ and $\lambda\psi$ that differ only by a complex factor λ with modulus $|\lambda| = 1$ define the same state since we have

$$\langle X\lambda\psi, \lambda\psi \rangle = |\lambda|^2 \langle X\psi, \psi \rangle = \langle X\psi, \psi \rangle.$$

As a consequence, up to the aforementioned equivalence, any vector state on $\mathbb{C}^2$ can be characterised by a vector of the form

$$\psi = \cos\frac{\theta}{2} e_1 + e^{i\phi} \sin\frac{\theta}{2} e_{-1} = \begin{bmatrix} \cos\frac{\theta}{2} \\ e^{i\phi} \sin\frac{\theta}{2} \end{bmatrix},$$

with $\theta \in [0, \pi)$, $\phi \in [0, 2\pi)$. In order to determine the distribution of $J(x)$ with respect to the state given by the vector ψ, we first compute the exponential of $t\xi(x)$. Note that we have

$$\xi(x)^2 = -\frac{1}{4} \begin{bmatrix} x_1^2 + x_2^2 + x_3^2 & 0 \\ 0 & x_1^2 + x_2^2 + x_3^2 \end{bmatrix} = -\frac{||x||^2}{2} I,$$

where $||x|| = \sqrt{x_1^2 + x_2^2 + x_3^2}$ denotes the norm of x. By induction we get

$$\xi(x)^k = \begin{cases} (-1)^\ell \dfrac{||x||^{2\ell}}{2^{2\ell}} \xi(x), & \text{if } k = 2\ell + 1 \text{ is odd,} \\[2ex] (-1)^\ell \dfrac{||x||^{2\ell}}{2^{2\ell}} I, & \text{if } k = 2\ell \text{ is even.} \end{cases}$$

Therefore we have

$$\begin{aligned} \exp\left(t\xi(x)\right) &= I + t\xi(x) - \frac{t^2}{2}\frac{||x||^2}{4} I - \frac{t^3}{3!}\frac{||x||^2}{4}\xi(x) + \frac{t^4}{4!}\frac{||x||^2}{16} \pm \cdots \\ &= \cos\left(\frac{t}{2}||x||\right) I + \frac{2}{||x||}\sin\left(\frac{t}{2}||x||\right)\xi(x). \end{aligned} \quad (4.6)$$

For the Fourier transform of the distribution of the quantum random variable $J(x)$ with respect to the state given by ψ, this yields

$$\begin{aligned} \langle \psi, \exp\left(itJ(x)\right)\psi \rangle &= \langle \psi, \exp\left(-t\xi(x)\right)\psi \rangle \\ &= \cos\left(\frac{t}{2}||x||\right)\langle \psi, I\psi \rangle - \frac{2}{||x||}\sin\left(\frac{t}{2}||x||\right)\langle \psi, \xi(x)\psi \rangle. \end{aligned}$$

But

$$
\langle \psi, \xi(x)\psi \rangle = \left\langle \begin{bmatrix} \cos\frac{\theta}{2} \\ e^{i\phi}\sin\frac{\theta}{2} \end{bmatrix}, -\frac{i}{2}\begin{bmatrix} x_3 & x_1 - ix_2 \\ x_1 + ix_2 & -x_3 \end{bmatrix}\begin{bmatrix} \cos\frac{\theta}{2} \\ e^{i\phi}\sin\frac{\theta}{2} \end{bmatrix}\right\rangle
$$

$$
= -\frac{i}{2}\left(x_1\left(e^{i\phi} + e^{-i\phi}\right)\sin\frac{\theta}{2}\cos\frac{\theta}{2} \right)
$$

$$
- \frac{i}{2}\left(ix_2\left(e^{-i\phi} - e^{i\phi}\right)\sin\frac{\theta}{2}\cos\frac{\theta}{2} + x_3\left(\cos^2\frac{\theta}{2} - \sin^2\frac{\theta}{2}\right)\right)
$$

$$
= -\frac{i}{2}\left(x_1 \sin\theta\cos\phi + x_2\sin\theta\sin\phi + x_3\cos\theta\right)
$$

$$
= -\frac{i}{2}\left\langle \begin{bmatrix} \cos\phi\sin\theta \\ \sin\phi\sin\theta \\ \cos\theta \end{bmatrix}, \begin{bmatrix} x_1 \\ x_2 \\ x_3 \end{bmatrix}\right\rangle
$$

$$
= -\frac{i}{2}\langle B(\psi), x\rangle,
$$

where the vector $\psi = e_1\cos\frac{\theta}{2} + e_{-1}e^{i\phi}\sin\frac{\theta}{2}$ is visualised as the point

$$
B(\psi) = \begin{bmatrix} \cos\phi\sin\theta \\ \sin\phi\sin\theta \\ \cos\theta \end{bmatrix}.
$$

on the unit sphere[1] with polar coordinates (θ, ϕ) in $\mathbb{R}^3$. Let us now denote by

$$
\gamma := \left\langle B(\psi), \frac{x}{||x||}\right\rangle \in [-1, 1]
$$

the cosine of the angle between $B(\psi)$ and x. We have

$$
\langle \psi, \exp\left(itJ(x)\right)\psi \rangle = \cos\left(\frac{t||x||}{2}\right) + i\gamma\sin\left(\frac{t||x||}{2}\right),
$$

which shows that the distribution of the Hermitian element $J(x)$ in the state associated to the vector ψ is given by

$$
\mathcal{L}_\Phi\left(J(x)\right) = \frac{1-\gamma}{2}\delta_{-\frac{||x||}{2}} + \frac{1+\gamma}{2}\delta_{\frac{||x||}{2}}.
$$

We find (again) a Bernoulli distribution with parameters $p = (1+\gamma)/2$ and $q = (1-\gamma)/2$.

More generally, a state defined from an n-dimensional representation will yield a discrete distribution with n equally distant points with distance $||x||/2$ and symmetric around 0.

---

[1] The unit sphere is also called the *Bloch sphere* in this case.

4.5.2 Three-dimensional representation of *so*(3) on $\mathfrak{h} = \mathbb{C}^3$

For $n = 2$, *i.e.*, in the three-dimensional representation of $so(3)$, Equations (2.5a), (2.5b), and (2.5c), give

$$\xi(x) = -i \begin{bmatrix} x_3 & \dfrac{x_1 - ix_2}{\sqrt{2}} & 0 \\ \dfrac{x_1 + ix_2}{\sqrt{2}} & 0 & \dfrac{x_1 - ix_2}{\sqrt{2}} \\ 0 & \dfrac{x_1 + ix_2}{\sqrt{2}} & -x_3 \end{bmatrix}$$

and

$$J(x) = \begin{bmatrix} x_3 & \dfrac{x_1 - ix_2}{\sqrt{2}} & 0 \\ \dfrac{x_1 + ix_2}{\sqrt{2}} & 0 & \dfrac{x_1 - ix_2}{\sqrt{2}} \\ 0 & \dfrac{x_1 + ix_2}{\sqrt{2}} & -x_3 \end{bmatrix}.$$

Therefore, we have

$$\xi(x)^2 = - \begin{bmatrix} x_3^2 + \dfrac{x_1^2 + x_2^2}{2} & \dfrac{(x_1 - ix_2)x_2}{\sqrt{2}} & \dfrac{(x_1 - ix_2)^2}{2} \\ \dfrac{(x_1 + ix_2)x_3}{\sqrt{2}} & x_1^2 + x_2^2 & \dfrac{(x_1 - ix_2)x_2}{\sqrt{2}} \\ \dfrac{(x_1 + ix_2)^2}{2} & \dfrac{(x_1 + ix_2)x_3}{\sqrt{2}} & x_3^2 + \dfrac{x_1^2 + x_2^2}{2} \end{bmatrix},$$

and

$$\xi(x)^3 = -(x_1^2 + x_2^2 + x_3^2)\xi(x),$$

which implies by induction

$$\xi(x)^n = \begin{cases} I & \text{if } n = 0, \\ (-||x||^2)^m \xi & \text{if } n \text{ is odd}, n = 2m + 1, \\ (-||x||^2)^m \xi^2 & \text{if } n \geq 2 \text{ is even}, n = 2m + 2. \end{cases}$$

The exponential of $\xi(x)$ is given by

$$\exp\left(t\xi(x)\right) = \sum_{n=0}^{\infty} \frac{t^n}{n!} \xi(x)^n$$

$$= I + \frac{1}{||x||} \sum_{m=0}^{\infty} (-1)^m \frac{(t||x||)^{2m+1}}{(2m+1)!} \xi(x)$$

$$+ \frac{1}{||x||^2} \sum_{m=0}^{\infty} (-1)^m \frac{(t||x||^2)^{2m+2}}{(2m+2)!} \xi(x)^2$$

$$= I + \frac{\sin(t||x||)}{||x||} \xi(x) + \frac{1 - \cos(t||x||)}{||x||^2} \xi(x)^2.$$

This formula is known as Rodrigues' rotation formula. We want to determine the law of $\xi(x)$ in the state given by

$$\psi = \begin{bmatrix} 1 \\ 0 \\ 0 \end{bmatrix},$$

for this we have to calculate the first two moments

$$\langle \psi, \xi(x)\psi \rangle = -ix_3, \qquad \langle \psi, \xi(x)^2 \psi \rangle = -\frac{x_1^2 + x_2^2 + 2x_3^2}{2}.$$

Thus we have

$$\langle \psi, \exp\left(itJ(x)\right)\psi \rangle = \langle \psi, \exp\left(-t\xi(x)\right)\psi \rangle$$

$$= I + ix_3 \frac{\sin(t||x||)}{||x||} - \frac{(x_1^2 + x_2^2 + 2x_3^2)(1 - \cos(t||x||))}{2||x||^2}$$

which shows that $J(x)$ has distribution

$$\mathcal{L}(J(x)) = \frac{(1-\gamma)^2}{4} \delta_{-||x||} + \frac{1-\gamma^2}{2} \delta_0 + \frac{(1+\gamma)^2}{4} \delta_{-||x||},$$

where $\gamma = x_3/||x||$ is the cosine of the angle between ψ and x. This is a binomial distribution with parameters $n = 2$ and $p = (1 + \gamma)/2$.

4.5.3 Two-dimensional representation of *so*(3)

Let us start by defining another model for the two-dimensional representation of *so*(3). We take the two-dimensional Hilbert space $L^2(\{-1, 1\}, b_p)$, where b_p denotes the Bernoulli distribution

$$b_q = p\delta_{+1} + q\delta_{-1},$$

with $0 < p < 1$, $q = 1 - p$. We define the representation on the basis vectors $\mathbf{1}_{\{+1\}}$ and $\mathbf{1}_{\{-1\}}$ by

$$\xi_0 \mathbf{1}_{\{x\}} = x \mathbf{1}_{\{x\}},$$

$$\xi_+ \mathbf{1}_{\{x\}} = \begin{cases} 0 & \text{if } x = +1, \\ \sqrt{\dfrac{q}{p}} \mathbf{1}_{\{+1\}} & \text{if } x = -1, \end{cases}$$

$$\xi_- \mathbf{1}_{\{x\}} = \begin{cases} \sqrt{\dfrac{p}{q}} \mathbf{1}_{\{-1\}} & \text{if } x = +1, \\ 0 & \text{if } x = -1, \end{cases}$$

for $x \in \{-1, +1\}$. Clearly, ξ_0 is Bernoulli distributed in the state given by the constant function $\mathbf{1}$, *i.e.*, $\mathcal{L}_1(\xi_0) = b_p$. More generally, let us consider the elements

$$X_\theta = \cos(\theta)\xi_0 + \sin(\theta)(\xi_+ + \xi_-) = 2i\big(\cos(\theta)\xi_3 + \sin(\theta)x_1\big)$$

with $\theta \in [0, 2\pi)$. By Lemma 2.6.1, X_θ can be obtained from $X_0 = 2i\xi_3$ by a rotation around the second axis, more precisely,

$$X_\theta = 2i\xi\big(R_\theta(e_3)\big) = 2i \exp\big(\mathrm{ad}(\theta\xi_2)\big)\xi_3 = e^{\theta\xi_2}\xi_3 e^{-\theta\xi_2},$$

where

$$R_\theta = \begin{bmatrix} \cos(\theta) & 0 & \sin(\theta) \\ 0 & 1 & 0 \\ -\sin(\theta) & 0 & \cos(\theta) \end{bmatrix}.$$

Therefore, we have

$$\langle \mathbf{1}, \exp(itX_\theta)\mathbf{1} \rangle = \langle \mathbf{1}, e^{\theta\xi_2} \exp(itX_0)e^{-\theta\xi_2}\mathbf{1} \rangle = \langle g_\theta, \exp(itX_0)g_\theta \rangle,$$

with

$$g_\theta = e^{-\theta\xi_2}\mathbf{1}$$

$$= \cos\left(\frac{\theta}{2}\right)\mathbf{1} - 2\sin\left(\frac{\theta}{2}\right)\xi_2\mathbf{1}$$

$$= \left(\cos\left(\frac{\theta}{2}\right) + \sqrt{\frac{q}{p}}\sin\left(\frac{\theta}{2}\right)\right)\mathbf{1}_{\{+1\}} + \left(\cos\left(\frac{\theta}{2}\right) - \sqrt{\frac{p}{q}}\sin\left(\frac{\theta}{2}\right)\right)\mathbf{1}_{\{+1\}},$$

where we could use Equation (4.6) to compute the exponential of

$$-\theta\xi_2 = \frac{1}{2}\begin{bmatrix} 0 & -\sqrt{q/p} \\ \sqrt{p/q} & 0 \end{bmatrix}.$$

We see that the law of X_θ has density $|g_\theta|^2$ with respect to the law of X_0, which gives

$$
\begin{aligned}
\mathcal{L}_1(X_\theta) &= p\left(\cos\left(\frac{\theta}{2}\right) + \sqrt{\frac{q}{p}}\sin\left(\frac{\theta}{2}\right)\right)^2 \delta_{+1} \\
&\quad + q\left(\cos\left(\frac{\theta}{2}\right) - \sqrt{\frac{p}{q}}\sin\left(\frac{\theta}{2}\right)\right)^2 \delta_{-1} \\
&= \frac{1}{2}\left(1 + (2p-1)\cos(\theta) + 2\sqrt{pq}\sin(\theta)\right)\delta_{+1} \\
&\quad + \frac{1}{2}\left(1 - (2p-1)\cos(\theta) - 2\sqrt{pq}\sin(\theta)\right)\delta_{-1}.
\end{aligned}
$$

4.6 Trace and density matrix

We now describe an important state called the *trace* on the algebra $M_n(\mathbb{C})$ of complex $n \times n$ matrices. As we shall see, the trace can be used to give a useful expression for arbitrary states.

Theorem 4.6.1 *There exists a unique linear functional on $M_n(\mathbb{C})$ that satisfies the two conditions*

$$\operatorname{tr}(I) = 1, \tag{4.7}$$

where I denotes the identity matrix, and

$$\operatorname{tr}(AB) = \operatorname{tr}(BA), \qquad A, B \in M_n(\mathbb{C}). \tag{4.8}$$

This unique functional $\operatorname{tr} : M_n(\mathbb{C}) \longrightarrow \mathbb{C}$ is called the trace *(or normalised trace) on $M_n(\mathbb{C})$ and is given by*

$$\operatorname{tr}(A) = \frac{1}{n}\sum_{j=1}^{n} a_{jj} \tag{4.9}$$

for $A = (a_{jk}) \in M_n(\mathbb{C})$. The trace is a state. We can compute the trace of a matrix $A \in M_n(\mathbb{C})$ also as

$$\operatorname{tr}(A) = \frac{1}{n}\sum_{j=1}^{n} \langle e_j, Ae_j \rangle, \tag{4.10}$$

where $\{e_1, \ldots, e_n\}$ is any orthonormal basis of $\mathbb{C}^n$.

Proof: *a*) Existence. It is straightforward to check that the functional defined in (4.9) does indeed satisfy the conditions (4.7) and (4.8). We have

$$\text{tr}(I) = \text{tr}\left((\delta_{jk})_{1 \le j, \ell \le n}\right) = \frac{1}{n}\sum_{j=1}^{n}\delta_{jj} = 1$$

and

$$\text{tr}(AB) = \text{tr}\left(\left(\sum_{k=1}^{n} A_{jk}B_{k\ell}\right)_{1 \le j, \ell \le n}\right)$$

$$= \sum_{j,\ell=1}^{n} a_{j\ell}b_{\ell j}$$

$$= \sum_{j,\ell=1}^{n} b_{j\ell}a_{\ell j} = \text{tr}(BA).$$

b) Uniqueness. Denote by e_{jk} with $1 \le j, k \le n$ the matrix units, *i.e.*, e_{jk} is the matrix with all coefficients equal to zero except for the coefficient in the *j*-th row and *k*-th column, which is equal to 1,

$$e_{jk} = (\delta_{jr}\delta_{ks})_{1 \le r,s \le n} = j \begin{pmatrix} 0 & \cdots & 0 & \overset{k}{0} & 0 & \cdots & 0 \\ \vdots & \ddots & \vdots & \vdots & \vdots & \ddots & \vdots \\ 0 & \cdots & 0 & 0 & 0 & \cdots & 0 \\ 0 & \cdots & 0 & 1 & 0 & \cdots & 0 \\ 0 & \cdots & 0 & 0 & 0 & \cdots & 0 \\ \vdots & \ddots & \vdots & \vdots & \vdots & \ddots & \vdots \\ 0 & \cdots & 0 & 0 & 0 & \cdots & 0 \end{pmatrix}.$$

The n^2 matrix units $\{e_{11}, \ldots, e_{1n}, e_{21}, \ldots, e_{nn}\}$ form a basis of $M_n(\mathbb{C})$. Therefore two linear functionals coincide, if they have the same values on all matrix units. For the trace tr we have

$$\text{tr}(e_{jk}) = \begin{cases} 0 & \text{if } j \ne k, \\ 1/n & \text{if } j = k. \end{cases}$$

Note that we have the following formula for the multiplication of the matrix units,

$$e_{jk}e_{\ell m} = \delta_{k\ell}e_{jm}.$$

Let $f \colon M_n(\mathbb{C}) \longrightarrow \mathbb{C}$ be a linear functional that satisfies conditions (4.7) and (4.8). We will show that f takes the same values as tr on the matrix units which then implies $f = $ tr and therefore establishes uniqueness.

For $j \neq k$, we have $e_{jk} = e_{j1}e_{1k}$, $e_{1k}e_{j1} = 0$ and therefore

$$f(e_{jk}) = f(e_{j1}e_{1k}) = f(e_{1k}e_{j1}) = f(0) = 0,$$

since f satisfies (4.8). We also have $e_{jk}e_{kj} = e_{jj}$, so (4.8) implies

$$f(e_{jj}) = f(e_{jk}e_{kj}) = f(e_{kj}e_{jk}) = f(e_{kk}),$$

for any $j, k \in \{1, \ldots, n\}$. This means that there exists a constant $c \in \mathbb{C}$ such that

$$f(e_{jj}) = c$$

for $j = 1, \ldots, n$. But it is easy to see that (4.7) implies $c = \frac{1}{n}$, and we have shown

$$f(e_{jk}) = \frac{1}{n}\delta_{jk} = \operatorname{tr}(e_{jk})$$

for $1 \leq j, k \leq n$.

c) Proof of (4.10). Let $e_1, \ldots, e_n$ be an orthonormal basis of $\mathbb{C}^n$. To prove formula (4.10) it is sufficient to prove that the functional

$$f \colon M_n(\mathbb{C}) \longrightarrow \mathbb{C}$$

defined by

$$f(A) = \frac{1}{n}\sum_{j=1}^{n}\langle e_j, Ae_j\rangle,$$

satisfies Equations (4.7) and (4.8). The first is obvious, we clearly have

$$f(A) = \frac{1}{n}\sum_{j=1}^{n}\langle e_j, Ie_j\rangle = \frac{1}{n}\sum_{j=1}^{n}||e_j||^2 = 1.$$

For (4.8) we use the identity

$$v = \sum_{j=1}^{v}\langle e_j, v\rangle e_j, \qquad v \in \mathbb{C}^n,$$

which develops a vector $v \in \mathbb{C}^n$ with respect to the basis $e_1, \ldots, e_n$. Let $A, B \in M_n(\mathbb{C})$, applying the formula to b_j, we get

$$f(AB) = \frac{1}{n} \sum_{j=1}^n \langle e_j, ABe_j \rangle$$

$$= \frac{1}{n} \sum_{j=1}^n \left\langle e_j, A \left(\sum_{\ell=1}^n \langle e_\ell, Be_j \rangle e_\ell \right) \right\rangle$$

$$= \frac{1}{n} \sum_{j,\ell=1}^n \langle e_j, Ae_\ell \rangle \langle e_\ell, Be_j \rangle$$

$$= \frac{1}{n} \sum_{j,\ell=1}^n \langle e_\ell, be_j \rangle \langle e_j, Ae_\ell \rangle$$

$$= \frac{1}{n} \sum_{\ell=1}^n \left\langle e_\ell, b \left(\sum_{j=1}^n \langle e_j, Ae_\ell \rangle e_j \right) \right\rangle$$

$$= \frac{1}{n} \sum_{\ell=1}^n \langle e_\ell, BAe_\ell \rangle$$

$$= f(BA).$$

This formula shows that the trace is a state. We have

$$\mathrm{tr}(I) = \frac{1}{n} \sum_{j=1}^n \langle e_j, e_j \rangle = 1,$$

and, if $A \in M_n(\mathbb{C})$ is a positive matrix, then there exists a matrix $B \in M_n(\mathbb{C})$ such that $A = B^*B$, and we have

$$\mathrm{tr}(B) = \frac{1}{n} \sum_{j=1}^n \langle e_j, Ae_j \rangle = \frac{1}{n} \langle e_j, B^*Be_j \rangle = \frac{1}{n} \|be_j\|^2 \geq 0.$$

$\square$

Let $\rho \in M_n(\mathbb{C})$ be a positive matrix with trace one. Then we can define a state on $M_n(\mathbb{C})$ on

$$\Phi(A) = \mathrm{tr}(\rho A)$$

for $A \in M_n(\mathbb{C})$. Indeed, since $\mathrm{tr}(\rho) = 1$ we have $\Phi(I) = \mathrm{tr}(\rho I) = 1$, and since ρ is positive, there exists a matrix $B \in M_n(\mathbb{C})$ such that $\rho = B^*B$ and therefore

$$\Phi(A) = \mathrm{tr}(\rho A) = \mathrm{tr}(B^*BA) = \mathrm{tr}(BAB^*) \geq 0$$

for any positive matrix $A \in M_n(\mathbb{C})$. Here we used the fact that A is of the form $A = C^*C$, since it is positive, and therefore $BAB^* = (CB^*)(CB^*)$ is also positive.

All states on $M_n(\mathbb{C})$ are of this form.

Theorem 4.6.2 *Let* $\Phi: M_n(\mathbb{C}) \longrightarrow \mathbb{C}$ *be a state. Then there exists a unique matrix* $\rho = (\rho_{jk}) \in M_n(\mathbb{C})$ *such that*

$$\Phi(A) = \mathrm{tr}(\rho A)$$

for all $A \in M_n(\mathbb{C})$*. The matrix* ρ *is positive and has trace equal to one. Its coefficients can be calculated as*

$$\rho_{jk} = n\Phi(e_{kj})$$

for $1 \leq j, k \leq n$*, where* e_{kj} *denotes the matrix unit, and*

$$e_{kj} := (\delta_{kr}\delta_{js})_{1 \leq r,s \leq n} = \begin{matrix} & & & j & & \\ & \\ k \end{matrix} \begin{pmatrix} 0 & \cdots & 0 & 0 & 0 & \cdots & 0 \\ \vdots & \ddots & \vdots & \vdots & \vdots & \ddots & \vdots \\ 0 & \cdots & 0 & 0 & 0 & \cdots & 0 \\ 0 & \cdots & 0 & 1 & 0 & \cdots & 0 \\ 0 & \cdots & 0 & 0 & 0 & \cdots & 0 \\ \vdots & \ddots & \vdots & \vdots & \vdots & \ddots & \vdots \\ 0 & \cdots & 0 & 0 & 0 & \cdots & 0 \end{pmatrix}.$$

The theorem can be deduced from the fact that $M_n(\mathbb{C})$ is a Hilbert space with the inner product

$$\langle A, B \rangle = \mathrm{tr}(A^*B) \qquad \text{for } A, B \in M_n(\mathbb{C}),$$

and from the observation that the matrices

$$\eta_{jk} = \sqrt{n}e_{jk}, \qquad j, k = 1, \ldots, n,$$

form an orthonormal basis for $M_n(\mathbb{C})$.

Definition 4.6.3 *A positive matrix* $\rho \in M_n(\mathbb{C})$ *with* $\mathrm{tr}(\rho) = 1$ *is called a density matrix.*

The expression "density matrix" is motivated by the observation that in quantum probability such matrices play the same role as probability densities do in classical probability.

4.7 Spin measurement and the Lie algebra *so*(3)

Since Stern and Gerlach's experiments in 1921–1922, it is known that many atoms and particles have a magnetic moment and that the measurement of this moment, in a chosen direction, will always product half-integer values. See, *e.g.*, [41, volume 3, chapters 5 and 6] or [82, 1.5.1 "The Stern–Gerlach experiment"] for an introduction to these experiments and their correct description in quantum physics.

By constructing the Stern–Gerlach device appropriately, one can cause the particle or atom to be deflected by an amount that depends upon the component of the particle or atom's magnetic dipole moment in a chosen direction. When the particle or atom hits a screen, this deflection can be measured and allows one to deduce the component of the particle or atom's magnetic dipole moment. In the most elementary case, when we observe, *e.g.*, the spin of an electron, we will obtain only two possible outcomes, $+1/2$ or $-1/2$, in appropriately chosen units. Similar experiments can also be conducted with photons, using their polarisation. We will here describe how such experiments can be modelled using our *so*(3)-quantum probabilty space.

In Section 2.6, we considered the random variable

$$J(x) = i(x_1\xi_1 + x_2\xi_2 + x_3\xi_3)$$

in several representations of *so*(3). The random variable $J(x)$ corresponds to the measurement of the component of the spin of our particle in the direction given by $x = \begin{bmatrix} x_1 \\ x_2 \\ x_3 \end{bmatrix}$. Let us assume that x is a unit vector, *i.e.*,

$$||x|| = \sqrt{x_1^2 + x_2^2 + x_3^2} = 1.$$

Let us begin with the two-dimensional representation, *i.e.*, with the representation with parameter $n = 1$. This corresponds to a spin 1/2-particle, as in Reference [41, volume 3, chapter 6 "Spin one-half"]. A general vector state in this representation is given by a vector of the form

$$\psi = \cos\frac{\theta}{2}e_1 + e^{i\phi}\sin\frac{\theta}{2}e_{-1}$$

with $\theta \in [0, \pi]$, $\phi \in [0, 2\pi)$. This corresponds to a particle whose spin points in the direction

$$B(\psi) = \begin{bmatrix} \cos\phi\sin\theta \\ \sin\phi\sin\theta \\ \cos\theta \end{bmatrix}.$$

Table 4.1 *Dictionary "Classical ↔ Quantum"*

	Classical	Quantum
Sample space	A set $\Omega = \{\omega_1, \ldots, \omega_n\}$	A Hilbert space $\mathfrak{h} = \mathbb{C}^n$
Events	Subsets of Ω that form a σ-algebra (also a Boolean algebra)	The orthogonal projections in $\mathfrak{h}$ that form a lattice, which is not Boolean (or distributive), *e.g.*, in general $E \wedge (F_1 \vee F_2) \neq (E \wedge F_1) \vee (E \wedge F_2)$
Random variables / observables	A measurable function $f : \Omega \longrightarrow \mathbb{R}$ forming form a commutative (von Neumann) algebra to each event $E \in \mathcal{F}$ to obtain an r.v. I_E	Self-adjoint operators $X : \mathfrak{h} \longrightarrow \mathfrak{h}, X^* = X$ spanning a noncommutative (von Neumann) algebra event are observables with values in $\{0, 1\}$. Note that $E_\lambda = I_{\{\lambda\}}(X)$.
Probability distribution/ state	A countably additive function $\mathbb{P} : \mathcal{F} \longrightarrow [0,1]$ determined by n pos. real numbers $p_k = \mathbb{P}(\{\omega_k\})$ such that $\sum_{k=1}^{n} p_k = 1$ $\mathbb{P}(E) = \sum_{\omega \in E} \mathbb{P}(\{\omega\})$	A density matrix, *i.e.* a pos. operator with $\mathrm{tr}(\rho) = 1$ $\mathbb{P}(X = \lambda) = \mathrm{tr}(\rho E_\lambda),$ $\mathbb{P}(X \in E) = \mathrm{tr}(\rho I_E(X)),$ $I_E(X) = \sum_{\lambda \in E \cap \sigma(X)} E_\lambda.$
Expectation	$\mathbf{E}[f] = \int_\Omega f d\mathbb{P}$	$\mathbf{E}[X] = \mathrm{tr}(\rho X) = \sum_{k=1}^{n} f(\omega)\mathbb{P}(\{\omega\})$
Variance	$\mathrm{Var}[f] = \mathbf{E}[f^2] - (\mathbf{E}[f])^2$	$\mathrm{Var}_\rho[X] = \mathrm{tr}(\rho X^2) - \left(\mathrm{tr}(\rho X)\right)^2$
Extreme points	The set of all probability distribution on Ω is a compact convex set exactly n extreme points $\delta_{\omega_k}, k = 1, \ldots, n$. If $\mathbb{P} = \delta_{\omega_k}$, then the distribution of any r.v. f is concentrated at one point (namely $f(\omega_k)$).	The extreme points of the set $\mathcal{S}(\mathfrak{h})$ of states on $\mathfrak{h}$ are exactly the one-dim. projections onto the rays $\mathbb{C}u, u \in \mathfrak{h}$ a unit vector. If $\rho = \mathbb{P}_u$ then $\mathrm{Var}[X]$ $= \|(X - \langle u, Xu \rangle)u\|^2$ Thus $\mathrm{Var}[X] = 0$ if and only if u is an eigenvector of X. *Degeneracy of the state does not kill the uncertainty of the observables!*
Product spaces systems	Given two systems described by $(\Omega_i, \mathcal{F}_i, \mathbb{P}_i), i = 1, 2$ then $(\Omega_1 \times \Omega_2, \mathcal{F}_1 \otimes \mathcal{F}_2, \mathbb{P}_1 \otimes \mathbb{P}_2)$ describes both independent systems as a single system	Given two systems described by $(\mathfrak{h}_i, \rho_i), i = 1, 2$ then $(\mathfrak{h}_1 \otimes \mathfrak{h}_2, \rho_1 \otimes \rho_2)$ describes both independent systems as a single system $\longrightarrow$ independence $\longrightarrow$ entanglement

We found that the law of $J(x)$ in the state with state vector ψ is given by

$$\mathcal{L}_\psi\big(J(x)\big) = \frac{1-\gamma}{2}\delta_{-1/2} + \frac{1+\gamma}{2}\delta_{1/2},$$

where γ is the cosine of the angle between $B(\psi)$ and $||x||$,

$$\gamma = \left\langle B(\psi), \frac{x}{||x||} \right\rangle.$$

So the measurement of the component of the spin in direction x of a spin $1/2$-particle whose spin points in the direction $b(\psi)$, will give $+1/2$ with probability $(1+\gamma)/2$, and $-1/2$ with probability $(1-\gamma)/2$.

The other representations correspond to particles with higher spin; the $n+1$-dimensional representation describes a particle with spin $\frac{n}{2}$. In particular, for $n = 2$, we have spin 1-particles, cf. Reference [41, volume 3, chapter 5 "Spin one"].

Notes

On $so(3)$, see [20, 21] for the Rotation Group $SO(3)$, its Lie algebra $so(3)$, and their applications to physics. See, *e.g.*, [36, 37, 100] for Krawtchouk polynomials and their relation to the binomial process (or Bernoulli random walk).

Table 4.1 presents an overview of the terminology used in classical and quantum probability, as in *e.g.,* [88].

Exercises

Exercise 4.1 Let $n \in \mathbb{N}$, $A \in M_n(\mathbb{C})$ be a Hermitian matrix. Let $f : \mathbb{R} \to \mathbb{C}$ be a function.

1. Find a polynomial $p(x) = \displaystyle\sum_{k=1}^{m} p_k x^k$ with

$$p(\lambda_i) = f(\lambda_i)$$

for all eigenvalues λ_i of A and set

$$f(A) = p(A) = \sum_{k=1}^{m} p_k A^k.$$

If A has m distinct eigenvalues $\lambda_1, \ldots, \lambda_m$ (counted without their multiplicities) then we can use *Lagrange interpolation* to find p as

$$p(x) = \sum_{k=1}^{m} f(\lambda_k) \prod_{i \neq k} \frac{x - \lambda_i}{\lambda_k - \lambda_i}$$

2. Find an invertible matrix $U \in M_n(\mathbb{C})$ that diagonalises A, *i.e.*, such that

$$U^{-1}AU = \begin{bmatrix} \lambda_1 & 0 & \cdots & 0 & 0 \\ 0 & \lambda_2 & \cdots & 0 & 0 \\ \vdots & \vdots & \ddots & \vdots & \vdots \\ 0 & 0 & \cdots & \lambda_{n-1} & 0 \\ 0 & 0 & \cdots & 0 & \lambda_n \end{bmatrix}$$

a diagonal matrix and let

$$f(A) = U \begin{bmatrix} f(\lambda_1) & 0 & \cdots & 0 & 0 \\ 0 & f(\lambda_2) & \cdots & 0 & 0 \\ \vdots & \vdots & \ddots & \vdots & \vdots \\ 0 & 0 & \cdots & f(\lambda_{n-1}) & 0 \\ 0 & 0 & \cdots & 0 & f(\lambda_n) \end{bmatrix} U^{-1}.$$

The numbers $\lambda_1, \ldots, \lambda_n$ are the eigenvalues counted with their multiplicities. By the spectral theorem we know that we can choose U unitary, so that $U^{-1} = U^*$.

3. Show that using the spectral theorem in the form stated earlier and writing X as

$$X = \sum_{\lambda \in \sigma(X)} \lambda E_\lambda,$$

we have

$$f(X) = \sum_{\lambda \in \sigma(X)} f(\lambda) E_\lambda.$$

Exercise 4.2 In the framework of the Examples of Section 4.4, define further

$$\Xi_0^n = \sum_{j=1}^{n} \xi_0^{(j)}, \quad \Xi_+^n = \sum_{j=1}^{n} \xi_+^{(j)}, \quad \Xi_-^n = \sum_{j=1}^{n} \xi_-^{(j)}.$$

1. Show that these operators define a representation of $so(3)$, *i.e.*, we have

$$[\Xi_+^n, \Xi_-^n] = \Xi_0^n, \qquad [\Xi_0^n, \Xi_\pm^n] = \pm 2\Xi_\pm,$$

and

$$(Xi_0^n)^* = \Xi_0^n, \qquad (\Xi_+^n)^* = \Xi_-^n.$$

2. Show that the indicator functions $\mathbf{1}_{\{x\}}$ are eigenvectors of Ξ_0^n, with eigenvalues equal to the difference of the number of $+1$s and -1s in x.
3. Show that Ξ_0^n has a binomial distribution on the set $\{-n, -n+2, \dots, n-2, n\}$ and compute the density of this distribution with respect to the constant function.
4. Compute the law of

$$\Xi_\theta^n = \Xi_0^n + \theta(\Xi_+^n + \Xi_+^n)$$

and discuss possible connections with the Krawtchouk polynomials.

Exercise 4.3 Calculate $\Phi(X^3)$ in (4.1).

5

Noncommutative stochastic integration

To invent, one must think aside.

(Paul Souriau, cited by Jacques Hadamard in "The Psychology
of Invention in the Mathematical Field," *1954.)*

In this chapter we revisit the construction of a Fock space over a given Hilbert space, which we already encountered in Section 1.3. In this book we will consider only symmetric (or boson) Fock spaces. In the previous chapters, we have already seen in many examples how they can be used to construct representations of Lie algebras and realisations of quantum probability spaces. In Sections 5.2 and 5.3 we will define the fundamental noise processes of creation, annihilation, and conservation operators on symmetric Fock space and define and study stochastic integrals for these processes. They provide a powerful and flexible tool for the construction of many other, more general processes.

5.1 Construction of the Fock space

We will present two constructions of the Fock space, one based on the theory of positive definite functions and one based on tensor products of Hilbert spaces.

5.1.1 Construction from a positive definite function

The following definition is fundamental for this paragraph.

Definition 5.1.1 *Let X be a set. A function $K : X \times X \longrightarrow \mathbb{C}$ is called a* positive definite kernel *on X, if for all $n \in \mathbb{N}$ and all choices of $x_1, \ldots, x_n \in X$ and $\lambda_1, \ldots, \lambda_n \in \mathbb{C}$ we have*

$$\sum_{j,k=1}^{n} \overline{\lambda_j}\lambda_k K(x_j, x_k) \geq 0.$$

Example 5.1.2

a) The inner product of a complex Hilbert space is a positive definite kernel. If H is a complex Hilbert space, $n \in \mathbb{N}$, $x_1, \ldots, x_n \in X$ and $\lambda_1, \ldots, \lambda_n \in \mathbb{C}$, then we clearly have

$$\sum_{j,k=1}^{n} \overline{\lambda_j}\lambda_k \langle x_j, x_k \rangle = \left\langle \sum_{j=1}^{n} \lambda_j x_j, \sum_{k=1}^{n} \lambda_k x_k \right\rangle = \left\| \sum_{j=1}^{n} \lambda_j x_j \right\|^2 \geq 0.$$

b) By the same argument the inner product of a complex Hilbert space H is also a positive definite kernel on any subset of H.

c) The inner product of a real Hilbert space is also a positive definite kernel, since a real Hilbert space can be viewed as a subset of its complexification.

The following theorem shows that all positive definite kernels are in a sense of the form of the examples above.

Theorem 5.1.3   *Let $K : X \times X \longrightarrow \mathbb{C}$ be a positive definite kernel. Then there exists a complex Hilbert space H_K and a map $\varphi : X \longrightarrow H_K$ such that*

i) $\varphi(K)$ *is total in* H_K,
ii) *we have* $K(x, y) = \langle \varphi(x), \varphi(y) \rangle$, *for all* $x, y \in X$.

The space H_K is unique up to unitary equivalence.

Recall that a subset M of a Hilbert space H is called *total*, if any vector $x \in H$ with

$$\forall y \in M, \quad \langle y, x \rangle = 0$$

is necessarily the zero vector. This is equivalent to the linear span of M being dense in H.

There are several constructions that allow to produce positive definite kernels. For example, if $K, L : X \times X \longrightarrow \mathbb{C}$ are positive definite kernels, then $K \cdot L : X \times X \longrightarrow \mathbb{C}$ with

$$(K \cdot L)(x, y) = K(x, y)L(x, y)$$

for $x, y \in X$ is again a positive definite kernel on X.

Lemma 5.1.4 *Let X be a set and $K, L : X \times X \longrightarrow \mathbb{C}$ two positive definite kernels on X. Then their pointwise product*

$$K \cdot L : X \times X \longrightarrow \mathbb{C}$$

$$(x, y) \longmapsto K(x, y)L(x, y)$$

is positive definite.

Proof: Let $n \in \mathbb{N}$, $x_1, \ldots, x_n \in X$. We have to show that the entrywise product (also called the *Hadamard product* or *Schur product*)

$$K^x \circ L^x = \left(K^x_{jk}L^x_{jk}\right)_{1 \leq j, k \leq n} = \left(K(x_j, x_k)L(x_j, x_k)\right)_{1 \leq j, k \leq n}$$

of the matrices

$$K^x = \left(K(x_j, x_k)\right)_{1 \leq j, k \leq n} \quad \text{and} \quad L^x = \left(L(x_j, x_k)\right)_{1 \leq j, k \leq n}$$

is positive. Let $\lambda_1, \ldots, \lambda_n \in \mathbb{C}$. We denote by diag$(x)$ the diagonal matrix $\text{diag}(x) = (x_j \delta_{jk})_{1 \leq j, k \leq n}$. We can write

$$\sum_{j,k=1}^{n} \overline{\lambda_j} \lambda_k (K \cdot L)(x_j, x_k) = \text{Tr}\left(\text{diag}(\lambda)^* K^x \text{diag}(\lambda)(L^x)^t\right),$$

where

$$\text{Tr}(A) = \sum_{j=1}^{n} a_{jj}, \qquad A = (a_{jk}) \in M_n(\mathbb{C}),$$

denotes the (non-normalised) *trace* on $M_n(\mathbb{C})$. Since L^x is a positive matrix, we can write it in the form $L^x = A^*A$ with a square matrix $A \in M_n(\mathbb{C})$. Substituting this expression into the aforementioned formula and using the trace property, we get

$$\sum_{j,k=1}^{n} \overline{\lambda_j} \lambda_k (K \cdot L)(x_j, x_k) = \text{Tr}\left((A^t)^* \text{diag}(\lambda)^* K^x \text{diag}(\lambda)A^t\right).$$

Since K^x is positive and since conjugation with another matrix preserves positivity, we see that this expression is positive. $\qquad\qquad\square$

Positive multiples and sums of positive definite kernels are clearly positive definite kernels. Since the constant function with a positive value is also a positive definite kernel, it follows that for a positive definite kernel K on X

$$\exp K = \sum_{n=0}^{\infty} K^n$$

is also a positive definite kernel on X. Therefore, we have the following result.

Proposition 5.1.5 *Let H be a Hilbert space. Then there exists a Hilbert space, denoted* $\exp(H)$ *spanned by the set*

$$\bigl\{ \mathcal{E}(h) \ : \ h \in H \bigr\},$$

with the inner product determined by

$$\langle \mathcal{E}(h_1), \mathcal{E}(h_2) \rangle = \exp\bigl(\langle h_1, h_2 \rangle \bigr),$$

for $h_1, h_2 \in H$.

Proof: Since the inner product of a Hilbert space is a positive definite kernel, the preceding discussion shows that

$$K(h_1, h_2) = \exp\bigl(\langle h_1, h_2 \rangle \bigr)$$

defines a positive definite kernel on H. Theorem 5.1.3 then gives the existence of the Hilbert space $\exp H$. □

Remark 5.1.6 *The Hilbert space* $\exp H$ *is called the* Fock space *over H, or the* symmetric *or* boson Fock space *over H. We will also use the notation* $\Gamma_s(H)$. *We will briefly see another kind of Fock space, the* full *or* free Fock *space $\mathcal{F}(H)$ over a given Hilbert space H in the next paragraph. But otherwise we will only use the symmetric Fock space and call it simply Fock space, when there is no danger of confusion.*

5.1.2 Construction via tensor products

We now consider another construction of the Fock space $\Gamma_s(H)$ over a given Hilbert space H, and refer to the appendix Section A.8 for background on the tensor products of Hilbert spaces. We set $H^{\otimes 0} = \mathbb{C}$, $H^{\otimes 1} = H$, and

$$H^{\otimes n} := \underbrace{H \otimes H \otimes \cdots \otimes H}_{n \text{ times}}, \qquad n \geq 2.$$

The (algebraic) direct sum

$$\mathcal{F}_{\mathrm{alg}}(H) = \bigoplus_{n=0}^{\infty} H^{\otimes n}$$

is a pre-Hilbert space with the "term-wise" inner product defined by

$$\langle (h_n)_{n \in \mathbb{N}}, (k_n)_{n \in \mathbb{N}} \rangle_{\mathcal{F}_{\mathrm{alg}}(H)} := \sum_{n=0}^{\infty} \langle h_n, k_n \rangle_{H^{\otimes n}}$$

for $(h_n)_{n \in \mathbb{N}}, (k_n)_{n \in \mathbb{N}} \in \mathcal{F}_{\mathrm{alg}}(H)$, *i.e.*, with $h_n, k_n \in H^{\otimes n}$ for $n \geq 0$. The inner product on $H^{\otimes 0} = \mathbb{C}$ is of course simply $\langle z_1, z_2 \rangle_{\mathbb{C}} = \overline{z_1} z_2$. Upon completion we get the Hilbert space

$$\mathcal{F}(H) = \overline{\bigoplus_{n=0}^{\infty}} H^{\otimes n} = \left\{ (h_n)_{n \in \mathbb{N}} \; : \; h_n \in H^{\otimes n}, \; \sum_{n=0}^{\infty} ||h_n||^2_{H^{\otimes n}} < \infty \right\},$$

which is the space of all sequences of tensors of increasing order whose squared norms are summable. This space is called the *free* or *full Fock space* over H and denoted by $\mathcal{F}(H)$, and it plays an important role in free probability theory.

We now turn to the definition of the *symmetric Fock space*, and for this purpose we start with the symmetrisation of the tensor powers of a Hilbert space. On $H^{\otimes n}$ we can define a *symmetrisation operator* $\mathcal{S}_n$ by

$$\mathcal{S}_n(h_1 \otimes \cdots \otimes h_n) = \frac{1}{n!} \sum_{\pi \in \Sigma_n} h_{\sigma(1)} \otimes \cdots \otimes h_{\sigma(n)}$$

for $h_1, \ldots, h_n \in H$, where Σ_n denotes the set of permutations of $\{1, \ldots, n\}$. One can check by direct calculation that we have $\mathcal{S}_n = (\mathcal{S}_n)^* = (\mathcal{S}_n)^2$, so the symmetrisation operator is an orthogonal projection. We define the symmetric tensor power of order n as the range of this projection,

$$H^{\circ n} = \mathcal{S}_n(H^{\otimes n}),$$

and the *symmetric Fock space* as the completed direct sum of the symmetric tensor powers,

$$\Gamma_s(H) = \overline{\bigoplus_{n=0}^{\infty}} H^{\circ n}.$$

If we denote by $\mathcal{S}$ the direct sum of the symmetrisation operators, *i.e.*,

$$\mathcal{S}\left((h_n)_{n=0}^{\infty} \right) = \left((\mathcal{S}_n(h_n))_{n=0}^{\infty} \right)$$

for $(h_n)_{n=0}^{\infty} \in \mathcal{F}(H)$, then we have

$$\Gamma_s(H) = \mathcal{S}\big(\mathcal{F}(H)\big).$$

One can show that the tensor powers

$$v^{\otimes n} = \underbrace{v \otimes \cdots \otimes v}_{n \text{ times}}$$

are total in the symmetric tensor power $H^{\circ n}$, see Exercise 5.1. Let us now use the *exponential vectors*

$$\mathcal{E}(h) = \left(\frac{h^{\otimes n}}{\sqrt{n!}}\right)_{n=0}^{\infty}$$

to show that the symmetric Fock space which we just constructed is the same as the space $\exp H$ that we obtained previously from the theory of positive definite kernels. We have

$$\sum_{n=0}^{\infty} \left\| \frac{h^{\otimes n}}{\sqrt{n!}} \right\|_{H^{\otimes n}}^{2} = \sum_{n=0}^{\infty} \frac{\|h\|^{2n}}{n!} = \exp\left(\|h\|^2\right) < \infty,$$

and therefore $\mathcal{E}(h) \in \mathcal{F}(H)$. Since each term is a product vector, we have furthermore

$$\mathcal{S}\big(\mathcal{E}(h)\big) = \mathcal{E}(h),$$

and therefore $\mathcal{E}(h) \in \Gamma_s(H)$. Another computation gives

$$\langle \mathcal{E}(h), \mathcal{E}(k) \rangle = \sum_{n=0}^{\infty} \frac{1}{n!} \langle h, k \rangle^n = \exp\left(\langle h, k \rangle\right), \qquad h, k \in H.$$

The totality of the exponential vectors $\{\mathcal{E}(h) : h \in H\}$ in the symmetric Fock space $\Gamma_s(H)$ follows from the totality of the tensor powers $\{v^{\otimes n} : v \in H\}$ in the symmetric tensor power $H^{\circ n}$, since

$$\left.\frac{d^n}{d^n t}\right|_{t=0} \langle \mathcal{E}(th), (k_n)_{n \in \mathbb{N}} \rangle = \frac{1}{\sqrt{n!}} \langle h^{\otimes n}, k_n \rangle,$$

for $n \geq 0$, $h \in H$, $(k_n)_{n \in \mathbb{N}} \in \Gamma_s(H)$.

We have checked that the map $H \ni h \longmapsto \mathcal{E}h \in \Gamma_s(H)$ satisfies all the conditions of Theorem 5.1.3 with respect to the positive definite kernel $\exp\langle \cdot, \cdot \rangle_H$, therefore $\Gamma_s(H)$ is indeed isomorphic to $\exp H$.

5.2 Creation, annihilation, and conservation operators

There is a special family of operators acting on the symmetric Fock space. We can use the symmetric tensor powers to define them. For $h \in H$ and $T \in B(H)$, we can define

$$\begin{cases} a_n^-(h) : H^{\circ n} \longrightarrow H^{\circ(n-1)} \\[2mm] a_n^+(h) : H^{\circ n} \longrightarrow H^{\circ(n+1)} \\[2mm] a_n^\circ(T) : H^{\circ n} \longrightarrow H^{\circ n}, \end{cases}$$

by setting

$$\begin{cases} a_n^-(h)v^{\otimes n} := \langle h, v \rangle v^{\otimes(n-1)}, \\[3mm] a_n^+(h)v^{\otimes n} := \sqrt{n+1}\,\mathcal{S}_{n+1}(h \otimes v^{\otimes n}) = \dfrac{1}{\sqrt{n+1}} \displaystyle\sum_{k=0}^{n} v^k \otimes h \otimes v^{n-k}, \\[3mm] a_n^{\circ}(T)v^{\otimes n} := \displaystyle\sum_{k=1}^{n} v^{k-1} \otimes Tv \otimes v^{n-k}, \end{cases}$$

on tensor powers $v^{\otimes n} \in H^{\circ n}$. These operators and their extensions to $\Gamma_s(H)$ are called the *annihilation operator*, the *creation operator*, and the *conservation operator*, respectively. The conservation operator with $T = I$ is also called the *number operator*, since it acts as

$$a_n^{\circ}(I)v^{\circ n} = nv^{\circ n},$$

i.e., it has the symmetric tensor powers as eigenspaces and the eigenvalues give exactly the order of the tensor power.

We set $H^{\circ(-1)} = \{0\}$, then the $0 - th$ order annihilation operator $a_0^-(h)$: $H^{\circ 0} = \mathbb{C} \longrightarrow H^{\circ(-1)} = \{0\}$ must clearly be the zero operator, *i.e.*, $a_0^-(h)(z) = 0$ for any $h \in H$ and $z \in \mathbb{C}$. The direct sums

$$a^-(h) = \bigoplus_{n=0}^{\infty} a_n^-(h), \quad a^+(h) = \bigoplus_{n=0}^{\infty} a_n^+(h), \quad \text{and} \quad a^{\circ}(T) = \bigoplus_{n=0}^{\infty} a_n^{\circ}(T)$$

are well-defined on the algebraic direct sum of the tensor powers. We have

$$\left(a^-(h)\right)^* = a^+(h) \quad \text{and} \quad \left(a^{\circ}(T)\right)^* = a^{\circ}(T^*),$$

so these operators have adjoints and therefore are closable. They extend to densely defined, closable, (in general) unbounded operators on $\Gamma_s(H)$.

There is another way to let operators $T \in B(H)$ act on the Fock spaces $\mathcal{F}(H)$ and $\Gamma_s(H)$, namely by setting

$$\Gamma(T)(v_1 \otimes \cdots \otimes v_n) := (Tv_1) \otimes \cdots \otimes (Tv_n)$$

for $v_1, \ldots, v_n \in H$. The operator $\Gamma(T)$ is called the *second quantisation* of T. It is easy to see that $\Gamma(T)$ leaves the symmetric Fock space invariant. The second quantisation operator $\Gamma(T)$ is bounded if and only if T is a contraction, *i.e.*, $\|T\| \leq 1$. The conservation operator $a^{\circ}(T)$ of an operator $T \in B(H)$ can be recovered from the operators $(e^{tT})_{t \in \mathbb{R}}$ via

$$a^{\circ}(T) = \frac{d}{dt}\bigg|_{t=0} \Gamma(e^{tT})$$

on some appropriate domain. On exponential vectors we have the following formulas for the annihilation, creation, and conservation operators

$$a^-(h)\mathcal{E}(k) = \langle h, k \rangle \mathcal{E}(k),$$

$$a^+(h)\mathcal{E}(k) = \left.\frac{d}{dt}\right|_{t=0} \mathcal{E}(k + th),$$

$$a^\circ(T)\mathcal{E}(k) = \left.\frac{d}{dt}\right|_{t=0} \mathcal{E}(e^{tT}k),$$

with $h, k \in H$, $T \in B(H)$. The creation, annihilation, and conservation operators satisfy the commutation relations

$$
\begin{cases}
[a^-(h), a^-(k)] = [a^+(h), a^+(k)] = 0, \\[2mm]
[a^-(h), a^+(k)] = \langle h, k \rangle I, \\[2mm]
[a^\circ(T), a^\circ(S)] = a^\circ([T, S]), \\[2mm]
[a^\circ(T), a^-(h)] = -a^-(Th), \\[2mm]
[a^\circ(T), a^+(h)] = a^+(Th),
\end{cases}
\tag{5.1}
$$

for $h, k \in H$, $S, T \in B(H)$, cf. [87, Proposition 20.12]. Since the operators are unbounded, these relations can only hold on some appropriate domain. One can take, *e.g.*, the algebraic direct sum of the symmetric tensor powers of H, since this is a common invariant domain for these operators. Another common way to give a meaning to products is to evaluate them between exponential vectors. The condition

$$\langle a^+(h)\mathcal{E}(\ell_1), a^+(k)\mathcal{E}(\ell_2) \rangle - \langle a^-(k)\mathcal{E}(\ell_1), a^-(h)\mathcal{E}(\ell_2) \rangle$$
$$= \langle h, k \rangle \langle \mathcal{E}(\ell_1), \mathcal{E}(\ell_2) \rangle$$

for all $h, k, \ell_1, \ell_2 \in H$ can be viewed as an alternative formulation of the second relation mentioned earlier. Instead of actually multiplying two unbounded operators, we take adjoints to let the left factor of the product act on the left vector of the scalar product. This technique will also be used in Section 5.4 to give a meaning to the Itô formula for products of quantum stochastic integrals.

An important property of the symmetric Fock space is its factorisation.

Theorem 5.2.1 *Let H_1 and H_2 be two Hilbert spaces. The map*

$$U : \{\mathcal{E}(h_1 + h_2) : h_1 \in H_1, h_2 \in H_2\} \longrightarrow \Gamma_s(H_1) \otimes \Gamma_s(H_2),$$
$$\mathcal{E}(h_1 + h_2) \longmapsto \mathcal{E}(h_1) \otimes \mathcal{E}(h_2),$$

$h_1 \in H_1$, $h_2 \in H_2$, *extends to a unique unitary isomorphism between* $\Gamma_s(H_1 \oplus H_2)$ *and* $\Gamma_s(H_1) \otimes \Gamma_s(H_2)$.

Proof: It easy to check that U preserves the inner product between exponential vectors. The theorem therefore follows from the totality of these vectors. □

5.3 Quantum stochastic integrals

For stochastic integrals and stochastic calculus we need a time parameter, so we choose

$$H = L^2(\mathbb{R}_+, \mathfrak{h}) \cong L^2(\mathbb{R}_+) \otimes \mathfrak{h}$$

with some Hilbert space $\mathfrak{h}$. Since we can write H as a direct sum

$$L^2(\mathbb{R}_+, \mathfrak{h}) = L^2([0, t], \mathfrak{h}) \oplus L^2([t, +\infty), \mathfrak{h})$$

of

$$H_{t]} = L^2([0, t], \mathfrak{h}) \quad \text{and} \quad H_{[t} = L^2([t, +\infty), \mathfrak{h})$$

by decomposing functions on $\mathbb{R}_+$ as $f = f 1_{[0,t]} + f 1_{[t,+\infty)}$, we get from Theorem 5.2.1 an isomorphism

$$U_t : \Gamma_s(H) \longrightarrow \Gamma_s(H_{t]}) \otimes \Gamma_s(H_{[t}) \tag{5.2}$$

which acts on exponential vectors as

$$U_t \mathcal{E}(k) = \mathcal{E}(k 1_{[0,t]}) \otimes \mathcal{E}(k 1_{[t,+\infty)})$$

for $k \in L^2(\mathbb{R}_+, \mathfrak{h})$, with its adjoint given by

$$U_t^* \big(\mathcal{E}(k_1) \otimes \mathcal{E}(k_2) \big) = \mathcal{E}(k_1 + k_2)$$

for $k_1 \in L^2([0, t], \mathfrak{h})$, $k_2 \in L^2([t, +\infty), \mathfrak{h})$. Let $m_t \in B(L^2(\mathbb{R}_+))$ denote multiplication by the indicator function $1_{[0,t]}$, *i.e.*,

$$m_t(f)(s) = \begin{cases} f(s), & \text{if } s \le t. \\ 0, & \text{else,} \end{cases}$$

Then the tensor product $m_t \otimes T$ of multiplication by the indicator function $1_{[0,t]}$ on $L^2(\mathbb{R}_t)$ and an operator $T \in B(\mathfrak{h})$ acts as

$$\big((m_t \otimes T)(f)\big)(s) = \begin{cases} Tf(s), & \text{if } s \le t, \\ 0, & \text{else,} \end{cases}$$

for $f \in L^2(\mathbb{R}_t, \mathfrak{h})$. We introduce the notation

$$
\begin{cases}
a_t^-(h) = a^-(\mathbf{1}_{[0,t]} \otimes h), \\[2mm]
a_t^+(h) = a^+(\mathbf{1}_{[0,t]} \otimes h), \\[2mm]
a_t^\circ(T) = a^\circ(m_t \otimes T), \quad t \in \mathbb{R}_+, \ h \in \mathfrak{h}, \ T \in B(\mathfrak{h}).
\end{cases}
$$

Note that the evaluation of these operators between a pair of exponential vectors is given by

$$
\begin{cases}
\langle \mathcal{E}(k_1), a_t^-(h)\mathcal{E}(k_2) \rangle = \displaystyle\int_0^t \langle h, k_2(s) \rangle_{\mathfrak{h}}\, ds \langle \mathcal{E}(k_1), \mathcal{E}(k_2) \rangle, \\[4mm]
\langle \mathcal{E}(k_1), a_t^+(h)\mathcal{E}(k_2) \rangle = \displaystyle\int_0^t \langle k_1(s), h \rangle_{\mathfrak{h}}\, ds \langle \mathcal{E}(k_1), \mathcal{E}(k_2) \rangle, \\[4mm]
\langle \mathcal{E}(k_1), a_t^\circ(T)\mathcal{E}(k_2) \rangle = \displaystyle\int_0^t \langle k_1(s), Tk_2(s) \rangle_{\mathfrak{h}}\, ds \langle \mathcal{E}(k_1), \mathcal{E}(k_2) \rangle,
\end{cases}
$$

$T \in B(\mathfrak{h})$, $h \in \mathfrak{h}$, $k_1, k_2 \in H$. An important notion in stochastic calculus is adaptedness. In our setting this is defined in the following way.

Definition 5.3.1 *Let $\mathfrak{h}$ be a Hilbert space and set $H = L^2(\mathbb{R}_+, \mathfrak{h})$. Let $t \in \mathbb{R}_+$. An operator $X \in B(\Gamma_s(H))$ is called t-adapted, if it can be written in the form*

$$
X = X_{t]} \otimes I
$$

with respect to the factorisation given in Equation (5.2) and with some operator $X_{t]} \in B(\Gamma_s(H_{t]}))$, a stochastic process $(X_t)_{t \in \mathbb{R}_+}$, i.e., a family $(X_t)_{t \in \mathbb{R}_+}$ of operators in $B(\Gamma_s(H))$, is called adapted, if X_t is t-adapted for all $t \in \mathbb{R}_+$.

We will further assume in the sequel that our processes are smooth, i.e., they are piece-wise continuous in the strong operator topology. This means that the maps

$$
\mathbb{R}_+ \ni t \longmapsto X_t v \in \Gamma_s(H)
$$

are piece-wise continuous for all $v \in \Gamma_s(H)$. For applications it is necessary to extend the notion of adaptedness to unbounded operators and processes, see the references mentioned at the end of this chapter.

Quantum stochastic integrals of smooth adapted processes with respect to the creation, annihilation, or conservation process can be defined as limits of Riemann–Stieltjes sums. Let

$$
\pi := \big\{ 0 = t_0 < t_1 < \cdots < t_n = t \big\}
$$

be a partition of the interval $[0, t]$, then the corresponding approximation of the stochastic integral

$$I(t) = \int_0^t X_s da_s^\varepsilon(h), \qquad t \in \mathbb{R}_+,$$

with $\varepsilon \in \{-, \circ, +\}$ and h a vector in $\mathfrak{h}$ (if $\varepsilon \in \{-, +\}$) or an operator on $\mathfrak{h}$ (if $\varepsilon = \circ$), is defined by

$$I_\pi(t) := \sum_{k=1}^n X_{t_{k-1}} \left(a_{t_k}^\varepsilon(h) - a_{t_{k-1}}^\varepsilon(h) \right), \qquad t \in \mathbb{R}_+.$$

We introduce the notation

$$a_{st}^\varepsilon(h) = a_t^\varepsilon(h) - a_s^\varepsilon(h), \qquad 0 \le s \le t,$$

for $\varepsilon \in \{-, \circ, +\}$. Since the creation and conservation operators are linear in h and the annihilation operator is conjugate linear in f, we can also write the increments as

$$a_{st}^\varepsilon(h) = a^\varepsilon(\mathbf{1}_{[s,t]} \otimes h),$$

where, in the case $\varepsilon = \circ$, $\mathbf{1}_{[s,t]}$ as to interpreted as an operator on $L^2(\mathbb{R}_+)$, acting by multiplication.

Under appropriate natural conditions one can show that these approximation do indeed to converge on appropriate domains (*e.g.*, in the strong operator topology) when the mesh

$$|\pi| = \max \{ t_k - t_{k-1} : k = 1, \ldots, n \}$$

of the partition π goes to zero, and define the quantum stochastic integral $\int_0^t X_s da_s^\varepsilon(h)$ as the limit. Evaluating a Riemann–Stieltjes sum over two exponential vectors yields

$$\langle \mathcal{E}(k_1), I_\pi(t) \mathcal{E}(k_2) \rangle =$$

$$\begin{cases} \sum_{k=1}^n \langle \mathcal{E}(k_1), X_{t_{k-1}} \mathcal{E}(k_2) \rangle \int_{t_{k-1}}^{t_k} \langle k_1(s), h \rangle_\mathfrak{h} ds & \text{if } \varepsilon = +, \\[2mm] \sum_{k=1}^n \langle \mathcal{E}(k_1), X_{t_{k-1}} \mathcal{E}(k_2) \rangle \int_{t_{k-1}}^{t_k} \langle h, k_2(s) \rangle_\mathfrak{h} ds & \text{if } \varepsilon = -, \\[2mm] \sum_{k=1}^n \langle \mathcal{E}(k_1), X_{t_{k-1}} \mathcal{E}(k_2) \rangle \int_{t_{k-1}}^{t_k} \langle k_1(s), h k_2(s) \rangle_\mathfrak{h} ds & \text{if } \varepsilon = \circ. \end{cases}$$

Note that for $\varepsilon \in \{-, +\}$, h is a vector in $\mathfrak{h}$, whereas for $\varepsilon = \circ$, it is an operator on $\mathfrak{h}$. These expressions can be derived from the action of the creation, annihilation, and conservation operators on exponential vectors. Note that the adaptedness condition insures that X_t commutes with increments of creation, annihilation, and conservation processes, *i.e.*, operators of the form $a_{t+s}^\varepsilon(h) - a_t^\varepsilon(h)$ for $s > 0$.

The values of quantum stochastic integrals between exponential vectors are given by the following First Fundamental Lemma, cf. [87, Proposition 25.9].

Theorem 5.3.2 *Let* $(X_t)_{t \geq 0}$ *be an adapted smooth quantum stochastic process,* $\varepsilon \in \{-, \circ, +\}$, $h \in \mathfrak{h}$ *if* $\varepsilon \in \{-, +\}$ *and* $h \in B(\mathfrak{h})$ *if* $\varepsilon = \circ$, *and* $k_1, k_2 \in H$. *Then the evaluation of*

$$I(t) = \int_0^t X_s da_s^\varepsilon(h)$$

between exponential vectors is given by

$$\langle \mathcal{E}(k_1), I(t)\mathcal{E}(k_2) \rangle =$$

$$\begin{cases} \int_0^t \langle \mathcal{E}(k_1), X_t\mathcal{E}(k_2) \rangle \langle k_1(s), h \rangle_\mathfrak{h} ds, & \text{if } \varepsilon = +, \\ \int_0^t \langle \mathcal{E}(k_1), X_t\mathcal{E}(k_2) \rangle \langle h, k_2(s) \rangle_\mathfrak{h} ds, & \text{if } \varepsilon = -, \\ \int_0^t \langle \mathcal{E}(k_1), X_t\mathcal{E}(k_2) \rangle \langle k_1(s), hk_2(s) \rangle_\mathfrak{h} ds, & \text{if } \varepsilon = \circ. \end{cases}$$

5.4 Quantum Itô table

A weak form of the analogue of Itô's formula for products of quantum stochastic integrals is given by the Second Fundamental Lemma.

Theorem 5.4.1 (Second Fundamental Lemma, [87, Proposition 25.10]) *Let* $(X_t)_{t \geq 0}$ *and* $(X_t)_{t \geq 0}$ *be adapted smooth quantum stochastic processes,* $\varepsilon, \delta \in \{-, \circ, +\}$, $h \in \mathfrak{h}$ *if* $\varepsilon \in \{-, +\}$, $h \in B(\mathfrak{h})$ *if* $\varepsilon = \circ$, $k \in \mathfrak{h}$ *if* $\delta \in \{-, +\}$, *and* $k \in B(\mathfrak{h})$ *if* $\delta = \circ$. *Set*

$$I_t = \int_0^t X_s da_s^\varepsilon(h), \quad \text{and} \quad J_t = \int_0^t Y_s da_s^\delta(k), \qquad t \in \mathbb{R}_+.$$

Then we have

$$\langle I_t\mathcal{E}(\ell_1), J_t\mathcal{E}(\ell_2) \rangle = \int_0^t \langle X_s\mathcal{E}(\ell_1), J_s\mathcal{E}(\ell_2) \rangle m_1(s) ds$$

$$+ \int_0^t \langle I_s\mathcal{E}(\ell_1), Y_s\mathcal{E}(\ell_2) \rangle m_2(s) ds + \int_0^t \langle X_s\mathcal{E}(\ell_1), Y_s\mathcal{E}(\ell_2) \rangle m_{12}(s) ds,$$

for all $\ell_1, \ell_2 \in H$, *where the functions* m_1, m_2, *and* m_{12} *are given by the following tables:*

ε	$-$	$\circ$	$+$
m_1	$\langle \ell_1(s), h \rangle_\mathfrak{h}$	$\langle h\ell_1(s), \ell_2(s) \rangle_\mathfrak{h}$	$\langle h, \ell_2(s) \rangle_\mathfrak{h}$

δ	$-$	$\circ$	$+$
m_2	$\langle k, \ell_2(s) \rangle_\mathfrak{h}$	$\langle \ell_1(s), k\ell_2(s) \rangle_\mathfrak{h}$	$\langle \ell_1(s), k \rangle_\mathfrak{h}$

$\varepsilon \backslash \delta$	$-$	$\circ$	$+$
$-$	0	0	0
$\circ$	0	$\langle h\ell_1, k\ell_2 \rangle_\mathfrak{h}$	$\langle h\ell_1, k \rangle_\mathfrak{h}$
$+$	0	$\langle h, k\ell_2 \rangle_\mathfrak{h}$	$\langle h, k \rangle_\mathfrak{h}$

A stronger form of the Itô formula, which holds on an appropriate domain and under appropriate conditions on the integrands, is

$$I_t J_t = \int_0^t I_s dJ_s + \int_0^t dI_s J_s + \int_0^t (dI \bullet dJ)_s,$$

where the product in the last term is computed according to the rule

$$\left(X_t da_t^\varepsilon(h) \right) \bullet \left(Y_t da_t^\delta(k) \right) = X_t Y_t \left(da_t^\varepsilon(h) \bullet da_t^\delta(k) \right)$$

and the Itô table

$\bullet$	$da_t^-(k)$	$da_t^\circ(k)$	$da_t^+(k)$
$da_t^+(h)$	0	0	0
$da_t^\circ(h)$	0	$da^\circ(hk)$	$da^+(hk)$
$da_t^-(h)$	0	$da^-(k^*h)$	$\langle h, k \rangle dt$

If one adds the differential dt and sets all products involving dt equal to zero, then

$$\text{span}\left(\{ da^+(h) : h \in \mathfrak{h} \} \cup \{ da^\circ(T) : T \in B(\mathfrak{h}) \} \cup \{ da^-(h) : h \in \mathfrak{h} \} \cup \{ dt \} \right)$$

becomes an associative algebra with the Itô product $\bullet$ called the *Itô algebra* over $\mathfrak{h}$. If $\dim \mathfrak{h} = n$, then the Itô algebra over $\mathfrak{h}$ has dimension $(n+1)^2$.

Example 5.4.2

To realise classical Brownian motion on a Fock space, we can take $\mathfrak{h} = \mathbb{C}$ and set

$$B_t := \int_0^t da_s^- + \int_0^t da_s^+,$$

where we wrote $a_s^\pm$ for $a_s^\pm(1)$. Then the quantum stochastic Itô formula given earlier shows that

$$B_t^2 = 2 \int_0^t B_s \, da_s^- + 2 \int_0^t B_s \, da_s^+ + \int_0^t ds = 2 \int_0^t B_s \, dB_s + t,$$

i.e., we recover the well-known result from classical Itô calculus. The integral for B_t can of course be computed explicitly, we get

$$B_t = a_t^- + a_t^+ = a^-(\mathbf{1}_{[0,t]}) + a^+(\mathbf{1}_{[0,t]}).$$

We have already shown in Section 3.1 that the sum of the creation and the annihilation operator are Gauss distributed.

Notes

See [16] for more information on positive definite functions and the proofs of the results quoted in Section 5.1. Guichardet [50] has given another construction of symmetric Fock space for the case where the Hilbert space H is the space of square integrable functions on a measurable space $(M, \mathcal{M}, m)$. This representation is used for another approach to quantum stochastic calculus, the so-called kernel calculus [69, 73].

P.-A. Meyer's book [79] gives an introduction to quantum probability and quantum stochastic calculus for readers who already have some familiarity with classical stochastic calculus. Other introductions to quantum stochastic calculus on the symmetric Fock space are [17, 70, 87]. For an abstract approach to Itô algebras, we refer to [15]. Noncausal quantum stochastic integrals, *i.e.*, integrals with integrands that are not adapted, were defined and studied by Belavkin and Lindsay, see [13, 14, 68]. The recent book by M.-H. Chang [26] focusses on the theory of quantum Markov processes.

We have not treated here the stochastic calculus on the free Fock space which was introduced by Speicher and Biane, see [19] and the reference therein. Free probability is intimately related to random matrix theory, cf. [9]. More information on the methods and applications of free probability can also be found in [81, 120, 121].

Exercises

Exercise 5.1 We want to show that the tensor powers $v^{\otimes n}$ do indeed span the symmetric tensor powers $H^{\circ n}$. For this purpose, prove the following *polarisation formulas*:

$$\mathcal{S}_n(v_1 \otimes \cdots \otimes v_n) = \frac{1}{n!2^n} \sum_{\epsilon \in \{\pm 1\}^n} \left(\prod_{k=1}^{n} \epsilon_k \right) (\epsilon_1 v_1 + \cdots + \epsilon_n v_n)^{\otimes n}$$

$$= \frac{1}{n!} \sum_{k=1}^{n} (-1)^{n-k} \sum_{l_1 < \cdots < l_k} (v_{l_1} + \cdots + v_{l_k})^{\otimes n},$$

for $n \geq 0$ and $v_1, \ldots, v_n \in H$, where the first summation runs over n-tuples $\epsilon = (\epsilon_1, \ldots, \epsilon_n)$ with coefficients -1 or $+1$.

6

Random variables on real Lie algebras

Any good theorem should have several proofs, the more the better.
For two reasons: usually, different proofs have different strengths and
weaknesses, and they generalise in different directions – they are not
just repetitions of each other.

(M. Atiyah, in Interview with M. Atiyah and I. Singer.*)*

In Chapter 3 we have considered the distribution of random variables on real
Lie algebras by specifying *ad hoc* Hilbert space representations. In this chapter
we revisit this construction via a more systematic approach based on the
framework of Chapter 4 and the splitting Lemma 6.1.3. We start as previously
from the Heisenberg–Weyl and oscillator Lie algebras, and then move on to
the Lie algebra $\mathfrak{sl}_2(\mathbb{R})$.

6.1 Gaussian and Poisson random variables on $\mathfrak{osc}$

Consider the oscillator algebra $\mathfrak{osc}$ with the relations

$$[N, a^-] = -a^-, \quad [N, a^+] = a^+,$$

and

$$(a^+)^* = a^-, \quad a^- e_0 = 0, \quad N e_0 = 0.$$

For $\alpha, \beta \in \mathbb{R}$ and $\zeta \in \mathbb{C}$, let $X_{\alpha,\zeta,\beta}$ denote the random variable

$$X_{\alpha,\zeta,\beta} := \alpha N + \zeta a^+ + \bar{\zeta} a^- + \beta E.$$

Proposition 6.1.1 *Let $\alpha, \beta \in \mathbb{R}$ and $\zeta \in \mathbb{C}$. The distribution of $X_{\alpha,\zeta,\beta}$ in the
vacuum vector e_0 has the characteristic function*

$$\langle e_0, e^{iX_{\alpha,\varsigma,\beta}} e_0 \rangle = \begin{cases} e^{i\beta - |\varsigma|^2/2}, & \text{for} \quad \alpha = 0, \\ e^{i\beta + |\varsigma|^2 (e^{i\alpha} - i\alpha - 1)/\alpha^2}, & \text{for} \quad \alpha \neq 0. \end{cases}$$

As a consequence of Proposition 6.1.1, $X_{0,\varsigma,\beta}$ is a Gaussian random variable with variance $|\varsigma|^2$ and mean β, while when $\alpha \neq 0$, $X_{\alpha,\varsigma,\beta}$ is a Poisson random variable with "jump size" α, intensity $|\varsigma|^2/\alpha^2$, and drift $\beta - |\varsigma|^2/\alpha$. The proof of Proposition 6.1.1 is a direct consequence of the splitting Lemma 6.1.3 which itself follows from the next Lemma 6.1.2 that gives the normally ordered form of the generalised Weyl operators.

Lemma 6.1.2   *Let $z \in \mathbb{C}$. We have the commutation relations*

$$e^{zN} a^- = e^{-z} a^- e^{zN}, \; e^{za^+} a^- = (a^- - zE) e^{za^+}, \; e^{za^+} N = (N - za^+) e^{za^+},$$

on the boson Fock space.

Proof: This can be deduced from the formula for the adjoint actions,

$$\begin{aligned} \mathrm{Ad}\, e^X Y :\; &= e^X Y e^{-X} \\ &= \sum_{n,m=0}^{\infty} \frac{(-1)^m}{n!m!} X^n Y X^m \\ &= \sum_{k=0}^{\infty} \frac{1}{k!} \sum_{m=0}^{k} \binom{k}{m} (-1)^m X^{k-m} Y X^m \\ &= Y + [X, Y] + \frac{1}{2}[X, [X, Y]] + \cdots \\ &= e^{\mathrm{ad}\, X} Y, \end{aligned}$$

cf. Section A.5 in appendix. For the first relation we get

$$\begin{aligned} \mathrm{Ad}\, e^{zN} a^- &= e^{zN} a^- e^{-zN} \\ &= e^{z\,\mathrm{ad}\, N} a^- \\ &= a^- + z[N, a^-] + \frac{z^2}{2}[N, [N, a^-]] + \frac{z^3}{3!}[N, [N, [N, a^-]]] + \cdots \\ &= a^- - za^- + \frac{z^2}{2} a^- - \frac{z^3}{3!} a^- + \cdots \\ &= \sum_{n=0}^{\infty} (-1)^n \frac{z^n}{n!} a^- \\ &= e^{-z} a^-. \end{aligned}$$

For the second relation we have

$$\mathrm{Ad}\, e^{za^+} a^- = e^{za^+} a^- e^{-za^+} = e^{z \,\mathrm{ad}\, a^+} a^-$$

$$= a^- + z[a^+, a^-] + \frac{z^2}{2}[a^+, [a^+, a^-]] + \cdots$$

$$= a^- - z[a^-, a^+]$$

$$= a^- - zE.$$

For the last relation we find

$$\mathrm{Ad}\, e^{za^+} N = e^{za^+} N e^{-za^+} = e^{z \,\mathrm{ad}\, a^+} N$$

$$= N + z[a^+, N] + \frac{z^2}{2}[a^+, [a^+, N]] + \cdots$$

$$= N + z[a^+, N]$$

$$= N - za^+.$$ □

The following formula, also known as the splitting lemma, cf. Proposition 4.2.1 in Chapter 1 of [38], provides the normally ordered form of the Weyl operators and is a key tool to calculate characteristic functions of elements of the oscillator algebra.

Lemma 6.1.3 *(Splitting lemma) Let $x, u, v, \alpha \in \mathbb{C}$. We have*

$$\exp\left(xN + ua^+ + va^- + \alpha E\right)$$
$$= \exp\left(\frac{u}{x}(e^x - 1)a^+\right) e^{\alpha + xN} \exp\left(\frac{v}{x}(e^x - 1)a^-\right) \exp\left(\frac{uv}{x^2}(e^x - x - 1)\right).$$

In particular, when $x = 0$ we have

$$\exp(ua^+ + va^- + \alpha E) = e^{ua^+} e^{va^-} e^{(\alpha + uv/2)E}, \qquad u, v, \alpha \in \mathbb{C}.$$

Proof: We will show that

$$\exp(xN + ua^+ + va^- + \alpha E) = e^{\tilde\alpha E} e^{\tilde u a^+} e^{xN} e^{\tilde v a^-}$$

on the boson Fock space, where

$$\begin{cases} \tilde u = \displaystyle\sum_{n=1}^{\infty} \frac{x^{n-1}}{n!} u = \frac{u}{x}(e^x - 1), \\[4ex] \tilde v = \displaystyle\sum_{n=1}^{\infty} \frac{x^{n-1}}{n!} v = \frac{v}{x}(e^x - 1), \\[4ex] \tilde\alpha = \alpha + uv \displaystyle\sum_{n=2}^{\infty} \frac{x^{n-2}}{n!} = \alpha + \frac{uv}{x^2}(e^x - x - 1). \end{cases}$$

Set

$$\omega_1(t) = \exp\left(t(xN + ua^+ + va^- + \alpha E)\right)$$

and

$$\omega_2(t) = e^{\tilde{u}(t)a^+} e^{txN} e^{\tilde{v}(t)a^-} e^{\tilde{\alpha}(t)E},$$

$t \in [0, 1]$, where

$$\begin{cases} \tilde{u}(t) := u \sum_{n=1}^{\infty} x^{n-1} \frac{t^n}{n!} = \frac{u}{x}(e^{tx} - 1), \\[3mm] \tilde{v}(t) := v \sum_{n=1}^{\infty} x^{n-1} \frac{t^n}{n!} = \frac{v}{x}(e^{tx} - 1), \\[3mm] \tilde{\alpha}(t) = t\alpha + uv \sum_{n=2}^{\infty} x^{n-2} \frac{t^n}{n!} = t\alpha + \frac{uv}{x^2}(e^{tx} - tx - 1). \end{cases}$$

We have

$$\omega_1'(t) = \left(xN + ua^+ + va^- + \alpha E\right) \exp(t(xN + ua^+ + va^- + \alpha E)),$$

and

$$\begin{aligned} \omega_2'(t) &= \tilde{u}'(t)a^+ e^{\tilde{u}(t)a^+} e^{txN} e^{\tilde{v}(t)a^-} e^{\tilde{\alpha}(t)E} + x e^{\tilde{u}(t)a^+} N e^{txN} e^{\tilde{v}(t)a^-} e^{\tilde{\alpha}(t)E} \\ &\quad + \tilde{v}'(t) e^{\tilde{u}(t)a^+} e^{txN} a^- e^{\tilde{v}(t)a^-} e^{\tilde{\alpha}(t)E} + \alpha'(t) e^{\tilde{u}(t)a^+} e^{txN} e^{\tilde{v}(t)a^-} E e^{\tilde{\alpha}(t)E} \\ &= u e^{tx} a^+ e^{\tilde{u}(t)a^+} e^{txN} e^{\tilde{v}(t)a^-} e^{\tilde{\alpha}(t)E} + x e^{\tilde{u}(t)a^+} N e^{txN} e^{\tilde{v}(t)a^-} e^{\tilde{\alpha}(t)E} \\ &\quad + v e^{tx} e^{\tilde{u}(t)a^+} e^{txN} a^- e^{\tilde{v}(t)a^-} e^{\tilde{\alpha}(t)E} \\ &\quad + \left(\alpha + \frac{uv}{x}(e^{tx} - 1)\right) e^{\tilde{u}(t)a^+} e^{txN} e^{\tilde{v}(t)a^-} E e^{\tilde{\alpha}(t)E} \\ &= u e^{tx} a^+ e^{\tilde{u}(t)a^+} e^{txN} e^{\tilde{v}(t)a^-} e^{\tilde{\alpha}(t)E} \\ &\quad + x(N - \tilde{u}(t)a^+) e^{\tilde{u}(t)a^+} e^{txN} e^{\tilde{v}(t)a^-} e^{\tilde{\alpha}(t)E} + v e^{\tilde{u}(t)a^+} a^- e^{txN} e^{\tilde{v}(t)a^-} e^{\tilde{\alpha}(t)E} \\ &\quad + \left(\alpha + \frac{uv}{x}(e^{tx} - 1)\right) E e^{\tilde{u}(t)a^+} e^{txN} e^{\tilde{v}(t)a^-} e^{\tilde{\alpha}(t)E} \\ &= u e^{tx} a^+ e^{\tilde{u}(t)a^+} e^{txN} e^{\tilde{v}(t)a^-} e^{\tilde{\alpha}(t)E} \\ &\quad + x(N - \tilde{u}(t)a^+) e^{\tilde{u}(t)a^+} e^{txN} e^{\tilde{v}(t)a^-} e^{\tilde{\alpha}(t)E} \\ &\quad + v(a^- - \tilde{u}(t)) e^{\tilde{u}(t)a^+} e^{txN} e^{\tilde{v}(t)a^-} e^{\tilde{\alpha}(t)E} \\ &\quad + \left(\alpha + \frac{uv}{x}(e^{tx} - 1)\right) E e^{\tilde{u}(t)a^+} e^{txN} e^{\tilde{v}(t)a^-} e^{\tilde{\alpha}(t)E} \end{aligned}$$

$$= ua^+ e^{\tilde{u}(t)a^+} e^{txN} e^{\tilde{v}(t)a^-} e^{\tilde{\alpha}(t)E} + xN e^{\tilde{u}(t)a^+} e^{txN} e^{\tilde{v}(t)a^-} e^{\tilde{\alpha}(t)E}$$

$$+ va^- e^{\tilde{u}(t)a^+} e^{txN} e^{\tilde{v}(t)a^-} e^{\tilde{\alpha}(t)E}$$

$$+ \left(\alpha + \frac{uv}{x}(e^{tx} - 1) - v\tilde{u}(t) \right) E e^{\tilde{u}(t)a^+} e^{txN} e^{\tilde{v}(t)a^-} e^{\tilde{\alpha}(t)E}$$

$$= ua^+ e^{\tilde{u}(t)a^+} e^{txN} e^{\tilde{v}(t)a^-} e^{\tilde{\alpha}(t)E} + xN e^{\tilde{u}(t)a^+} e^{txN} e^{\tilde{v}(t)a^-} e^{\tilde{\alpha}(t)E}$$

$$+ va^- e^{\tilde{u}(t)a^+} e^{txN} e^{\tilde{v}(t)a^-} e^{\tilde{\alpha}(t)E} + \alpha E e^{\tilde{u}(t)a^+} e^{txN} e^{\tilde{v}(t)a^-} e^{\tilde{\alpha}(t)E}$$

$$= ua^+ e^{\tilde{u}(t)a^+} e^{txN} e^{\tilde{v}(t)a^-} e^{\tilde{\alpha}(t)E} + xN e^{\tilde{u}(t)a^+} e^{txN} e^{\tilde{v}(t)a^-} e^{\tilde{\alpha}(t)E}$$

$$+ va^- e^{\tilde{u}(t)a^+} e^{txN} e^{\tilde{v}(t)a^-} e^{\tilde{\alpha}(t)E} + \alpha E e^{\tilde{u}(t)a^+} e^{txN} e^{\tilde{v}(t)a^-} e^{\tilde{\alpha}(t)E}$$

$$= \left(xN + ua^+ + va^- + \alpha E \right) \exp(t(xN + ua^+ + va^- + \alpha E)),$$

where, using Lemma 6.1.2, we checked that both expressions coincide for all $t \in [0, 1]$ since $\omega_1(0) = \omega_2(0) = 1$. Therefore, we have $\omega_1(t) = \omega_2(t)$, $t \in \mathbb{R}_+$, which yields the conclusion. $\qquad\square$

We find in particular

$$\exp(uQ) = \exp(ua^- + ua^+) = e^{ua^+} e^{ua^-} e^{uvE/2}.$$

Proof of Proposition 6.1.1: This is now a consequence of the splitting Lemma 6.1.3, using the relations $(a^+)^* = a^-$ and $a^- e_0 = N e_0 = 0$. $\qquad\square$

We close this section with a lemma on cyclicity of the vacuum state e_0 and will be used for the Girsanov theorem in Chapter 10.

Lemma 6.1.4 *For any $\zeta \neq 0$, the vacuum vector e_0 is cyclic for $X_{\alpha,\zeta,\beta}$, i.e.,*

$$\overline{\mathrm{span}\, \{X_{\alpha,\zeta,\beta}^k e_0 \ : \ k = 0, 1, \ldots\}} = \ell^2.$$

Proof: Due to the creation operator a^+ in the definition of $X_{\alpha,\zeta,\beta}$ we have

$$X_{\alpha,\zeta,\beta}^k e_0 = \zeta^k \sqrt{k!} e_k + \sum_{\ell=0}^{k-1} c_\ell e_\ell,$$

for some coefficients $c_\ell \in \mathbb{C}$, therefore we have

$$\mathrm{span}\, \{e_0, X_{\alpha,\zeta,\beta} e_0, \ldots, X_{\alpha,\zeta,\beta}^k e_0\} = \mathrm{span}\, \{e_0, \ldots, e_k\},$$

for all $k \in \mathbb{N}$, if $\zeta \neq 0$. $\qquad\square$

6.2 Meixner, gamma, and Pascal random variables on $\mathfrak{sl}_2(\mathbb{R})$

As in the case of the oscillator algebra $\mathfrak{osc}$ we will need the following version of the splitting lemma, cf. [38, Chapter 1, Proposition 4.3.1].

Lemma 6.2.1   *([38]) For any* $x, u, v \in \mathbb{R}$ *we have*

$$\exp(uB^+ + xM + vB^-) = \exp(\tilde{u}B^+)\exp(\tilde{x}M)\exp(\tilde{v}B^-)$$

on the boson Fock space, where

$$\begin{cases} \tilde{u} = \dfrac{\tanh(\delta u)}{\delta - (x/u)\tanh(\delta u)}, \\[3mm] \tilde{x} = \log\left(\dfrac{\delta\,\mathrm{sech}\,(\delta u)}{\delta - (x/u)\tanh(\delta u)}\right), \\[3mm] \tilde{v} = v\dfrac{\tilde{u}}{u} = \dfrac{v\tanh(\delta u)}{u(\delta - (x/u)\tanh(\delta u))}, \end{cases}$$

with $\delta = \sqrt{x^2 - v^2}$ *and* $\mathrm{sech}\, x = 1/\cosh x$.

Using the above splitting Lemma 6.2.1 for $\mathfrak{sl}_2(\mathbb{R})$ we write e^{X_β} with

$$X_\beta = B^+ + B^- + \beta M, \qquad \beta \in \mathbb{R},$$

as a product

$$e^{X_\beta} = e^{v_+ B^+}e^{v_0 M}e^{v_- B^-}.$$

It is straightforward to compute the distribution of $\rho_\lambda(X_\beta)$ in the state vector e_0 as in the following proposition in which we consider the representation $\rho_\lambda(X_\beta)$ of X_β, which satisfies

$$\rho_\lambda(M)e_0 = \lambda e_0 \quad \text{and} \quad \rho_\lambda(B^-)e_0 = 0.$$

Using the splitting lemma to write $e^{\lambda X_\beta}$ as a product $e^{v_+ B^+}e^{v_0 M}e^{v_- B^-}$, the distribution of $\rho_\lambda(X_\beta)$ in the state vector e_0 is computed in the next proposition which will also be used for the Girsanov theorem in Chapter 10.

Proposition 6.2.2   *The Fourier–Laplace transform of the distribution of* $\rho_\lambda(X_\beta)$ *with respect to* e_0 *is given by*

$$\langle e_0, e^{\rho_\lambda(X_\beta)}e_0\rangle = \left(\frac{\sqrt{\beta^2 - 1}}{\sqrt{\beta^2 - 1}\cosh\left(\sqrt{\beta^2 - 1}\right) - \beta\sinh\left(\sqrt{\beta^2 - 1}\right)}\right)^\lambda.$$

For $\beta = 0$ we find the Fourier transform $(\cosh \lambda)^{-\lambda}$ which corresponds to the hyperbolic secant distribution.

More generally when $|\beta| < 1$, the above distribution is called the Meixner distribution. It is absolutely continuous with respect to the Lebesgue measure and the density is given by

$$C \exp\left(\frac{(\pi - 2\arccos\beta)x}{2\sqrt{1-\beta^2}}\right)\left|\Gamma\left(\frac{\lambda}{2} + \frac{ix}{2\sqrt{1-\beta^2}}\right)\right|^2,$$

where C is a normalisation constant, see also [1]. When $\beta = 0$ and $\lambda = 1$ we find the density

$$\xi_1 \mapsto C\left|\Gamma\left(\frac{1}{2} + \frac{i}{2}\xi_1\right)\right|^2 = \frac{1}{2\cosh(\pi\xi_1/2)},$$

of the continuous binomial distribution, with $C = 1/\pi$. For $\beta = \pm 1$ we get the gamma distribution, which has the density

$$\frac{1}{\Gamma(\lambda)}|x|^{\lambda-1}e^{-\beta x}\mathbf{1}_{\beta\mathbb{R}_+}.$$

Finally, for $|\beta| > 1$, we get a discrete measure, the negative binomial distribution (also called Pascal distribution). The next cyclicity lemma will also be used for the Girsanov theorem in Chapter 10.

Lemma 6.2.3 *The lowest weight vector e_0 is cyclic for $\rho_\lambda(X_\beta)$ for all $\beta \in \mathbb{R}$, $\lambda > 0$.*

Proof: On e_0, we get

$$\rho_\lambda(X_\beta)^k e_0 = \sqrt{k!\lambda(\lambda+1)\cdots(\lambda+k-1)}e_k + \sum_{\ell=0}^{k-1}c_\ell e_\ell$$

for some coefficients $c_\ell \in \mathbb{C}$. Therefore,

$$\text{span}\{e_0, \rho_\lambda(X_\beta)e_0, \ldots, \rho_\lambda(X_\beta)^k e_0\} = \text{span}\{e_0, \ldots, e_k\},$$

for all $k \in \mathbb{N}$, if $\lambda > 0$. $\qquad\square$

6.3 Discrete distributions on $so(2)$ and $so(3)$

6.3.1 The Bernoulli distribution on $so(2)$

Recall that the Lie algebra $so(2)$ is generated by

$$\xi_1 = \begin{bmatrix} 0 & i \\ -i & 0 \end{bmatrix}.$$

We can check that the Bernoulli distribution is generated by $so(2)$. In other words taking $a, b \in \mathbb{C}$ such that $|a|^2 + |b|^2 = 1$ we can compute the characteristic function

$$\left\langle \begin{bmatrix} a \\ b \end{bmatrix}, \exp\left(\theta \begin{bmatrix} 0 & i \\ -i & 0 \end{bmatrix}\right) \begin{bmatrix} a \\ b \end{bmatrix}\right\rangle = \left\langle \begin{bmatrix} a \\ b \end{bmatrix}, \exp\left(\begin{bmatrix} 0 & -i\theta \\ -i\theta & 0 \end{bmatrix}\right) \begin{bmatrix} a \\ b \end{bmatrix}\right\rangle$$

$$= \left\langle \begin{bmatrix} \cosh\theta & \sinh\theta \\ -\sinh\theta & \cosh\theta \end{bmatrix} \begin{bmatrix} a \\ b \end{bmatrix}, \begin{bmatrix} a \\ b \end{bmatrix}\right\rangle$$

$$= (|a|^2 + |b|^2)\cosh\theta - 2i\Im(\bar{a}b)\sinh\theta$$

$$= \cosh\theta - i(p-q)\sinh\theta$$

$$= (p+q)\cosh\theta - i(p-q)\sinh\theta$$

$$= pe^\theta + qe^{-\theta},$$

with $\bar{a}b - \bar{b}a = 2i\Im(ab) = 2i(p-q)$ and $|a|^2 + |b|^2 = p+q = 1$. This yields the characteristic function of the Bernoulli distribution

$$q\delta_{-1} + p\delta_1$$

with

$$p = \frac{1}{4i}(\bar{a}b - \bar{b}a + 2i|a|^2 + i|b|^2) \text{ and } q = \frac{1}{4i}(-\bar{a}b + \bar{b}a + 2i|a|^2 + 2i|b|^2),$$

which is supported by $\{-1, 1\}$.

6.3.2 The three-point distribution on $so(3)$

Recall that Lie algebra $so(3)$ is generated by

$$\xi_1 = \begin{bmatrix} 0 & -1 & 0 \\ 1 & 0 & 0 \\ 0 & 0 & 0 \end{bmatrix}, \quad \xi_2 = \begin{bmatrix} 0 & 0 & 1 \\ 0 & 0 & 0 \\ -1 & 0 & 0 \end{bmatrix}, \quad \xi_3 = \begin{bmatrix} 0 & 0 & 0 \\ 0 & 0 & -1 \\ 0 & 1 & 0 \end{bmatrix}.$$

The probability distribution associated to $so(3)$ is supported by three points.

6.4 The Lie algebra $e(2)$

Consider the representation of the Euclidean Lie algebra $e(2)$ on $\ell^2(\mathbb{Z})$ given by

$$Me_k = 2ke_k, \qquad E^+e_k = e_{k+1}, \qquad E^-e_k = e_{k-1},$$

where $(e_k)_{k\in\mathbb{Z}}$ denotes the standard basis of $\ell^2(\mathbb{Z})$. Let us now give a probabilistic interpretation of $e(2)$, *i.e.*, compute the distribution of the quantum random variable

$$E_{\alpha,\zeta} = \alpha M + \zeta E^+ + \bar{\zeta}E^-$$

with $\alpha \in \mathbb{R}, \zeta \in \mathbb{C}$.

By the version (6.1) of the splitting lemma proved in Exercise 6.1 it is easy to compute the moment generating function of the law of $E_{\alpha,\zeta}$ in the state given by the vector e_0. We have

$$\exp(zE^-)e_0 = \sum_{k=0}^{\infty} \frac{z^k}{k!} e_{-k}$$

and

$$\langle e_0, \exp(\lambda E_{\alpha,\zeta})e_0 \rangle$$

$$= \left\langle \exp\left(\frac{\zeta}{2\alpha}\left(e^{2\lambda\alpha}-1\right)E^-\right)e_0, \exp(\lambda\alpha M)\exp\left(\frac{\overline{\zeta}}{2\alpha}\left(e^{2\lambda\alpha}-1\right)E^-\right)e_0 \right\rangle$$

$$= \sum_{k=0}^{\infty} e^{-2\alpha\lambda k}\frac{|\zeta|^{2k}}{(2\alpha k!)^2}\left(e^{2\lambda\alpha}-1\right)^{2k}$$

$$= \sum_{k=0}^{\infty} \frac{|\zeta|^{2k}}{\alpha^2 k!^2}\sinh(\alpha\lambda)^{2k}$$

$$= J_0\left(\frac{2|\zeta|}{\alpha}\sinh(\alpha\lambda)\right),$$

where J_0 denotes the modified Bessel (or hyperbolic) Bessel function of the first kind,

$$J_0(x) = \sum_{m=0}^{\infty} \frac{(-1)^m}{(m!)^2}\left(\frac{x}{2}\right)^{2m}.$$

See also Section 5.V "e2 and Lommel polynomials" in Feinsilver and Schott's, *Algebraic Structures and Operator Calculus*, Volume III: *Representations of Lie Groups*.

6.4.1 The case $\alpha = 0$

The operator E^+ is the shift on $\mathbb{Z}$ in our representation, note that it is normal, since E^+ and $E^- = (E^+)^*$ commute, and even unitary. In the state given by the vector e_0 the distribution of E^+ is the uniform distribution on the unit circle,

$$\langle e_0, E_n^+ E_m^- e_0 \rangle = \delta_{n,m}.$$

Therefore, the distribution of $E_{0,1} = E^+ + E^-$ is the image measure of the uniform distribution on the unit circle under the map $S^1 \ni z \mapsto z + \overline{z} \in [-2,2]$, which is the arcsine distribution,

$$\mathcal{L}_{e_0}(E^+ + E^-) \simeq \frac{1}{2\pi\sqrt{4-x^2}} \mathbf{1}_{(-2,2)}(x)dx.$$

6.4.2 The case $\alpha \neq 0$

For $\alpha \neq 0$, $E_{\alpha,\zeta}$ had moment generating function

$$\langle e_0, \exp(\lambda E_{\alpha,\zeta})e_0 \rangle = J_0\left(\frac{2|\zeta|\sinh(\alpha\lambda)}{\alpha}\right)$$

and characteristic function

$$\langle e_0, \exp(i\lambda E_{\alpha,\zeta})e_0 \rangle = J_0\left(\frac{2|\zeta|\sin(\alpha\lambda)}{\alpha}\right).$$

The distribution of $E_{\alpha,\zeta}$ is a discrete measure supported on $\mathbb{Z}$, we have

$$\mathcal{L}_{e_0}(E_{\alpha,\zeta}) = \sum_{m\in\mathbb{Z}} \left(J_m\left(\frac{|\zeta|}{2\alpha}\right)\right)^2 \delta_{2m\alpha},$$

cf. Theorem 5.9.2 in [40]. Here,

$$J_m(x) = \left(\frac{x}{2}\right)^m \sum_{k=0}^{\infty} \frac{(-1)^k}{k!\Gamma(k+m+1)} \left(\frac{x}{2}\right)^{2k},$$

are the Bessel functions of the first kind of order $m \geq 0$.

Notes

We also refer the reader to [118], [115], [117], [116] for additional noncommutative relations on real Lie algebras, and to [10] and references therein for further discussion of "noncommutative (or quantum) mathematics".

Exercises

Exercise 6.1 Splitting lemma for the two-dimensional Euclidean group.

The goal of this exercise is to prove the following version

$$\exp\left(xM + yE^+ + zE^-\right) \qquad (6.1)$$
$$= \exp\left(\frac{y}{2x}\left(e^{2x}-1\right)E^+\right)\exp(xM)\exp\left(\frac{z}{2x}\left(e^{2x}-1\right)E^-\right)$$

of the splitting lemma, for $x, y, z \in \mathbb{C}$, where M, E^+, E^- denote a basis of the Lie algebra of the group of rigid motions in two dimensions (*i.e.*, the

Euclidean Lie algebra). Denote by $e(2)$ the Lie algebra with basis R, T_x, T_y and the relations

$$[T_x, T_y] = 0, \quad [R, T_x] = T_y, \quad [R, T_y] = -T_x,$$

and $R^* = -R$, $T_x^* = -T_x$, $T_y^* = -T_y$.

1. Show that

$$\tilde{R} = \begin{bmatrix} 0 & -1 & 0 \\ 1 & 0 & 0 \\ 0 & 0 & 0 \end{bmatrix}, \quad \tilde{T}_x = \begin{bmatrix} 0 & 0 & 1 \\ 0 & 0 & 0 \\ 0 & 0 & 0 \end{bmatrix}, \quad \tilde{T}_y = \begin{bmatrix} 0 & 0 & 0 \\ 0 & 0 & 1 \\ 0 & 0 & 0 \end{bmatrix},$$

satisfy the same commutation relations as R, T_x, T_y.

2. Consider the affine subspace

$$K_2 = \left\{ \begin{bmatrix} x \\ y \\ 1 \end{bmatrix} : x, y \in \mathbb{R} \right\} \subseteq \mathbb{R}^3$$

and show that

$$\exp(\theta \tilde{R}), \quad \exp(v \tilde{T}_x), \quad \exp(w \tilde{T}_y)$$

act as rigid motions on K_2 (*i.e.*, maps that preserve distances).

3. Show that we can find a basis M, E^+, E^- for $e(2)$ that satisfies the relations

$$M^* = M, \quad (E^+)^* = E^-,$$

and

$$[E^+, E^-] = 0, \quad [M, E^\pm] = \pm 2E^\pm.$$

4. Show that we have

$$\begin{cases} \exp(uE^+)M = (M - 2E^+)\exp(uE^+), \\[2mm] \exp(uM)E^- = e^{-2u}E^- \exp(uM), \\[2mm] \exp(uE^+)E^- = E^- \exp(uB^+). \end{cases}$$

5. Define

$$\begin{cases} \omega_1(t) = \exp\left(t(xM + yE^+ + zE^-)\right), \\[2mm] \omega_2(t) = \exp\left(\tilde{y}(t)E^+\right)\exp\left(\tilde{x}(t)M\right)\exp\left(\tilde{z}(t)E^-\right), \end{cases}$$

with

$$\tilde{x}(t) = tx, \quad \tilde{y}(t) = \frac{y}{2x}\left(e^{2tx} - 1\right), \quad \tilde{z}(t) = \frac{z}{2x}\left(e^{2tx} - 1\right).$$

Show that we have $\omega_1(0) = \omega_2(0)$ and that $\omega_1(t)$ and $\omega_2(t)$ satisfy the same differential equation

$$\omega_j'(t) = (xM + yE^+ + zE^-)\omega_j(t), \qquad j = 1, 2.$$

6. Conclude that Equation (6.1) holds.

Exercise 6.2 Splitting lemma on the Heisenberg–Weyl algebra $\mathfrak{hw}$.

Consider the Heisenberg–Weyl algebra $\mathfrak{hw}$ generated by a^-, a^+, E with the commutation relations

$$[a^-, a^+] = E, \qquad [E, a^-] = [E, a^+] = 0.$$

The goal of this question is to prove the splitting lemma

$$\exp\left(ua^+ + va^- + wE\right) = e^{ua^+} e^{va^-} e^{(w+uv/2)E}, \qquad u, v, w \in \mathbb{C}.$$

1. Using the relation

$$e^{zX} Y e^{-zX} = Y + z[X, Y] + \frac{z^2}{2}[X, [X, Y]] + \frac{z^3}{3!}[X[X, [X, Y]]] + \cdots$$

$$= Y + \sum_{n=1}^{\infty} \frac{z^n}{n!} \underbrace{[X, [X, \ldots [X, Y] \cdots]]}_{n \text{ times}}, \qquad (6.2)$$

show that for all $z \in \mathbb{C}$ we have

   i) $e^{za^+} a^- e^{-za^+} = a^- - zE$ and $e^{za^+} a^- = (a^- - zE)e^{za^+}$,

   ii) $e^{za^+} E e^{-za^+} = E$ and $e^{za^+} E = E e^{za^+}$,

   iii) $e^{za^-} E e^{-za^-} = E$ and $e^{za^-} E = E e^{za^-}$.

2. Given $u, v, w \in \mathbb{C}$, let

$$\omega_1(t) := \exp\left(t(ua^+ + va^- + wE)\right), \qquad t \in \mathbb{R}_+.$$

Show that

$$\omega_1'(t) = (ua^+ + va^- + wE)\omega_1(t), \qquad t \in \mathbb{R}_+. \qquad (6.3)$$

3. Given $u, v, w \in \mathbb{C}$, let now

$$\omega_2(t) := e^{uta^+} e^{vta^-} e^{(tw+t^2uv/2)E}, \qquad t \in \mathbb{R}_+.$$

Show that

$$\omega_2'(t) = ua^+ e^{uta^+} e^{vta^-} e^{(tw+t^2uv/2)E}$$
$$+ ve^{uta^+} a^- e^{vta^-} e^{(tw+t^2uv/2)E}$$
$$+ (w + tuv)e^{uta^+} e^{vta^-} E e^{(tw+t^2uv/2)E}. \qquad (6.4)$$

4. Using Relations (6.3) and (6.4) and the result of Question (1), show that
$$\omega_2'(t) = (ua^+ + va^- + wE)\omega_2(t), \qquad t \in \mathbb{R}_+.$$
Show from (6.3) that, as a consequence, we have $\omega_1(t) = \omega_2(t), t \in \mathbb{R}_+$, and
$$e^{ua^+ + va^- + wE} = e^{ua^+} e^{va^-} e^{(w+uv/2)E}. \tag{6.5}$$

5. Using the splitting lemma Relation (6.5), show that when $E = \sigma^2 I$ we have
$$\langle e_0, e^{u(a^- + a^+)} e_0 \rangle = \langle e_0, e^{iu(a^- - a^+)} e_0 \rangle = e^{u^2 \sigma^2/2},$$
where e_0 is a unit vector in a Hilbert space H with inner product $\langle \cdot, \cdot \rangle$, such that $\langle e_0, e_0 \rangle = 1$ and $a^- e_0 = 0$.

6. From the result of Question (5) show that $a^- + a^+$ and $i(a^- - a^+)$ have centered Gaussian distribution with variance σ^2.

Exercise 6.3 Consider the differential operators $\tilde{a}^+, \tilde{a}^-, \tilde{a}^\circ$ defined in Section 3.3 by $\tilde{a}^- = -\tau \partial_\tau$, $\tilde{a}^+ = \tau - 1 - \tau \partial_\tau$, and $\tilde{a}^\circ = -(1 - \tau)\partial_\tau - \tau \partial_\tau^2$.

1. Show that for $s \in \mathbb{R}$ we have
$$\exp\left(-is\tilde{Q}\right) \tilde{a}^\circ \exp\left(is\tilde{Q}\right) = \tilde{a}^\circ + is\tilde{a}^+ - is\tilde{a}^- + s^2\tau,$$
and
$$\exp\left(is\tilde{P}\right) \tilde{a}^\circ \exp\left(-is\tilde{P}\right) = e^{-2s}\tilde{a}^\circ - \frac{1}{2}\sinh(2s)\left(\tilde{a}^+ + \tilde{a}^-\right) + \sinh^2(s). \tag{6.6}$$

2. Show that for any $s \in \mathbb{R}$ the operator
$$\tilde{a}^\circ + is\tilde{a}^+ - is\tilde{a}^- + s^2 x$$
has a geometric distribution with parameter $s^2/(1 + s^2)$ in the vacuum state $\mathbf{1}$, and that the operator
$$e^{-2s}\tilde{a}^\circ - \frac{1}{2}\sinh(2s)\left(\tilde{a}^+ + \tilde{a}^-\right) + \sinh^2(s)$$
has a geometric distribution with parameter $\tanh^2(s)$ in the vacuum state.

3. Conclude that the distribution of $\tilde{P}$ has the Fourier transform
$$\mathbf{E}[\exp(it\tilde{P})] = \frac{1}{\cosh(t)}, \qquad t \in \mathbb{R}.$$

7

Weyl calculus on real Lie algebras

Couples are wholes and not wholes, what agrees disagrees, the
concordant is discordant. From all things one and from one all things.
(Heraclitus, On the Universe 59.)

This chapter introduces the notion of joint (Wigner) density of random
variables, for future use in quantum Malliavin calculus. For this we will rely
on functional calculus on general Lie algebras, starting with the Heisenberg–
Weyl algebra. We also consider some applications to quantum optics and time-
frequency analysis.

7.1 Joint moments of noncommuting random variables

The notion of moment $\langle \phi, P^n \psi \rangle$ of order n of P in the state $\langle \phi, \cdot \psi \rangle$ is well
understood as the noncommutative analogue of $\mathbb{E}[X^n]$. Indeed, in the single
variable case, various functional calculi are available for elements of an
algebra. In particular,

- polynomials can be applied to any element in an algebra;
- holomorphic functions can be applied to elements of a Banach algebra
  provided the function is holomorphic on a neighborhood of the spectrum of
  the element;
- continuous functions can be applied to normal elements of C*-algebras
  provided they are continuous on the spectrum of the element;
- measurable (*i.e.*, Borel) functions can be applied to normal elements in a
  von Neumann algebra (*e.g.*, $B(\mathfrak{h})$), provided they are measurable on the
  spectrum of the operator.

In addition, the Borel functional calculus can be extended to unbounded normal operators (where normal means that XX^* and X^*X have the same domain and coincide on this domain).

All the above functional calculi work due to a restriction to a *commutative* subalgebra. In that sense, *Weyl calculus* is fundamentally different because it applies to functions of *noncommuting* operators.

We now need to focus on the meaning of joint moments. For example, which is the noncommutative analogue of $\mathbf{E}[XY]$? A possibility is to choose $\langle \phi, PQ\psi \rangle$ as the the noncommutative analogue of $\mathbf{E}[XY]$, however this would lead to $\langle \phi, QP\psi \rangle$ as the the noncommutative analogue of $\mathbf{E}[YX]$ and we have $\mathbf{E}[XY] = \mathbf{E}[YX]$ while

$$\langle \phi, PQ\psi \rangle \neq \langle \phi, QP\psi \rangle$$

due to noncommutativity. A way to solve this contradiction can be to define the first joint moment as

$$\frac{1}{2} \left(\langle \phi, QP\psi \rangle + \langle \phi, PQ\psi \rangle \right).$$

Next, how can we define the analogue of $\mathbf{E}[XY^2]$? Proceeding similarly, a natural extension could be

$$\frac{1}{3} \left(\langle \phi, PQ^2\psi \rangle + \langle \phi, QPQ\psi \rangle + \langle \phi, Q^2P\psi \rangle \right).$$

Then what is the correct way to construct $\mathbf{E}[X^2Y^3]$? Clearly there is a combinatorial pattern which should be made explicit.

We note that the exponentials $e^{iby+iax} = e^{iby}e^{iax} = e^{iby}e^{iax}$ commute, hence

$$e^{iby+iax} = e^{iby}e^{iax}$$

$$= \sum_{n,m=0}^{\infty} \frac{i^{n+m}}{n!m!} a^n b^m x^n y^m$$

$$= 1 + iax + iby - \frac{(ax+by)^2}{2} - i\frac{(ax+by)^3}{3!} + \sum_{n=4}^{\infty} \frac{(i)q^n}{n!}(ax+by)$$

$$= 1 + iax + iby - \frac{1}{2}(a^2x^2 + b^2y^2 + 2abxy)$$

$$- \frac{i}{3!}(a^3x^3 + b^3y^3 + 3ab^2xy^2 + 3a^2bx^2y) + \sum_{n=4}^{\infty} \frac{(i)^n}{n!}(ax+by)$$

$$= \sum_{n,m=0}^{\infty} \frac{i^{n+m}}{n!m!} a^n b^m x^n y^m.$$

On the other hand, the exponentials e^{ibQ} and e^{iaP} do not commute, and expanding the exponential series we get

$$e^{ibQ+iaP} = \sum_{n=0}^{\infty} \frac{1}{n!}(ibQ + iaP)^n$$

$$= I + ibQ + iaP + \frac{(ibQ + iaP)^2}{2} + \frac{(ibQ + iaP)^3}{3!} + \sum_{n=4}^{\infty} \frac{(ibQ + iaP)^n}{n!}$$

$$= I + ibQ + iaP - \frac{b^2Q^2 + a^2P^2 + abQP + abPQ}{2}$$

$$- \frac{i}{3!}\left(b^3Q^3 + a^3P^3 + ba^2(QP^2 + P^2Q + PQP) + ab^2(Q^2P + PQ^2 + QPQ)\right)$$

$$+ \sum_{n=4}^{\infty} \frac{i^n}{n!}(bQ + aP)^n,$$

and by identifying the above with the exponential series of $e^{iby+iax}$ we get

i) from the terms in ab with $k = l = 1$,

$$xy \longleftrightarrow \frac{QP + PQ}{2}.$$

ii) from the terms in ab^2 and a^2b we recover

$$\frac{y^2}{2!}x \longleftrightarrow \frac{QP^2 + P^2Q + PQP}{3!}$$

and

$$\frac{x^2}{2!}y \longleftrightarrow \frac{Q^2P + PQ^2 + QPQ}{3!}.$$

More generally, by identifying the terms in $a^k b^m$, we map

$$\frac{y^k}{k!}\frac{x^m}{m!}$$

to the coefficient of order $a^k b^m$ in

$$\frac{1}{n!}(bQ + aP)^{k+m} = \frac{1}{n!}\sum_{A \subset \{1,\dots,n\}} a^{|A|} b^{k+m-|A|} \prod_{l=1}^{n} (Q)^{\mathbf{1}_{\{l \in A\}}} (P)^{\mathbf{1}_{\{l \notin A\}}},$$

hence we map

$$\binom{k+m}{k} y^k x^m \quad \text{to} \quad \sum_{\substack{A \subset \{1,\dots,k+m\} \\ |A|=k}} \prod_{l=1}^{k+m} Q^{\mathbf{1}_{\{l \in A\}}} P^{\mathbf{1}_{\{l \notin A\}}},$$

with the correspondence

$$y^k x^m \quad \longleftrightarrow \quad \frac{k!m!}{(k+m)!} \sum_{\substack{A \subset \{1,\ldots,k+m\} \\ |A|=k}} \prod_{l=1}^{k+m} Q^{\mathbf{1}_{\{l \in A\}}} P^{\mathbf{1}_{\{l \notin A\}}}.$$

7.2 Combinatorial Weyl calculus

The above arguments can be generalised as a combinatorial Weyl calculus. Denote by $W(X_1,\ldots,X_k:j_1,\ldots,j_k)$ the set of all words in the letters $X_1,\ldots,X_k$ such that X_ℓ occurs j_ℓ times for $\ell = 1,\ldots,k$. For example, denoting the empty word by I, we have

$$\begin{cases} W(X_1,\ldots,X_k;0,\ldots,0) = \{I\}, \\[2mm] W(X_1,X_2,1,1) = \{X_1X_2, X_2X_1\} \\[2mm] W(X_1,X_2,2,1) = \{X_1^2 X_2, X_1 X_2 X_1, X_2 X_1^2\}. \end{cases}$$

The next lemma is a noncommutative extension of the multinomial identity (A.5).

Lemma 7.2.1

$$(X_1 + \cdots + X_k)^n = \sum_{\substack{j_1 \cdots j_k \geq 0 \\ j_1 + \cdots + j_k = n}} \sum_{X \in W(X_1,\ldots,X_k;j_1,\ldots,j_k)} X.$$

More generaly we can build a *combinatorial Weyl calculus* as in the next definition.

Definition 7.2.2 *Let $X_1,\ldots,X_k$ be elements in some algebra. We define a linear map from the algebra of polynomials in k variables $x_1,\ldots,x_k$ to the algebra generated by $X_1,\ldots,X_k$, by*

$$x_1^{j_1} \cdots x_j^{j_k} \quad \longleftrightarrow \quad \frac{1}{\dbinom{n}{j_1,\ldots,j_k}} \sum_{X \in W(X_1,\ldots,X_k;j_1,\ldots,j_k)} X.$$

We can also summarise the above definition as the one-to-one correspondence

$$e^{t_1 x_1 + \cdots + t_k x_k} \quad \longleftrightarrow \quad \exp(t_1 X_1 + \cdots + t_k X_k),$$

where we interpret the exponentials as formal power series. As an example, the above combinatorial Weyl calculus recovers the identifications

$$x_1 x_2 \quad \longleftrightarrow \quad \frac{1}{2}(X_1 X_2 + X_2 X_1)$$

and

$$x_1^2 x_2 \quad \longleftrightarrow \quad \frac{1}{3}(X_1^2 X_2 + X_1 X_2 X_1 + X_2 X_1^2).$$

The combinatorial Weyl calculus is a homomorphism when restricted to a subalgebra generated by a linear combination $\alpha_1 x_1 + \cdots + \alpha_k x_k$, *i.e.*, we have

$$P(\alpha_1 x_1 + \cdots + \alpha_k x_k) \quad \longleftrightarrow \quad P(\alpha_1 X_1 + \cdots + \alpha_k X_k)$$

for all $\alpha_1, \ldots, \alpha_k \in \mathbb{C}$ and all polynomials P.

7.2.1 Lie-theoretic Weyl calculus

In the "Lie-theoretic" Weyl calculus the definition of the "combinatorial" Weyl calculus is modified (condition of square integrability, appearance of the modular function σ and the Duflo-Moore operator C, see Theorem 7.4.1) to achieve some "nice" properties, such as invariance or isometry. These modifications are rather moderate in the unimodular case, but they lead to more complicated formulas in the non-unimodular case. They also destroy the nice relation we have for the marginal distributions of the Wigner function.

7.3 Heisenberg–Weyl algebra

7.3.1 Functional calculus on the Heisenberg–Weyl algebra

Let $B_2(\mathfrak{h})$ denote the space of Hilbert–Schmidt operators equipped with the scalar product

$$\langle \rho_1, \rho_2 \rangle_B := \mathrm{Tr}[\rho_1^* \rho_2], \quad \rho_1, \rho_2 \in B_2(\mathfrak{h}).$$

The operator $f(P, Q)$ is defined by

$$f(P, Q) := \int_{\mathbb{R}^2} e^{ixP + iyQ} (\mathcal{F}f)(x, y)\, dx\, dy,$$

and is also denoted by $O(f) = f(P, Q)$ with

$$O(e^{-iux - ivy}) = \exp(iuP + ivQ), \quad u, v \in \mathbb{R},$$

and the bound

$$\|f(P,Q)\|_{B_2(\mathfrak{h})} = \|O(f)\|_{B_2(\mathfrak{h})} \le C_p\|f\|_{L^p(\mathbb{R}^2)}$$

for $1 \le p \le 2$, see Lemma 7.3.2.

Definition 7.3.1 *The domain* $\mathrm{Dom}(O)$ *of the mapping* O *can be extended to the set of measurable functions* $\varphi\colon \mathbb{R}^2 \longrightarrow \mathbb{C}$ *for which there exists a bounded operator* $M \in B(L^2(\mathbb{R}))$ *on* $L^2(\mathbb{R})$ *such that*

$$\langle f, Mg \rangle = \frac{1}{2\pi}\int_{\mathbb{R}^2} \langle f, e^{iuP+ivQ}g \rangle \mathcal{F}^{-1}\varphi(u,v)dudv, \qquad (7.1)$$

for all $f,g \in \mathcal{S}(\mathbb{R})$, *and for* $\varphi \in \mathrm{Dom}\,O$ *we define* $O(\varphi)$ *to be the bounded operator* M *appearing in Equation (7.1), it is uniquely determined due to the totality of* $\mathcal{S}(\mathbb{R})$.

If $\varphi \in \mathcal{S}(\mathbb{R})$ is a Schwartz function on $\mathbb{R}^2$ one can check that

$$\varphi(P,Q) = O(\varphi) = \frac{1}{2\pi}\int_{\mathbb{R}^2} e^{iuP+ivQ}\mathcal{F}^{-1}\varphi(u,v)dudv$$

defines a bounded operator, and the map

$$O\colon \mathcal{S}(\mathbb{R}^2) \longrightarrow B(\mathfrak{h})$$

defined in this way extends to a continuous map from $L^p(\mathbb{R})$ to $B(\mathfrak{h})$ for all $p \in [1,2]$, as shown in the next lemma.

Lemma 7.3.2 *Let* $1 \le p \le 2$. *Then we have* $L^p(\mathbb{R}^2) \subseteq \mathrm{Dom}\,O$ *and there exists a constant* C_p *such that*

$$\|O(\varphi)\| = \|f(P,Q)\| \le C_p\|\varphi\|_p$$

for all $\varphi \in L^p(\mathbb{R}^2)$.

Proof: This follows immediately from [125, Theorem 11.1], where it is stated for the irreducible unitary representation with parameter $\hbar = 1$ of the Heisenberg–Weyl group. □

In case $p > 2$ there exist functions in $L^p(\mathbb{R}^2)$ for which we cannot define a bounded operator in this way, see, e.g., [125] and references therein. Nevertheless, since

$$\frac{1}{2\pi}\mathcal{F}^{-1}e^{ix_0u+iy_0v} = \delta_{(x_0,y_0)},$$

the map O can be extended to exponential functions, with the relation

$$O\!\left(e^{ix_0u+iy_0v}\right) = \exp\left(ix_0P + iy_0Q\right).$$

7.3.2 Wigner functions on the Heisenberg–Weyl algebra

Our goal is to define a noncommutative analogue of the joint probability density of a couple of noncommutative random variables. Here we are concerned with the notion of joint distribution for the couple (P, Q). Namely we define a noncommutative analogue of the joint probability density of the couple (P, Q) on the Heisenberg–Weyl algebra $\mathfrak{hw}$ with $\{P, Q, I\}$ with $[P, Q] = 2iI$. The "joint density" of the pair (P, Q) will be represented by its Wigner distribution.

Definition 7.3.3 *Let* Φ *be a state on* $B(\mathfrak{h})$. *We will call* $W_\Phi(dx)$ *the* Wigner distribution *of* (P, Q) *in the state* Φ, *if the relation*

$$\int_{-\infty}^{\infty} \varphi(x) W_\Phi(dx) = \Phi\big(O(\varphi)\big)$$

is satisfied for all Schwartz functions $\varphi \in \mathcal{S}(\mathbb{R})$.

In general, $W_\Phi(dx)$ may not be positive and it is a signed measure, since O does not map positive functions to positive operators. However, we can show in the next proposition that it admits a density.

Proposition 7.3.4 *Let* Φ *be a state on* $B(\mathfrak{h})$. *Then there exists a function* $w_\Phi \in \bigcap_{2 \leq p \leq \infty} L^p(\mathbb{R}^2)$ *such that*

$$W_\Phi(dx, dy) = w_\Phi(x, y) dx dy.$$

We call w_Φ the *Wigner function* of (P, Q) in the state Φ.

Proof: It is sufficient to observe that Lemma 7.3.2 implies that the map $\varphi \longmapsto \Phi\big(O(\varphi)\big)$ defines a continuous linear functional on $L^p(\mathbb{R}^2)$ for $p \in [1, 2]$. $\qquad\square$

By analogy with (A.10) we let

$$W_{|\phi\rangle\langle\psi|}(u, v) := \frac{1}{(2\pi)^2} \int_{\mathbb{R}^2} e^{-ixu-iyv} \langle \psi, e^{ixP/2-iyQ}\phi\rangle_{\mathfrak{h}} \, dx dy \qquad (7.2)$$

denote the Wigner density of $(P/2, -Q)$, defined by the Fourier inversion of the characteristic function

$$(x, y) \longmapsto \langle \psi, e^{ixP/2-iyQ}\phi\rangle_{\mathfrak{h}}.$$

Here, $|\phi\rangle\langle\psi|$, $\phi, \psi \in \mathfrak{h}$ denotes the state defined by

$$\rho \longmapsto \langle \psi, \rho\phi\rangle, \qquad \rho \in B(\mathfrak{h}).$$

In other words we have

$$\langle \psi, e^{ixP/2-iyQ}\phi\rangle_{\mathfrak{h}} = \int_{\mathbb{R}^2} e^{iux+ivy} W_{|\phi\rangle\langle\psi|}(u, v) du dv, \qquad (7.3)$$

$x, y \in \mathbb{R}$, *i.e.*, $W_{|\phi\rangle\langle\psi|}(u, v)$ represents the Wigner density of (P, Q) in the state $|\phi\rangle\langle\psi|$, and we can also write

$$\int_{\mathbb{R}^2} W_{|\phi\rangle\langle\psi|}(u,v)\varphi(u,v)dudv = \langle\phi,\big(O(\varphi)\big)\psi\rangle$$

for all Schwartz functions φ, or

$$\int_{\mathbb{R}^2} W_{|\phi\rangle\langle\psi|}(u,v)\varphi(u,v)dudv$$
$$= \frac{1}{(2\pi)^2}\int_{\mathbb{R}^2}\int_{\mathbb{R}^2} e^{-ixu-iyv}\langle\psi,e^{ixP/2-iyQ}\phi\rangle_{\mathfrak{h}}\,dxdy\,\varphi(u,v)dudv$$

and

$$\int_{\mathbb{R}^2} W_{|\phi\rangle\langle\psi|}(u,v)e^{iau+ibv}dudv$$
$$= \frac{1}{(2\pi)^2}\int_{\mathbb{R}^2}\int_{\mathbb{R}^2} e^{-ixu-iyv}\langle\psi,e^{ixP/2-iyQ}\phi\rangle_{\mathfrak{h}}\,dxdy\,e^{iau+ibv}dudv$$
$$= \frac{1}{(2\pi)^2}\int_{\mathbb{R}^2}\int_{\mathbb{R}^2} e^{-i(x-a)u-i(y-b)v}\langle\psi,e^{ixP/2-iyQ}\phi\rangle_{\mathfrak{h}}\,dxdydudv$$
$$= \langle\psi,e^{iaP/2-ibQ}\phi\rangle_{\mathfrak{h}},$$

where we applied (A.9), and this recovers (7.3). In particular we have

$$\int_{\mathbb{R}^2} uv W_{|\phi\rangle\langle\psi|}(u,v)dudv = -\left\langle\phi,\frac{QP/2+PQ/2}{2}\psi\right\rangle.$$

Next, by identification of the terms in xy^2 in (7.3) we recover

$$\int_{\mathbb{R}^2}\frac{uv^2}{2!}W_{|\phi\rangle\langle\psi|}(u,v)dudv = \frac{1}{3!}\left\langle\phi,Q^2\frac{P}{2}+P\frac{Q^2}{2}+QP\frac{Q}{2}\right\rangle,$$

while the identification of the terms in x^2y yields

$$\frac{1}{2!}\int_{\mathbb{R}^2} u^2v W_{|\phi\rangle\langle\psi|}(u,v)dudv$$
$$= -\frac{1}{3!}\left\langle\phi,Q\left(\frac{P}{2}\right)^2+\left(\frac{P}{2}\right)^2 Q+\left(\frac{P}{2}\right)Q\left(\frac{P}{2}\right)\right\rangle.$$

Note that letting

$$P\phi(t) = -2i\phi'(t) \quad\text{and}\quad Q\phi(t) = t\phi(t), \qquad \phi\in\mathcal{S}(\mathbb{R}),$$

also defines a representation based on $\mathfrak{h}=L^2(\mathbb{R};\mathbb{C},dx)$ of (P,Q) on $\mathfrak{h}\mathfrak{w}$, with the involutions $P^*=P$ and $Q^*=Q$, as follows from the integration by parts

$$\langle f, Pg \rangle_{\mathfrak{h}} = \int_{-\infty}^{\infty} \bar{f}(t) Pg(t) dt$$

$$= -2i \int_{-\infty}^{\infty} \bar{f}(t) g'(t) dt$$

$$= 2i \int_{-\infty}^{\infty} \bar{f}'(t) g(t) dt$$

$$= \int_{-\infty}^{\infty} \overline{(-2if')}(t) g(t) dt$$

$$= \langle Pf, g \rangle_{\mathfrak{h}}.$$

On the other hand, by the following version

$$\exp\left(-ix\frac{P}{2} + iyQ\right) = e^{-ixy/2} \exp(iyQ) \exp\left(-ix\frac{P}{2}\right)$$

of the splitting lemma on the Heisenberg–Weyl algebra $\mathfrak{hw}$, cf. Exercise 6.2, we have

$$\exp\left(-ix\frac{P}{2} + iyQ\right)\psi(t) = e^{-ixy/2} \exp(iyQ) \exp\left(-ix\frac{P}{2}\right)\psi(t)$$

$$= e^{iyt - ixy/2} \sum_{n=0}^{\infty} (-1)^n \frac{x^n}{n!} \frac{\partial^n \psi}{\partial t^n}(t)$$

$$= e^{iyt - ixy/2} \psi(t - x), \tag{7.4}$$

$x, y, t \in \mathbb{R}$, $\psi \in \mathcal{S}(\mathbb{R})$. As a consequence, by (7.2) we can write

$$W_{|\phi\rangle\langle\psi|}(u, v) = \frac{1}{(2\pi)^2} \int_{\mathbb{R}^2} e^{-ixu - iyv} \langle \psi, e^{ixP/2 - iyQ}\phi \rangle_{\mathfrak{h}} dxdy$$

$$= \frac{1}{(2\pi)^2} \int_{\mathbb{R}^2} e^{-ixu - iyv} \langle e^{-ixP/2 + iyQ}\psi, \phi \rangle_{\mathfrak{h}} dxdy$$

$$= \frac{1}{(2\pi)^2} \int_{\mathbb{R}^2} e^{-ixu - iyv} \int_{-\infty}^{\infty} e^{iyt - ixy/2} \overline{\psi}(t - x)\phi(t) dtdxdy$$

$$= \frac{1}{(2\pi)^2} \int_{\mathbb{R}^2} \int_{-\infty}^{\infty} e^{iy(-2v + 2t - x)/2 - ixu} \overline{\psi}(t - x)\phi(t) dtdxdy$$

$$= \frac{2}{(2\pi)^2} \int_{\mathbb{R}^2} \int_{-\infty}^{\infty} e^{iy(-2v + 2t - x) - ixu} \overline{\psi}(t - x)\phi(t) dtdxdy$$

$$= \frac{1}{\pi} \int_{-\infty}^{\infty} e^{-2i(t - v)u} \overline{\psi}(2v - t)\phi(t) dt$$

$$= \frac{1}{\pi} \int_{-\infty}^{\infty} e^{-2itu} \overline{\psi}(v - t)\phi(t + v) dt$$

$$= \frac{1}{2\pi} \int_{-\infty}^{\infty} e^{-itu} \overline{\psi}(v - t/2)\phi(v + t/2) dt.$$

Consequently the second marginal is given by

$$\int_{-\infty}^{\infty} W_{|\phi\rangle\langle\psi|}(u,v)du = \frac{1}{2\pi}\int_{-\infty}^{\infty}\int_{-\infty}^{\infty} e^{-itu}\overline{\psi}(v-t/2)\phi(v+t/2)dtdu$$

$$= \phi(v)\bar{\psi}(v), \qquad v \in \mathbb{R},$$

cf. (A.9). For the first marginal we have

$$\int_{-\infty}^{\infty} W_{|\phi\rangle\langle\psi|}(u,v)dv = (\mathcal{F}\phi)(u)(\overline{\mathcal{F}\psi})(u), \qquad u \in \mathbb{R},$$

since

$$\mathcal{F}\left(e^{-ixP/2+iyQ}\phi\right)(t) = \frac{1}{\sqrt{2\pi}}\int_{-\infty}^{\infty} e^{izt+iyz-ixy/2}\phi(z-x)dz$$

$$= \frac{1}{\sqrt{2\pi}}\int_{-\infty}^{\infty} e^{it(x+z)}e^{iy(x+z)-ixy/2}\phi(z)dz$$

$$= \frac{1}{\sqrt{2\pi}}\int_{-\infty}^{\infty} e^{i(x+z)t}e^{iyz+ixy/2}\phi(z)dz$$

$$= \frac{1}{\sqrt{2\pi}}e^{ixy/2+ixt}\int_{-\infty}^{\infty} e^{iz(y+t)}\phi(z)dz$$

$$= e^{ixy/2+ixt}\mathcal{F}\phi(t+y)$$

$$= \exp\left(ixQ+iy\frac{P}{2}\right)\mathcal{F}\phi(t),$$

$x,y,t \in \mathbb{R}$, by the splitting lemma (7.4), which shows that

$$(\mathcal{F}\phi)(v)(\overline{\mathcal{F}\psi})(v) = \int_{-\infty}^{\infty} W_{|\mathcal{F}\phi\rangle\langle\mathcal{F}\psi|}(u,v)du$$

$$= \frac{1}{(2\pi)^2}\int_{-\infty}^{\infty}\int_{\mathbb{R}^2} e^{-ixu-iyv}\langle\mathcal{F}\psi, e^{ixP/2-iyQ}\mathcal{F}\phi\rangle_{\mathfrak{h}}dxdydu$$

$$= \frac{1}{(2\pi)^2}\int_{-\infty}^{\infty}\int_{\mathbb{R}^2} e^{-ixu-iyv}\langle\mathcal{F}\psi, \mathcal{F}e^{ixQ+iyP/2}\phi\rangle_{\mathfrak{h}}dxdydu$$

$$= \frac{1}{(2\pi)^2}\int_{-\infty}^{\infty}\int_{\mathbb{R}^2} e^{-ixu-iyv}\langle\psi, e^{ixQ+iyP/2}\phi\rangle_{\mathfrak{h}}dxdydu$$

$$= \frac{1}{(2\pi)^2}\int_{-\infty}^{\infty}\int_{\mathbb{R}^2} e^{-iyu-ixv}\langle\psi, e^{ixP/2+iyQ}\phi\rangle_{\mathfrak{h}}dxdydu$$

$$= -\frac{1}{(2\pi)^2}\int_{-\infty}^{\infty}\int_{\mathbb{R}^2} e^{-ixv+iyu}\langle\psi, e^{ixP/2-iyQ}\phi\rangle_{\mathfrak{h}}dxdydu$$

$$= -\int_{-\infty}^{\infty} W_{|\phi\rangle\langle\psi|}(v,-u)du$$

$$= \int_{-\infty}^{\infty} W_{|\phi\rangle\langle\psi|}(v,u)du.$$

Figure 7.1 presents a graph of $W_{|\phi\rangle\langle\psi|}$ with

$$\phi(x) = \psi(x) = \frac{1}{\sqrt{2}}(H_0 + H_1(x))e^{-x^2/4} = \frac{1}{\sqrt{2}}(1 + x)e^{-x^2/4},$$

drawn with the QuTiP package, cf. [61], [62].

Figure 7.2 represents the colour map of the aforementioned Wigner function, also drawn using QuTiP.

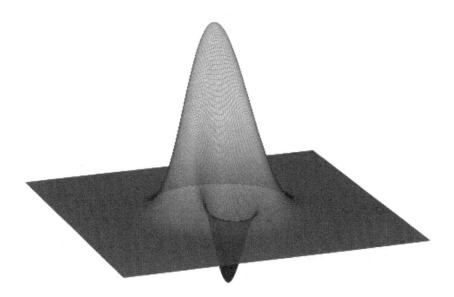

Figure 7.1 Wigner function.

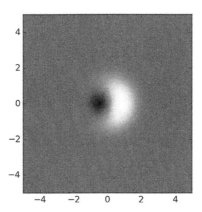

Figure 7.2 Wigner function colour map.

7.4 Functional calculus on real Lie algebras

In this section we extend the aforementioned functional calculus to more general Lie algebras via the approach of [7], and we also include some results and proofs not explicitly stated therein. For this we start by introducing Wigner functions for square-integrable representations of general Lie algebras.

Let $\langle \cdot, \cdot \rangle_{\mathcal{G}^*, \mathcal{G}}$ denote the pairing between the Lie algebra $\mathcal{G}$ and its dual $\mathcal{G}^*$. Let G be a Lie group with Lie algebra $\mathcal{G}$ and let U be a unitary representation of G on some Hilbert space $\mathfrak{h}$ with inner product $\langle \cdot, \cdot \rangle_{\mathfrak{h}}$. We assume that U is irreducible, and square integrable, *i.e.*, there exists a non-zero vector $\psi \in \mathfrak{h}$ such that

$$\int_G |\langle \psi, U(g)\psi \rangle_{\mathfrak{h}}|^2 d\mu(g) < \infty,$$

where μ denotes the left Haar measure on G. The following theorem defines the Duflo-Moore operator C.

Theorem 7.4.1 *([34]) There exists a positive self-adjoint operator C on $\mathfrak{h}$ such that*

$$\int_G \overline{\langle \psi_1, U(g)\phi_1 \rangle_{\mathfrak{h}}} \langle \psi_2, U(g)\phi_2 \rangle_{\mathfrak{h}} d\mu(g) = \langle C\psi_2, C\psi_1 \rangle_{\mathfrak{h}} \langle \phi_1, \phi_2 \rangle_{\mathfrak{h}}. \quad (7.5)$$

Moreover, the operator C is the identity if and only if G is unimodular, and $\mathrm{Dom}\, C^{-1}$ is dense in $\mathfrak{h}$. We assume the existence of an open subset N_0 of $\mathcal{G}$, symmetric around the origin, whose image $\exp(N_0)$ by $\exp: \mathcal{G} \longrightarrow G$ is dense in G with $\mu(G \setminus \exp(N_0)) = 0$. The image measure of μ on N_0 by $\exp^{-1}: \exp(N_0) \longrightarrow N_0$ is called the Haar measure on $\mathcal{G}$, and we denote by $m(x)$ its density with respect to the Lebesgue measure dx on $\mathcal{G}$.

Let now $\sigma(\xi)$ denote the density in the decomposition of the Lebesgue measure $d\xi$ on $\mathcal{G}^*$:

$$d\xi = dk(\lambda)\sigma(\xi)d\boldsymbol{\Omega}_\lambda(\xi),$$

where $dk(\lambda)$ is a measure on the parameter space of the co-adjoint orbits in $\mathcal{G}^*$ and $d\boldsymbol{\Omega}_\lambda(\xi)$ is the invariant measure on the orbit $\mathcal{O}_\lambda^*$. Let also $(X_1, \ldots, X_n)$, resp. $(X_1^*, \ldots, X_n^*)$, denote a basis of $\mathcal{G}$, resp. $\mathcal{G}^*$.

Definition 7.4.2 *([7]) Given $(\phi, \psi) \in \mathfrak{h} \times \mathrm{Dom}\, C^{-1}$ the Wigner function $W_{|\phi\rangle\langle\psi|}$ is defined as*

$$W_{|\phi\rangle\langle\psi|}(\xi) = \frac{\sqrt{\sigma(\xi)}}{(2\pi)^{n/2}} \int_{N_0} e^{-i\langle\xi, x\rangle_{\mathcal{G}^*, \mathcal{G}}} \langle U\left(e^{x_1 X_1 + \cdots + x_n X_n}\right) C^{-1}\psi, \phi \rangle_{\mathfrak{h}} \sqrt{m(x)} dx,$$

$\xi \in \mathcal{G}^*$.

The following proposition extends the definition of W_ρ in $L^2_{\mathbb{C}}(\mathcal{G}^*; d\xi/\sigma(\xi))$ to $\rho \in B_2(\mathfrak{h})$.

Proposition 7.4.3 *([7]) The mapping*

$$\mathfrak{h} \times \operatorname{Dom} C^{-1} \longrightarrow L^2_{\mathbb{C}}(\mathcal{G}^*; d\xi/\sigma(\xi))$$
$$\rho \longmapsto W_\rho$$

extends to an isometry on $B_2(\mathfrak{h})$:

$$\langle W_{\rho_1}, W_{\rho_2} \rangle_{L^2_{\mathbb{C}}(\mathcal{G}^*; d\xi/\sigma(\xi))} = \langle \rho_1, \rho_2 \rangle_{B_2(\mathfrak{h})}, \qquad \rho_1, \rho_2 \in B_2(\mathfrak{h}).$$

Proof: By a density argument it suffices to consider

$$\rho_1 = |\phi_1\rangle\langle\psi_1| \quad \text{and} \quad \rho_2 = |\phi_2\rangle\langle\psi_2|,$$

with $(\phi_1, \psi_1), (\phi_2, \psi_2) \in \mathfrak{h} \times \operatorname{Dom} C^{-1}$. From the identity (7.5) and since

$$\frac{1}{(2\pi)^n} \int_{-\infty}^{\infty} e^{i\langle \xi, x-x' \rangle_{\mathcal{G}^*, \mathcal{G}}} d\xi\, dx' = \delta_x(dx'), \tag{7.6}$$

we have

$$\langle W_{\rho_1}, W_{\rho_2} \rangle_{L^2_{\mathbb{C}}(\mathcal{G}^*; d\xi/\sigma(\xi))}$$

$$= \frac{1}{(2\pi)^n} \int_{\mathcal{G}^*} \left(\int_{N_0} e^{-i\langle \xi, x \rangle_{\mathcal{G}^*, \mathcal{G}}} \overline{\operatorname{Tr}[U\left(e^{-(x_1 X_1 + \cdots + x_n X_n)}\right) \rho_1 C^{-1}]} \sqrt{m(x)} dx \right.$$

$$\left. \times \int_{N_0} e^{-i\langle \xi, x' \rangle_{\mathcal{G}^*, \mathcal{G}}} \operatorname{Tr}\left[U\left(e^{-(x_1' X_1 + \cdots + x_n' X_n)}\right) \rho_2 C^{-1}\right] \sqrt{m(x')} dx' \right) d\xi$$

$$= \int_{N_0} \overline{\operatorname{Tr}\left[U\left(e^{-(x_1 X_1 + \cdots + x_n X_n)}\right) \rho_1 C^{-1}\right]} \operatorname{Tr}\left[U\left(e^{-(x_1 X_1 + \cdots + x_n X_n)}\right) \rho_2 C^{-1}\right] m(x) dx$$

$$= \int_{N_0} \overline{\langle U\left(e^{x_1 X_1 + \cdots + x_n X_n}\right) C^{-1} \psi_1, \phi_1 \rangle_{\mathfrak{h}}} \langle U\left(e^{x_1 X_1 + \cdots + x_n X_n}\right) C^{-1} \psi_2, \phi_2 \rangle_{\mathfrak{h}} m(x) dx$$

$$= \int_G \overline{\langle U(g) C^{-1} \psi_1, \phi_1 \rangle_{\mathfrak{h}}} \langle U(g) C^{-1} \psi_2, \phi_2 \rangle_{\mathfrak{h}} d\mu(g)$$

$$= \langle \psi_2, \psi_1 \rangle_{\mathfrak{h}} \langle \phi_1, \phi_2 \rangle_{\mathfrak{h}}$$

$$= \langle \rho_2, \rho_1 \rangle_{B_2(\mathfrak{h})},$$

where we used the relation

$$\operatorname{Tr}(U(g)^* \rho C^{-1}) = \operatorname{Tr}(U(g)^* |\phi\rangle\langle\psi| C^{-1}) = \operatorname{Tr}(C^{-1} U(g)^* |\phi\rangle\langle\psi|)$$
$$= \langle \psi, C^{-1} U(g)^* \phi \rangle_{\mathfrak{h}} = \langle U(g) C^{-1} \psi, \phi \rangle_{\mathfrak{h}}.$$

$\square$

As a result, the definition of $W_\rho(\xi)$ extends to $\rho \in B_2(\mathfrak{h})$ as:

$$W_\rho(\xi) = \frac{\sqrt{\sigma(\xi)}}{(2\pi)^{n/2}} \int_{N_0} e^{-i\langle \xi, x \rangle_{\mathcal{G}^*, \mathcal{G}}} \operatorname{Tr}\left[U\left(e^{-(x_1 X_1 + \cdots + x_n X_n)} \right) \rho C^{-1} \right] \sqrt{m(x)} dx,$$

$\xi \in \mathcal{G}^*$.

Definition 7.4.4 *Let* $O \colon L^2_{\mathbb{C}}(\mathcal{G}^*; d\xi/\sigma(\xi)) \longrightarrow B_2(\mathfrak{h})$ *denote the dual of* $\rho \longmapsto W_\rho$, *i.e.,*

$$\langle \rho | O(f) \rangle_{B_2(\mathfrak{h})} = \int_{\mathcal{G}^*} \overline{W}_\rho(\xi) f(\xi) \frac{d\xi}{\sigma(\xi)},$$

$f \in L^2_{\mathbb{C}}(\mathcal{G}^*; d\xi/\sigma(\xi))$, $\rho \in B_2(\mathfrak{h})$.

Note that for $\rho = |\phi\rangle\langle\psi|$ we have

$$\begin{aligned}
\langle \rho | O(f) \rangle_{B_2(\mathfrak{h})} &= \operatorname{Tr} |\phi\rangle\langle\psi|^* O(f) \\
&= \langle \phi | O(f) \psi \rangle_{\mathfrak{h}} \\
&= \langle W_{|\phi\rangle\langle\psi|}, f \rangle_{L^2_{\mathbb{C}}(\mathcal{G}^*; d\xi/\sigma(\xi))} \\
&= \int_{\mathcal{G}^*} \overline{W}_{|\phi\rangle\langle\psi|}(\xi) f(\xi) \frac{d\xi}{\sigma(\xi)}.
\end{aligned}$$

The next proposition allows to extend O as a bounded operator from $L^2_{\mathbb{C}}(\mathcal{G}^*; d\xi/\sigma(\xi))$ to $B_2(\mathfrak{h})$.

Proposition 7.4.5 *We have the bound*

$$\|O(f)\|_{B_2(\mathfrak{h})} \leq \|f\|_{L^2_{\mathbb{C}}(\mathcal{G}^*; d\xi/\sigma(\xi))}, \qquad f \in L^2_{\mathbb{C}}(\mathcal{G}^*; d\xi/\sigma(\xi)),$$

and the expression

$$O(f) = \int_{N_0} \sqrt{m(x)} \mathcal{F}\left(\frac{f}{\sqrt{\sigma}} \right)(x) U\left(e^{x_1 X_1 + \cdots + x_n X_n} \right) C^{-1} dx.$$

Proof: We have

$$\begin{aligned}
|\langle O(f), \rho \rangle_{B_2(\mathfrak{h})}| &= |\langle f, W_\rho \rangle_{L^2_{\mathbb{C}}(\mathcal{G}^*; d\xi/\sigma(\xi))}| \\
&\leq \|f\|_{L^2_{\mathbb{C}}(\mathcal{G}^*; d\xi/\sigma(\xi))} \|W_\rho\|_{L^2_{\mathbb{C}}(\mathcal{G}^*; d\xi/\sigma(\xi))} \\
&\leq \|f\|_{L^2_{\mathbb{C}}(\mathcal{G}^*; d\xi/\sigma(\xi))} \|\rho\|_{B_2(\mathfrak{h})},
\end{aligned}$$

and

$$\langle \phi, O(f)\psi \rangle_{\mathfrak{h}} = \mathrm{Tr}|\phi\rangle\langle\psi|^* O(f) = \int_{\mathcal{G}^*} \overline{W}_{|\phi\rangle\langle\psi|}(\xi)f(\xi)d\xi/\sigma(\xi)$$

$$= \frac{1}{(2\pi)^{n/2}} \int_{\mathcal{G}^*} \int_{N_0} e^{i\langle\xi,x\rangle_{\mathcal{G}^*,\mathcal{G}}} \overline{\mathrm{Tr} U\left(e^{-(x_1 X_1 + \cdots + x_n X_n)}\right) |\phi\rangle\langle\psi|C^{-1}} \sqrt{\frac{m(x)}{\sigma(\xi)}} dx f(\xi)d\xi$$

$$= \int_{N_0} \mathcal{F}\left(\frac{f}{\sqrt{\sigma}}\right)(x)\langle\phi|U\left(e^{x_1 X_1 + \cdots + x_n X_n}\right)C^{-1}\psi\rangle_{\mathfrak{h}}\sqrt{m(x)}dx$$

$$= \left\langle\phi \, \Big| \, \int_{N_0} \mathcal{F}\left(\frac{f}{\sqrt{\sigma}}\right)(x)U\left(e^{x_1 X_1 + \cdots + x_n X_n}\right)C^{-1}\sqrt{m(x)}dx\psi\right\rangle_{\mathfrak{h}}.$$

$\square$

In other words, we have

$$O\left(e^{-i\langle\cdot,x\rangle_{\mathcal{G}^*,\mathcal{G}}}\sqrt{\sigma(\cdot)}\right) = (2\pi)^{n/2}\sqrt{m(x)}U\left(e^{x_1 X_1 + \cdots + x_n X_n}\right)C^{-1}, \qquad (7.7)$$

and

$$O\left(f\sqrt{\sigma}\right) = \frac{1}{(2\pi)^{n/2}}\int_{N_0} O\left(e^{-i\langle\cdot,x\rangle_{\mathcal{G}^*,\mathcal{G}}}\sqrt{\sigma}\right)(\mathcal{F}f)(x)dx,$$

$f \in L^2_{\mathbb{C}}(\mathcal{G}^*; d\xi)$.

7.5 Functional calculus on the affine algebra

In this section we study in detail the particular case of the affine algebra generated by

$$X_1 = \begin{bmatrix} 1 & 0 \\ 0 & 0 \end{bmatrix}, \qquad X_2 = \begin{bmatrix} 0 & 1 \\ 0 & 0 \end{bmatrix},$$

with $[X_1, X_2] = X_2$. The affine group can be constructed as the group of 2×2 matrices of the form

$$g = \begin{bmatrix} a & b \\ 0 & 1 \end{bmatrix} = \begin{bmatrix} e^{x_1} & x_2 e^{\frac{x_1}{2}}\mathrm{sinch}\frac{x_1}{2} \\ 0 & 1 \end{bmatrix} = \exp\left(x_1 X_1 + x_2 X_2\right), \qquad (7.8)$$

$a > 0, b \in \mathbb{R}$, where

$$\mathrm{sinch}\,x = \frac{\sinh x}{x}, \qquad x \in \mathbb{R}.$$

We will work with

$$X_1 = -i\frac{P}{2} \quad \text{and} \quad X_2 = i(Q + M),$$

which form a representation of the affine algebra, as

$$[X_1, X_2] = X_2.$$

Here, $N_0 = \mathcal{G}$ is identified to $\mathbb{R}^2$ and

$$m(x_1, x_2) = e^{-x_1/2}\operatorname{sinch}\frac{x_1}{2}, \qquad x_1, x_2 \in \mathbb{R},$$

moreover from (92) in [7],

$$d\Omega_{\pm}(\xi_1, \xi_2) = \frac{1}{2\pi|\xi_2|}d\xi_1 d\xi_2,$$

hence

$$\sigma(\xi_1, \xi_2) = 2\pi|\xi_2|, \qquad \xi_1, \xi_2 \in \mathbb{R},$$

and the operator C is given by

$$Cf(\tau) = \sqrt{\frac{2\pi}{|\tau|}}f(\tau), \qquad \tau \in \mathbb{R}.$$

In order to construct the Malliavin calculus on the affine algebra we will have to use the functional calculus presented in Section 11.1. Letting $B_2(\mathfrak{h})$ denote the space of Hilbert–Schmidt operators on $\mathfrak{h}$, the results of Section 11.1 allow us to define a continuous map

$$O: L^2_{\mathbb{C}}\left(\mathbb{R}^2, \frac{1}{|\xi_2|}d\xi_1 d\xi_2\right) \longrightarrow B_2(\mathfrak{h})$$

as

$$O(f) := \int_{\mathbb{R}^2} (\mathcal{F}f)(x_1, x_2)e^{-ix_1 P/2 + iv x_2(Q+M)}dx_1 dx_2.$$

Relation (7.7), *i.e.*,

$$O\left(e^{-i\langle \cdot, x\rangle_{\mathcal{G}^*, \mathcal{G}}}\sqrt{\sigma(\cdot)}\right) = (2\pi)^{n/2}\sqrt{m(x)}U\left(e^{x_1 X_1 + \cdots + x_n X_n}\right)C^{-1},$$

shows that

$$e^{-iuP/2 + iv(Q+M)} = \frac{1}{\sqrt{2\pi}}\left(e^{-u/2}\operatorname{sinch}\frac{u}{2}\right)^{-1/2}O\left(e^{-iu\xi_1 - iv\xi_2}\sqrt{|\xi_2|}\right)C.$$

The next proposition shows that these relations can be simplified, and that the Wigner function is directly related to the density of the couple $(P, Q+M)$, with the property (7.9).

Proposition 7.5.1 *We have*

$$O\left(e^{iu\xi_1 + iv\xi_2}\right) = \exp\left(iu\frac{P}{2} - iv(Q+M)\right), \tag{7.9}$$

$u, v \in \mathbb{R}$.

Proof: We have for all $\phi, \psi \in \mathfrak{h}$:

$$\langle \phi, e^{-iuP/2 + iv(Q+M)} \psi \rangle_{\mathfrak{h}}$$

$$= \frac{\left(e^{-u/2}\operatorname{sinch}\frac{u}{2}\right)^{-1/2}}{\sqrt{2\pi}} \langle \phi, O\left(e^{-iu\xi_1 - iv\xi_2}\sqrt{|\xi_2|}\right) C\psi \rangle_{\mathfrak{h}}$$

$$= \frac{\left(e^{-u/2}\operatorname{sinch}\frac{u}{2}\right)^{-1/2}}{\sqrt{2\pi}} \langle W_{|\phi\rangle\langle C\psi|}(\xi_1, \xi_2), e^{-iu\xi_1 - iv\xi_2}\sqrt{|\xi_2|}\rangle_{L^2_{\mathbb{C}}(\mathcal{G}^*; \frac{d\xi_1 d\xi_2}{2\pi|\xi_2|})}$$

$$= \frac{1}{2\pi}\int_{\mathbb{R}^3} e^{-iu\xi_1 - iv\xi_2}\overline{\phi}\left(\frac{e^{-\frac{x}{2}}}{\operatorname{sinch}\frac{x}{2}}\right)\frac{e^{ix\xi_1}}{\operatorname{sinch}\frac{x}{2}}\sqrt{\frac{e^{-x/2}\operatorname{sinch}\frac{x}{2}}{e^{-u/2}\operatorname{sinch}\frac{u}{2}}}$$

$$\times \psi\left(\frac{\xi_2 e^{-x/2}}{\operatorname{sinch}\frac{x}{2}}\right)e^{-|\xi_2|\frac{\cosh\frac{x}{2}}{\operatorname{sinch}\frac{x}{2}}}\left(\frac{|\xi_2|}{\operatorname{sinch}\frac{x}{2}}\right)^{\beta-1}\frac{dx}{\Gamma(\beta)}d\xi_1 d\xi_2$$

$$= \frac{1}{2\pi}\int_{\mathbb{R}^3} e^{-iu\xi_1 - iv\xi_2}\overline{\phi}\left(\frac{\xi_2 e^{x/2}}{\operatorname{sinch}\frac{x}{2}}\right)\frac{e^{ix\xi_1}}{\operatorname{sinch}\frac{x}{2}}\psi\left(\frac{\xi_2 e^{x/2}}{\operatorname{sinch}\frac{x}{2}}\right)$$

$$\times e^{-|\xi_2|\frac{\cosh\frac{x}{2}}{\operatorname{sinch}\frac{x}{2}}}\left(\frac{|\xi_2|}{\operatorname{sinch}\frac{x}{2}}\right)^{\beta-1}\frac{dx}{\Gamma(\beta)}d\xi_1 d\xi_2$$

$$= \langle W_{|\phi\rangle\langle\psi|}, e^{-iu\xi_1 - iv\xi_2}\rangle_{L^2_{\mathbb{C}}(\mathcal{G}^*; \frac{d\xi_1 d\xi_2}{2\pi|\xi_2|})}$$

$$= \langle \phi, O\left(e^{-iu\xi_1 - iv\xi_2}\right)\psi \rangle_{\mathfrak{h}}.$$

$\square$

As a consequence of (7.9), the operator $O(f)$ has the natural expression

$$O(f) = O\left(\int_{\mathbb{R}^2}(\mathcal{F}f)(x_1, x_2)e^{-ix_1\xi_1 - ix_2\xi_2}dx_1 dx_2\right)$$

$$= \int_{\mathbb{R}^2}(\mathcal{F}f)(x_1, x_2)O(e^{-ix_1\xi_1 - ix_2\xi_2})dx_1 dx_2$$

$$= \int_{\mathbb{R}^2}(\mathcal{F}f)(x_1, x_2)e^{-ix_1 P/2 + ix_2(Q+M)}dx_1 dx_2.$$

As a consequence of Proposition 7.4.5 we find the bound

$$\|O(f)\|_{B_2(\mathfrak{h})} \le \|f\|_{L^2_{\mathbb{C}}(\mathcal{G}^*; \frac{d\xi_1 d\xi_2}{2\pi|\xi_2|})}.$$

This allows us to define the Wigner density $\tilde{W}_{|\phi\rangle\langle\psi|}(\xi_1, \xi_2)$ as the joint density of $(-P/2, Q+M)$,

$$\langle\psi|e^{iuP/2-iv(Q+M)}\phi\rangle_{\mathfrak{h}} = \int_{\mathbb{R}^2} e^{iu\xi_1+iv\xi_2}\tilde{W}_{|\phi\rangle\langle\psi|}(\xi_1, \xi_2)d\xi_1 d\xi_2,$$

$\phi, \psi \in \mathfrak{h}$, and this density has continuous binomial and gamma laws as marginals. Using a noncommutative integration by parts formula, we will be able to prove the smoothness of the joint density of $(P, Q+M)$. We also have the relations

$$\langle\psi|O(f)\phi\rangle_{\mathfrak{h}} = \int_{\mathcal{G}^*} \overline{W}_{|\psi\rangle\langle\phi|}(\xi_1, \xi_2)f(\xi_1, \xi_2)\frac{d\xi_1 d\xi_2}{2\pi|\xi_2|}$$

$$= \int_{\mathcal{G}^*} W_{|\phi\rangle\langle\psi|}(\xi_1, \xi_2)f(\xi_1, \xi_2)\frac{d\xi_1 d\xi_2}{2\pi|\xi_2|},$$

and

$$\langle\psi|e^{iuP/2-iv(Q+M)}\phi\rangle_{\mathfrak{h}} = \frac{1}{2\pi}\int_{\mathcal{G}^*} e^{iu\xi_1+iv\xi_2}W_{|\phi\rangle\langle\psi|}(\xi_1, \xi_2)\frac{d\xi_1 d\xi_2}{|\xi_2|},$$

which show that the density $\tilde{W}_{|\phi\rangle\langle\psi|}$ of $(P/2, -(Q+M))$ in the state $|\phi\rangle\langle\psi|$ has the expression

$$\tilde{W}_{|\phi\rangle\langle\psi|}(\xi_1, \xi_2) = \frac{1}{2\pi|\xi_2|}W_{|\phi\rangle\langle\psi|}(\xi_1, \xi_2) \tag{7.10}$$

$$= \frac{1}{2\pi}\int_{\mathbb{R}} \phi\left(\frac{\xi_2 e^{-x/2}}{\sinh\frac{x}{2}}\right)\frac{e^{-ix\xi_1}}{\sinh\frac{x}{2}}\overline{\psi}\left(\frac{\xi_2 e^{x/2}}{\sinh\frac{x}{2}}\right)e^{-|\xi_2|\frac{\cosh\frac{x}{2}}{\sinh\frac{x}{2}}}\left(\frac{|\xi_2|}{\sinh\frac{x}{2}}\right)^{\beta-1}\frac{dx}{\Gamma(\beta)},$$

as in Relation (102) of [7]. Note that $\tilde{W}_{|\phi\rangle\langle\psi|}$ has the correct marginals since integrating in $d\xi_1$ in (7.10) we have, using (7.6),

$$\frac{1}{2\pi|\xi_2|}\int_{\mathbb{R}} W_{|\phi\rangle\langle\psi|}(\xi_1, \xi_2)d\xi_1 = \gamma_\beta(|\xi_2|)\overline{\phi}(\xi_2)\psi(\xi_2),$$

and

$$\frac{1}{2\pi}\int_{\mathbb{R}} W_{|\phi\rangle\langle\psi|}(\xi_1, \xi_2)\frac{d\xi_2}{|\xi_2|}$$

$$= \frac{1}{2\pi}\int_{\mathbb{R}^2} e^{-i\xi_1 x}\overline{\phi}(\omega e^{x/2})\psi(\omega e^{-x/2})e^{-|\omega|\cosh\frac{x}{2}}\frac{|\omega|^{\beta-1}}{\Gamma(\beta)}dx d\omega.$$

In the vacuum state, *i.e.*, for $\phi = \psi = \Omega = \mathbb{1}_{\mathbb{R}_+}$, we have

$$\frac{1}{2\pi} \int_{\mathbb{R}} W_{|\Omega\rangle\langle\Omega|}(\xi_1, \xi_2) \frac{d\xi_2}{\xi_2} = \frac{1}{2\pi} \int_{\mathbb{R}} \int_0^\infty e^{-i\xi_1 x} \frac{\tau^{\beta-1}}{\Gamma(\beta)} e^{-\tau \cosh \frac{x}{2}} d\tau\, dx$$

$$= \frac{1}{2\pi} \int_{\mathbb{R}} e^{-i\xi_1 x} \frac{1}{\left(\cosh \frac{x}{2}\right)^\beta} dx$$

$$= c \left| \Gamma\left(\frac{\beta}{2} + \frac{i}{2}\xi_1\right) \right|^2,$$

where c is a normalisation constant and Γ is the gamma function. When $\beta = 1$ we have $c = 1/\pi$ and P has the hyperbolic cosine density in the vacuum state $\Omega = \mathbb{1}_{\mathbb{R}_+}$:

$$\xi_1 \longmapsto \frac{1}{2\cosh(\pi\xi_1/2)}.$$

Proposition 7.5.2 *The characteristic function of $(P, Q + M)$ in the state $|\phi\rangle\langle\psi|$ is given by*

$$\langle\psi, e^{iuP+iv(Q+M)}\phi\rangle_{\mathfrak{h}} = \int_{\mathbb{R}} e^{iv\omega\mathrm{sinch}\, u} \overline{\psi}(\omega e^u)\phi(\omega e^{-u}) e^{-|\omega|\cosh u} \frac{|\omega|^{\beta-1}}{\Gamma(\beta)} d\omega.$$

In the vacuum state $\Omega = \mathbb{1}_{\mathbb{R}_+}$ we find

$$\langle\Omega, e^{iuP+iv(Q+M)}\Omega\rangle_{\mathfrak{h}} = \frac{1}{(\cosh u - iv\mathrm{sinch}\, u)^\beta}, \qquad u, v \in \mathbb{R}.$$

Proof: Using the modified representation $\hat{U}$ defined in (2.4) on $\mathfrak{h} = L_{\mathbb{C}}^2(\mathbb{R}, \gamma_\beta(|\tau|)d\tau)$, we have

$$\langle\psi, e^{-iuP/2+iv(Q+M)}\phi\rangle_{\mathfrak{h}} = \left\langle \psi, \hat{U}\left(e^u, ve^{u/2}\mathrm{sinch}\, \frac{u}{2}\right)\phi \right\rangle_{\mathfrak{h}}$$

$$= \int_{\mathbb{R}} \overline{\psi}(\tau)\phi\left(\tau e^u\right) e^{iv\tau e^{u/2}\mathrm{sinch}\, \frac{u}{2}} e^{-(e^u-1)|\tau|/2} e^{\beta u/2} \frac{|\tau|^{\beta-1}}{\Gamma(\beta)} e^{-|\tau|} d\tau$$

$$= \int_{\mathbb{R}} e^{iv\omega\mathrm{sinch}\, \frac{u}{2}} \overline{\psi}\left(\omega e^{-u/2}\right) \phi\left(\omega e^{u/2}\right) e^{-|\omega|\cosh \frac{u}{2}} \frac{|\omega|^{\beta-1}}{\Gamma(\beta)} d\omega.$$

In the vacuum state $|\Omega\rangle\langle\Omega|$ we have

$$\langle\Omega, e^{-iuP/2+iv(Q+M)}\Omega\rangle_{\mathfrak{h}} = \int_0^\infty e^{i\omega\mathrm{sinch}\, \frac{u}{2}-|\omega|\cosh \frac{u}{2}} \frac{|\omega|^{\beta-1}}{\Gamma(\beta)} d\omega$$

$$= \frac{1}{(\cosh \frac{u}{2} - iv\mathrm{sinch}\, \frac{u}{2})^\beta}. \qquad \square$$

In particular, we have

$$\langle \psi, e^{iv(Q+M)} \phi \rangle_{\mathfrak{h}} = \frac{1}{\Gamma(\beta)} \int_{\mathbb{R}} e^{iv\omega} \overline{\psi}(\omega) \phi(\omega) e^{-|\omega|} |\omega|^{\beta-1} d\omega$$

hence as expected, $Q+M$ has density $\overline{\psi}(\omega)\phi(\omega)\gamma_\beta(|\omega|)$, in particular a gamma law in the vacuum state. On the other hand we have

$$\langle \psi, e^{iuP} \phi \rangle_{\mathfrak{h}} = \frac{1}{\Gamma(\beta)} \int_{\mathbb{R}} \overline{\psi}(\omega e^u) \phi(\omega e^{-u}) e^{-|\omega| \cosh u} |\omega|^{\beta-1} d\omega,$$

which recovers the density of P:

$$\xi_1 \longmapsto \frac{1}{2\pi \Gamma(\beta)} \int_{\mathbb{R}^2} e^{-i\xi_1 x} \overline{\psi}(\omega e^x) \phi(\omega e^{-x}) e^{-|\omega| \cosh x} |\omega|^{\beta-1} dx d\omega.$$

In the vacuum state we find

$$\langle \Omega, e^{iuP} \Omega \rangle_{\mathfrak{h}} = \frac{1}{(\cosh u)^\beta}, \qquad u \in \mathbb{R}.$$

7.6 Wigner functions on $so(3)$

In this section we consider Wigner functions on the Lie algebra $su(2) \equiv so(3)$. We consider the Lie group $SU(2)$ of unitary 2×2 matrices with determinant one. This example illustrates the importance of the choice of the set N_0 in the definition of the Wigner function given by [7]. We use the notation and results of Section 4.5.

A basis of the Lie algebra $\mathfrak{g} = su(2) \equiv so(3)$ of $SU(2)$ is given by X_1, X_2, X_3 with the relations

$$[X_1, X_2] = X_3 \quad \text{and cyclic permutations,}$$

and $X_j^* = -X_j$. For the adjoint action $\mathrm{ad} \colon \mathfrak{g} \longrightarrow \mathrm{Lin}(\mathfrak{g})$, $\mathrm{ad}(X)Z = [X, Z]$, we have

$$\begin{cases} \mathrm{ad}(X_1) = \begin{bmatrix} 0 & 0 & 0 \\ 0 & 0 & -1 \\ 0 & 1 & 0 \end{bmatrix}, \\[2em] \mathrm{ad}(X_2) = \begin{bmatrix} 0 & 0 & 1 \\ 0 & 0 & 0 \\ -1 & 0 & 0 \end{bmatrix}, \\[2em] \mathrm{ad}(X_3) = \begin{bmatrix} 0 & -1 & 0 \\ 1 & 0 & 0 \\ 0 & 0 & 0 \end{bmatrix}, \end{cases}$$

or

$$\mathrm{ad}(X) = \begin{bmatrix} 0 & -x_3 & x_2 \\ x_3 & 0 & -x_1 \\ -x_2 & x_1 & 0 \end{bmatrix}$$

for $X = x_1 X_1 + x_2 X_2 + x_3 X_3$. The dual adjoint action

$$\mathrm{ad}^* : \mathfrak{g} \longrightarrow \mathrm{Lin}(\mathfrak{g}^*)$$

is given by

$$\mathrm{ad}^*(X) = -\mathrm{ad}(X)^T = \begin{bmatrix} 0 & -x_3 & x_2 \\ x_3 & 0 & -x_1 \\ -x_2 & x_1 & 0 \end{bmatrix}$$

if we use the dual basis e_1, e_2, e_3,

$$\langle e_j, X_k \rangle = \delta_{jk}$$

for $\mathfrak{g}^*$. Similarly for $\mathrm{ad}^*(X_1), \mathrm{ad}^*(X_2), \mathrm{ad}^*(X_3)$. By exponentiation we get

$$\begin{cases} \mathrm{Ad}^*(e^{tX_1}) = e^{\mathrm{ad}(X_1)} = \begin{bmatrix} 0 & 0 & 0 \\ 0 & \cos 1 & -\sin t \\ 0 & \sin t & \cos t \end{bmatrix}, \\[2em] \mathrm{Ad}^*(e^{tX_2}) = e^{\mathrm{ad}(X_2)} = \begin{bmatrix} \cos t & 0 & \sin t \\ 0 & 0 & 0 \\ -\sin t & 0 & \cos t \end{bmatrix}, \\[2em] \mathrm{Ad}^*(e^{tX_3}) = e^{\mathrm{ad}(X_3)} = \begin{bmatrix} \cos t & -\sin t & 0 \\ \sin t & \cos t & 0 \\ 0 & 0 & 0 \end{bmatrix}, \end{cases}$$

and we see that $\mathfrak{g}$ acts on its dual as rotations. The orbits are therefore the spheres

$$\mathcal{O}_r = \{\xi = \xi_1 e_1 + \xi_2 e_2 + \xi_3 e_3 : \xi_1^2 + \xi_2^2 + \xi_3^2 = r\}, \quad r \geq 0.$$

The invariant measure on these orbits is just the uniform distribution, in polar coordinates $\sin \vartheta \, d\vartheta \, d\phi$. The Lebesgue measure on $\mathfrak{g}^*$ can be written as

$$r^2 dr \sin \vartheta_r d\vartheta_r d\varphi_r$$

so that we get $\sigma_r(\vartheta_r, \varphi_r) = 1$, cf. [7, Equation (19)]. The transfer of the Haar measure μ on the group

$$d\mu(e^X) = m(X)dX$$

gives

$$m(X) = \left| \det \left(\sum_{n=0}^{\infty} \frac{\operatorname{ad}(X)^n}{(n+1)!} \right) \right|$$

see Equation (27) in [8]. Since $\operatorname{ad}(X)$ is normal and has simple eigenvalues $\pm i\sqrt{x_1^2 + x_2^2 + x_3^2}$ and 0, we can use the spectral decomposition to get

$$m(X) = \frac{2 - 2\cos\sqrt{x_1^2 + x_2^2 + x_3^2}}{x_1^2 + x_2^2 + x_3^2} = 4\frac{\sin^2 t/2}{t^2},$$

where $t = \sqrt{x_1^2 + x_2^2 + x_3^2}$. For N_0 we take the ball

$$N_0 = \{X = (x_1, x_2, x_3) : x_1^2 + x_2^2 + x_3^2 \le 2\pi\}.$$

7.6.1 Group-theoretical Wigner function

If $U \colon SU(2) \longrightarrow \mathcal{U}(\mathfrak{h})$ is a unitary representation of $SU(2)$ on some Hilbert space $\mathfrak{h}$ and $\rho \colon B(\mathfrak{h}) \longrightarrow \mathbb{C}$ a state, then the associated Wigner function is given by

$$W_\rho(\xi) = \frac{2}{(2\pi)^{3/2}} \int_{N_0} e^{-i\langle \xi, X \rangle} \rho(U(e^X)) \frac{\sin\left(\sqrt{x_1^2 + x_2^2 + x_3^2}/2\right)}{\sqrt{x_1^2 + x_2^2 + x_3^2}} dX$$

see [7, Equation (48)]. We compute this for the irreducible $n + 1$-dimensional representations $D_{n/2} = \operatorname{span}\{e_n, e_{n-2}, \ldots, e_{-n}\}$, given by

$$U(X_+)e_k = \begin{cases} 0 & \text{if } k = n, \\ \dfrac{i}{2}\sqrt{(n-k)(n+k+2)}e_{k+2} & \text{else.} \end{cases},$$

$$U(X_-)e_k = \begin{cases} 0 & \text{if } k = -n, \\ \dfrac{i}{2}\sqrt{(n+k)(n-k+2)}e_{k-2} & \text{else.} \end{cases},$$

$$U(X_3)e_k = \frac{ik}{2}e_k,$$

in terms of $X_+ = X_1 + iX_2$, $X_- = X_1 - iX_2$, and X_3, cf. Equation (2.5), and $\rho = (n+1)^{-1}\,\mathrm{tr}$. The operator $U(X)$ has the eigenvalues

$$-i\frac{n}{2}\sqrt{x_1^2 + x_2^2 + x_3^2}, \quad -i\frac{n-2}{2}\sqrt{x_1^2 + x_2^2 + x_3^2}, \quad \ldots, \quad i\frac{n}{2}\sqrt{x_1^2 + x_2^2 + x_3^2},$$

which yields

$$W_\rho(\xi)$$

$$= \frac{2}{(2n+1)(2\pi)^{3/2}} \int_{N_0} e^{-i\langle\xi,X\rangle} \sum_{k=-n/2}^{n/2} e^{ik\sqrt{x_1^2 + x_2^2 + x_3^2}} \frac{\sin\sqrt{x_1^2 + x_2^2 + x_3^2}/2}{\sqrt{x_1^2 + x_2^2 + x_3^2}}\,dX$$

$$= \frac{2}{(2n+1)(2\pi)^{3/2}} \sum_{k=-n/2}^{n/2} \int_0^{2\pi} \int_0^\pi \int_0^{r\pi} e^{-i\langle\xi,X\rangle} e^{ikt} t\sin(t/2)\sin\theta\, dt\, d\theta\, d\phi$$

by using polar coordinates on N_0. Since the integrand is rotationally invariant in the variable X, the Wigner function will again be rotationally invariant and it is sufficient to consider $\xi = (0, 0, z)$,

$$W_\rho(0, 0, z)$$

$$= \frac{2}{(2n+1)(2\pi)^{3/2}} \sum_{k=-n/2}^{n/2} \int_0^{2\pi} \int_0^\pi \int_0^{2\pi} e^{-itz\cos\theta} e^{ikt} t\sin(t/2)\sin\theta\, dt\, d\theta\, d\phi$$

$$= \frac{2}{(2n+1)(2\pi)^{1/2}} \sum_{k=-n/2}^{n/2} \int_0^{2\pi} \int_0^\pi e^{-itz\cos\theta} e^{ikt} t\sin(t/2)\sin\theta\, dt\, d\theta$$

$$= \frac{4}{z(2n+1)(2\pi)^{1/2}} \sum_{k=-n/2}^{n/2} \int_0^{2\pi} e^{ikt}\sin(zt)\sin(t/2)\, dt$$

$$= -\frac{1}{z(2n+1)(2\pi)^{1/2}}$$

$$\times \sum_{k=-n/2}^{n/2} \int_0^{2\pi} \left(e^{it(k+z+1/2)} - e^{it(k-z+1/2)} - e^{it(k+z-1/2)} + e^{it(k-z-1/2)} \right) dt$$

$$= \frac{2(e^{2i\pi} - 1)e^{2\pi i(k-1/2)}}{iz(2n+1)(2\pi)^{1/2}} \sum_{k=-n/2}^{n/2} \left(\frac{e^{2\pi iz}}{2k + 2z + 1} - \frac{e^{-2\pi iz}}{2k - 2z - 1} \right)$$

$$= \frac{4(e^{2i\pi} - 1)e^{2\pi i(k-1/2)}}{z(2n+1)(2\pi)^{1/2}} \sum_{k=-n/2}^{n/2} \frac{2k\sin(2\pi z) + (2z+1)i\cos(2\pi z)}{4k^2 - (2z+1)^2},$$

cf. also [12, 28]. Note that even for the trivial representation with $n = 0$ and $U(e^X) = 1$, this doesn't give the Dirac measure at the origin.

7.6.2 Probabilistic Wigner density

We define the function W_ρ^{pr} as

$$W_\rho^{pr}(\xi) = \frac{1}{(2\pi)^{3/2}} \int_{\mathfrak{g}} e^{-i\langle \xi, X \rangle} \rho(U(e^X)) dX$$

and this will always have the right marginals. It is again rotationally invariant for the representation $D_{n/2}$ and the state $\rho = (2n+1)^{-1} \operatorname{tr}$, and we get

$$W_\rho^{pr}(\xi_1, \xi_2, \xi_3) = \frac{1}{(2n+1)(2\pi)^{3/2}} \sum_{k=-n/2}^{n/2} \int_{\mathfrak{g}} e^{-i\langle \xi, X \rangle} e^{ik\|X\|} dX.$$

Using polar coordinates (t, θ, ψ) in $\mathfrak{g}$, with the north pole in direction $\xi = (\xi_1, \xi_2, \xi_3)$, we get

$$W_\rho^{pr}(\xi) = \frac{1}{(2n+1)(2\pi)^{3/2}} \sum_{k=-n/2}^{n/2} \int_0^\infty \int_0^\pi \int_0^{2\pi} e^{-it\|\xi\|\cos\theta} e^{ikt} t^2 \sin\theta \, dt d\theta d\phi$$

$$= \frac{1}{(2n+1)(2\pi)^{1/2}} \sum_{k=-n/2}^{n/2} \int_0^\infty \int_0^\pi e^{-it\|\xi\|\cos\theta} e^{ikt} t^2 \sin\theta \, dt d\theta$$

$$= \frac{1}{\|\xi\|(2n+1)(2\pi)^{1/2}} \sum_{k=-n/2}^{n/2} \int_0^\infty t \sin(t\|\xi\|) e^{ikt} dt$$

$$= \frac{1}{2i\|\xi\|(2n+1)(2\pi)^{1/2}} \sum_{k=-n/2}^{n/2} \int_{-\infty}^\infty t e^{it\|\xi\|} \cos(kt) dt,$$

so that we get

$$W_\rho^{pr}(\xi) = \frac{1}{2\|\xi\|(2n+1)(2\pi)^{1/2}} \sum_{k=-n/2}^{n/2} \delta'_{\|\xi\|-k},$$

which is clearly not a measure, but a distribution. However, there should be a better description of W_ρ^{pr} ...

$$\int_{\|X\|^2 \le R} \frac{\Delta e^{-i\langle \xi, X \rangle}}{\|X\|} dX = \int_0^R \int_0^\pi \int_0^{2\pi} \frac{-\|\xi\|^2 e^{-i\|\xi\| r \cos\theta}}{r} r^2 \sin\theta \, dr d\theta d\phi$$

$$= -2\pi \|\xi\|^2 \int_0^R \int_0^\pi e^{-i\|\xi\| r \cos\theta} r \sin\theta \, dr d\theta$$

$$= -2\pi \|\xi\|^2 \int_0^R \int_{-1}^1 e^{i\|\xi\| r z} r dz dr$$

$$= -2\pi||\xi||^2 \int_0^R \frac{2\sin(r||\xi||)}{r||\xi||} r\,dr$$

$$= -4\pi||\xi|| \int_0^R \sin(r||\xi||)\,dr$$

$$= 4\pi\left(\cos(R||\xi||) - 1\right),$$

i.e., the Fourier transform of the distribution

$$f \longmapsto \int_{||X||^2 \le R} \frac{\Delta f}{||X||}\,dX$$

is almost what we are looking for. *E.g.*, for the spin 1/2 representation we have

$$\rho(U(e^X)) = \cos(||X||/2)$$

and therefore the associated "probabilistic" Wigner distribution is the distribution

$$W_\rho^{pr} : f \longmapsto \frac{1}{4\pi} \int_{||X||^2 \le 1/2} \frac{\Delta f}{||X||}\,dX + f(0).$$

Assuming that f depends only on z we can check that we get the right marginals, as

$$\frac{1}{4\pi} \int_{||X||^2 \le R} \frac{\Delta f}{||X||}\,dX + f(0)$$

$$= \frac{1}{4\pi} \int_0^R \int_0^\pi \int_0^{2\pi} \frac{f''(r\cos\theta)}{r} r^2 \sin\theta\,dr\,d\theta\,d\phi + f(0)$$

$$= \frac{1}{2} \int_0^R \int_0^\pi f''(r\cos\theta) r\sin\theta\,dr\,d\theta + f(0)$$

$$= \frac{1}{2} \int_0^R \int_{-r}^r f''(z)\,dz\,dr + f(0)$$

$$= \frac{1}{2} \int_0^R (f'(r) - f'(-r))\,dr + f(0)$$

$$= \frac{1}{2}\left(f(R) - f(0) + f(-R) - f(0)\right) + f(0)$$

$$= \frac{1}{2}(f(R) + f(-R)),$$

which gives the correct marginal distribution $\frac{1}{2}(\delta_{-1/2} + \delta_{1/2})$, if we set $R = 1/2$. Since $\Delta(fg) = f\Delta(g) + 2\vec{\nabla}(f)\cdot\vec{\nabla}(g) + g\Delta(f)$, and therefore

$$\Delta\left(\frac{f}{r}\right) = \frac{\Delta f}{r} - \frac{2}{r^2}\frac{\partial f}{\partial r},$$

we can rewrite W_ρ^{pr} also as

$$W_\rho^{pr}(f) = f(0) + \frac{1}{4\pi} \int_{||x|| \leq 1/2} \left(\Delta \left(\frac{f}{r} \right) + \frac{2}{r^2} \frac{\partial f}{\partial r} \right) dX.$$

Using Gauss' integral theorem, we can transform the first part of the integral into a surface integral,

$$W_\rho^{pr}(f) = f(0) + \frac{1}{4\pi} \int_{||x||=1/2} \vec{\nabla} \left(\frac{f}{r} \right) \cdot d\vec{n} + \frac{1}{2\pi} \int_{||x|| \leq 1/2} \frac{1}{r^2} \frac{\partial f}{\partial r} dX.$$

7.7 Some applications

7.7.1 Quantum optics

We refer to [48] for details on the material presented in this section. In optics one may consider the Maxwell equations for an electric field on the form

$$E(x, t) = C_E q(t) \sin(kx)$$

polarised along the z-axis, where C_E is a normalisation constant, and the magnetic field

$$B(x, t) = C_B q'(t) \cos(kx),$$

polarised along the y-axis, where C_B is a normalisation constant.

The classical field energy at time t is given by

$$\frac{1}{2} \int (E^2(x, t) + B^2(x, t)) dx,$$

and this system is quantised using the operators (Q, P) and the harmonic oscillator Hamiltonian

$$H = \frac{1}{2}(P^2 + \omega^2 Q^2)$$

where $\omega = kc$ is the frequency and c is the speed of light. Here the operator $N = a_+ a_-$ is called the photon number operator.

Each number eigenstate e_n is mapped to the wave function

$$\Psi(x) = (2^n n!)^{-1/2} e^{-x^2/2} H_n(x), \qquad x \in \mathbb{R},$$

while the coherent state $\epsilon(\alpha)$ corresponds to

$$\epsilon(\alpha) = e^{-|\alpha|^2/2}e^{-x^2/2}\sum_{n=0}^{\infty}(2^n n!)^{-1/2}\frac{\alpha^n}{\sqrt{n!}}H_n(x)$$

$$= e^{-|\alpha|^2/2}e^{-x^2/2}\sum_{n=0}^{\infty}\frac{\alpha^n}{n!}\frac{H_n(x)}{\sqrt{2}^n}$$

$$= e^{-|\alpha|^2/2}e^{-x^2/2}e^{x\alpha/\sqrt{2}-\alpha^2x^2/4}$$

$$= \exp\left(-\frac{|\alpha|^2}{2}-\left(x-\frac{\alpha}{\sqrt{2}}\right)^2/2\right).$$

The Wigner phase-space (quasi)-probability density function in the quasi-state $|\phi\rangle\langle\psi|$ is then given by

$$W_{|\phi\rangle\langle\psi|}(x,y) = \frac{1}{2\pi}\int_{-\infty}^{\infty}\bar{\phi}(x-t)\psi(x+t)e^{iyt}dt,$$

while the probability density in a pure state $|\phi\rangle\langle\phi|$ is

$$W_{|\phi\rangle\langle\phi|}(x,y) = \frac{1}{2\pi}\int_{-\infty}^{\infty}\bar{\phi}(x-t)\phi(x+t)e^{iyt}dt.$$

7.7.2 Time-frequency analysis

The aim of time-frequency analysis is to understand signals with changing frequencies. Say we have recorded a signal, *e.g.*, an opera by Mozart or the noise of a gearbox, which can be represented as a function $f: \mathbb{R} \longrightarrow \mathbb{C}$ that depends on time. Looking at the square of the modulus $|f(t)|^2$, we can see at what time the signal was strong or noisy, but it is more difficult to determine the frequency. An analysis of the frequencies present in the signal can be done by passing to the Fourier transform,

$$\tilde{f}(\omega) = \frac{1}{\sqrt{2\pi}}\int_{-\infty}^{+\infty}f(t)e^{it\omega}dt.$$

The square of the modulus $|\tilde{f}(\omega)|^2$ of the Fourier transform gives us a good idea to what extent the frequency ω is present in the signal. But it will not tell us at what time the frequency ω was played, the important information about how the frequency changed with time is not visible in $|\tilde{f}(\omega)|^2$, *e.g.*, if we play Mozart's opera backwards we still get the same function $|\tilde{f}(\omega)|^2$.

On the other hand, although the Wigner function

$$W_f(\omega,t) = \frac{1}{\pi}\int_{-\infty}^{\infty}\bar{f}(t+s)f(t-s)e^{2is\omega}ds,$$

cf. Section 7.4, does not have a clear interpretation as a joint probability distribution since it can take negative values, it can give approximate information on which frequency was present in the signal at what time. The use of Wigner functions in time-frequency analysis, where they are often called Wigner–Ville functions after Ville [119], is carefully explained in Cohen's book [29]. In [85], Wigner functions are used to analyse the sound of a gearbox in order to predict when it will break down.

Note

We also refer the reader to [3] for more background on Wigner functions and their use in quantum optics.

Exercises

Exercise 7.1 Quantum optics.

1. Compute the distribution of the photon number operator $N = a_+ a_-$ in the coherent state

$$\epsilon(\alpha) = e^{-|\alpha|^2/2} \sum_{n=0}^{\infty} \frac{\alpha^n}{\sqrt{n!}} e_n.$$

2. Show that in the quasi-state $|\phi\rangle\langle\phi|$ with

$$\phi(z) = \frac{1}{(2\pi)^{1/4}} e^{-z^2/4},$$

the Wigner phase-space (quasi)-probability density function is a standard two-dimensional joint Gaussian density.

8

Lévy processes on real Lie algebras

I really have long been of the mind that the quantity of noise that anyone can carefreely tolerate is inversely proportional to his mental powers, and can therefore be considered as an approximate measure of the same.

(A. Schopenhauer, in The World as Will and Representation.*)*

In this chapter we present the definition and basic theory of Lévy processes on real Lie algebras with several examples. We use the theories of factorisable current representations of Lie algebras and Lévy processes on ∗-bialgebras to provide an elegant and efficient formalism for defining and studying quantum stochastic calculi with respect to additive operator processes satisfying Lie algebraic relations. The theory of Lévy processes on ∗-bialgebras can also handle processes whose increments are not simply additive, but are composed by more complicated formulas, the main restriction is that they are independent (in the tensor sense).

8.1 Definition

Lévy processes, *i.e.*, stochastic processes with independent and stationary increments, are used as models for random fluctuations, in physics, finance, etc. In quantum physics so-called quantum noises or quantum Lévy processes occur, *e.g.*, in the description of quantum systems coupled to a heat bath [47] or in the theory of continuous measurement [53]. Motivated by a model introduced for lasers [122], Schürmann et al. [2, 106] have developed the theory of Lévy processes on involutive bialgebras. This theory generalises, in a sense, the theory of factorisable representations of current groups and current algebras as well as the theory of classical Lévy processes with values in Euclidean space or, more generally, semigroups. Note that many interesting classical stochastic processes arise as components of these quantum Lévy processes, cf. [1, 18, 42, 105].

Let D be a complex pre-Hilbert space with inner product $\langle \cdot, \cdot \rangle$. We denote by $\mathcal{L}(D)$ the algebra of linear operators on D having an adjoint defined everywhere on D, *i.e.*,

$$\mathcal{L}(D) := \big\{ A \colon D \longrightarrow D \text{ linear} : \exists A^* \colon D \longrightarrow D \text{ linear operator} \qquad (8.1)$$
$$\text{such that } \langle x, Ay \rangle = \langle A^* x, y \rangle \text{ for all } x, y \in D \big\}.$$

By $\mathcal{L}_{\mathrm{AH}}(D)$ we mean the anti-Hermitian linear operators on D, *i.e.*,

$$\mathcal{L}_{\mathrm{AH}}(D) = \{ A \colon D \longrightarrow D \text{ linear} : \langle x, Ay \rangle = -\langle Ax, y \rangle \text{ for all } x, y \in D \}.$$

In the sequel, $\mathfrak{g}$ denotes a Lie algebra over $\mathbb{R}$, D is a complex pre-Hilbert space, and $\Omega \in D$ is a unit vector.

Definition 8.1.1 *Any family*

$$\big(j_{st} \colon \mathfrak{g} \longrightarrow \mathcal{L}_{\mathrm{AH}}(D) \big)_{0 \leq s \leq t}$$

of representations of $\mathfrak{g}$ is called a Lévy *process on $\mathfrak{g}$ over D (with respect to the unit vector $\Omega \in D$) provided the following conditions are satisfied:*

i) (Increment property). *We have*

$$j_{st}(X) + j_{tu}(X) = j_{su}(X)$$

for all $0 \leq s \leq t \leq u$ and all $X \in \mathfrak{g}$.
ii) (Boson independence). *We have*

$$[j_{st}(X), j_{s't'}(Y)] = 0, \qquad X, Y \in \mathfrak{g},$$

$0 \leq s \leq t \leq s' \leq t'$, *and*

$$\langle \Omega, j_{s_1 t_1}(X_1)^{k_1} \cdots j_{s_n t_n}(X_n)^{k_n} \Omega \rangle$$
$$= \langle \Omega, j_{s_1 t_1}(X_1)^{k_1} \Omega \rangle \cdots \langle \Omega, j_{s_n t_n}(X_n)^{k_n} \Omega \rangle,$$

for all $n, k_1, \ldots, k_n \in \mathbb{N}$, $0 \leq s_1 \leq t_1 \leq s_2 \leq \cdots \leq t_n$, $X_1, \ldots, X_n \in \mathfrak{g}$.
iii) (Stationarity). *For all $n \in \mathbb{N}$ and all $X \in \mathfrak{g}$, the moments*

$$m_n(X; s, t) = \langle \Omega, j_{st}(X)^n \Omega \rangle$$

depend only on the difference $t - s$.
iv) (Weak continuity). *We have*

$$\lim_{t \searrow s} \langle \Omega, j_{st}(X)^n \Omega \rangle = 0, \qquad n \in \mathbb{N}, \quad X \in \mathfrak{g}.$$

Such a process extends to a family of ∗-representations of the complexification $\mathfrak{g}_{\mathbb{C}} = \mathfrak{g} \oplus i\mathfrak{g}$ with the involution

$$(X + iY)^* = -X + iY, \qquad X, Y \in \mathfrak{g},$$

by setting

$$j_{st}(X + iY) = j_{st}(X) + ij_{st}(Y).$$

We denote by $\mathcal{U}(\mathfrak{g})$ the *universal enveloping algebra* of $\mathfrak{g}_{\mathbb{C}}$, i.e., the unital associative algebra over $\mathbb{C}$ generated by the elements of $\mathfrak{g}_{\mathbb{C}}$ with the relation $XY - YX = [X, Y]$ for $X, Y \in \mathfrak{g}$. If $X_1, \ldots, X_d$ is a basis of $\mathfrak{g}$, then

$$\{X_1^{n_1} \cdots X_d^{n_d} : n_1, \ldots, n_d \in \mathbb{N},\}$$

is a basis of $\mathcal{U}(\mathfrak{g})$. The set

$$\{X_1^{n_1} \cdots X_d^{n_d} : n_1, \ldots, n_d \in \mathbb{N}, \ n_1 + \cdots + n_d \geq 1\}$$

is a basis of a nonunital subalgebra $\mathcal{U}_0(\mathfrak{g})$ of $\mathcal{U}$ generated by $\mathfrak{g}$, and we have $\mathcal{U}(\mathfrak{g}) = \mathbb{C}\mathbf{1} \oplus \mathcal{U}_0(\mathfrak{g})$.

Furthermore, we extend the involution on $\mathfrak{g}_{\mathbb{C}}$ defined in Remark 2.1.3 as an antilinear antihomomorphism to $\mathcal{U}(\mathfrak{g})$ and $\mathcal{U}_0(\mathfrak{g})$. It acts on the basis given earlier by

$$\left(X_1^{n_1} \cdots X_d^{n_d}\right)^* = (-1)^{n_1 + \cdots + n_d} X_d^{n_d} \cdots X_1^{n_1}.$$

A Lévy process $(j_{st})_{0 \leq s \leq t}$ on $\mathfrak{g}$ extends to a family of ∗-representations of $\mathcal{U}(\mathfrak{g})$, and the functionals

$$\varphi_t = \langle \mathbf{\Omega}, j_{0t}(\cdot)\mathbf{\Omega} \rangle : \mathcal{U}(\mathfrak{g}) \longrightarrow \mathbb{C}$$

are states, *i.e.*, unital positive linear functionals. Furthermore, they are differentiable with respect to t and

$$L(u) = \lim_{t \searrow 0} \frac{1}{t} \varphi_t(u), \quad u \in \mathcal{U}_0(\mathfrak{g}),$$

defines a positive Hermitian linear functional on $\mathcal{U}_0(\mathfrak{g})$. In fact, one can prove that the family $(\varphi_t)_{t \in \mathbb{R}_+}$ is a convolution semigroup of states on $\mathcal{U}_0(\mathfrak{g})$. The functional L is also called the *generating functional* of the process. It satisfies the conditions of the following definition.

Definition 8.1.2 *A linear functional* $L : \mathcal{U}_0 \longrightarrow \mathbb{C}$ *on a (non-unital) ∗-algebra* $\mathcal{U}_0$ *is called a* generating functional *if*

i) *L is Hermitian, i.e.,* $L(u^*) = \overline{L(u)}$ *for all* $u \in \mathcal{U}_0$;
ii) *L is positive, i.e.,* $L(u^*u) \geq 0$ *for all* $u \in \mathcal{U}_0$.

Schürmann has shown that there exists indeed a Lévy process for any generating functional on $\mathcal{U}_0(\mathfrak{g})$, cf. [106]. Let

$$\left(j_{st}^{(1)} \colon \mathfrak{g} \longrightarrow \mathcal{L}_{AH}(D^{(1)})\right)_{0 \leq s \leq t} \quad \text{and} \quad \left(j^{(2)} \colon \mathfrak{g} \longrightarrow \mathcal{L}_{AH}(D^{(2)})\right)_{0 \leq s \leq t}$$

be two Lévy processes on $\mathfrak{g}$ with respect to the state vectors $\mathbf{\Omega}^{(1)}$ and $\mathbf{\Omega}^{(2)}$, respectively. We call them *equivalent*, if all their moments agree, *i.e.*, if

$$\langle \mathbf{\Omega}^{(1)}, j_{st}^{(1)}(X)^k \mathbf{\Omega}^{(1)}\rangle = \langle \mathbf{\Omega}^{(2)}, j_{st}^{(2)}(X)^k \mathbf{\Omega}^{(2)}\rangle,$$

for all $k \in \mathbb{N}$, $0 \leq s \leq t$, $X \in \mathfrak{g}$. This implies that all joint moments also agree on $\mathcal{U}(\mathfrak{g})$, *i.e.*,

$$\langle \mathbf{\Omega}^{(1)}, j_{s_1 t_1}^{(1)}(u_1) \cdots j_{s_n t_n}^{(1)}(u_1) \mathbf{\Omega}^{(1)}\rangle$$
$$= \langle \mathbf{\Omega}^{(2)}, j_{s_1 t_1}^{(2)}(u_1) \cdots j_{s_n t_n}^{(2)}(u_n) \mathbf{\Omega}^{(2)}\rangle,$$

for all $0 \leq s_1 \leq t_1 \leq s_2 \leq \cdots \leq t_n$, $u_1, \ldots, u_n \in \mathcal{U}(\mathfrak{g})$, $n \geq 1$.

8.2 Schürmann triples

By a Gelfand–Naimark–Segal (GNS)-type construction, one can associate to every generating functional a Schürmann triple.

Definition 8.2.1 *A Schürmann triple on $\mathfrak{g}$ is a triple (ρ, η, ψ), where*

a) $\rho \colon \mathfrak{g} \longrightarrow \mathcal{L}_{AH}(D)$ *is a representation on some pre-Hilbert space D, i.e.,*

$$\rho([X, Y]) = \rho(X)\rho(Y) - \rho(Y)\rho(X) \quad \text{and} \quad \rho(X)^* = -\rho(X),$$

for all $X, Y \in \mathfrak{g}$,

b) $\eta \colon \mathfrak{g} \longrightarrow D$ *is a ρ-1-cocycle, i.e., it satisfies*

$$\eta([X, Y]) = \rho(X)\eta(Y) - \rho(X)\eta(Y), \qquad X, Y \in \mathfrak{g},$$

and

c) $\psi \colon \mathfrak{g} \longrightarrow \mathbb{C}$ *is a linear functional with imaginary values such that the bilinear map $(X, Y) \longmapsto \langle \eta(X), \eta(Y)\rangle$ is the 2-coboundary of ψ (with respect to the trivial representation), i.e.,*

$$\psi([X, Y]) = \langle \eta(Y), \eta(X)\rangle - \langle \eta(X), \eta(Y)\rangle, \qquad X, Y \in \mathfrak{g}.$$

The functional ψ in the Schürmann triple associated to a generating functional

$$L \colon \mathcal{U}_0(\mathfrak{g}) \longrightarrow \mathbb{C}$$

is the restriction of the generating functional L to $\mathfrak{g}$. Conversely, given a Schürmann triple (ρ, η, ψ), we can reconstruct a generating functional $L: \mathcal{U}_0(\mathfrak{g}) \longrightarrow \mathbb{C}$ from it by setting

$$\begin{cases} L(X_1) = \psi(X_1), & n = 1, \\[2ex] L(X_1 X_2) = -\langle \eta(X_1), \eta(X_2) \rangle, & n = 2, \\[2ex] L(X_1 \cdots X_n) = -\langle \eta(X_1), \rho(X_2) \cdots \rho(X_{n-1}) \eta(X_n) \rangle, & n \geq 3, \end{cases}$$

for $X_1, \ldots, X_n \in \mathfrak{g}$. We will now see how a Lévy processes can be reconstructed from its Schürmann triple.

Let (ρ, η, ψ) be a Schürmann triple on $\mathfrak{g}$, acting on a pre-Hilbert space D. We can define a Lévy process on the symmetric Fock space

$$\Gamma\left(L^2(\mathbb{R}_+, D)\right) = \bigoplus_{n=0}^{\infty} L^2(\mathbb{R}_+, D)^{\circ n}$$

by setting

$$j_{st}(X) = a_{st}^{\circ}\left(\rho(X)\right) + a_{st}^{+}\left(\eta(X)\right) - a_{st}^{-}\left(\eta(X)\right) + \psi(X)(t-s)I_d, \tag{8.2}$$

for $X \in \mathfrak{g}$, where a_{st}°, a_{st}^{+}, and a_{st}^{-} denote increments of the conservation, creation, and annihilation processes on $\Gamma\left(L^2(\mathbb{R}_+, D)\right)$ defined in Chapter 5, see also [79, 87]. Using the commutation relations (5.1) satisfied by the conservation, creation, and annihilation operators, it is straightforward to check that we have

$$\left[j_{st}(X), j_{st}(Y)\right] = j_{st}\left([X, Y]\right), \quad \text{and} \quad j_{st}(X)^* = -j_{st}(X),$$

$0 \leq s \leq t$, $X, Y \in \mathfrak{g}$. One also checks that the moments of j_{st} satisfy the stationarity and continuity properties of Definition 8.1.1. Furthermore, using the identification

$$\Gamma\left(L^2(\mathbb{R}_+, D)\right) \cong \Gamma\left(L^2([0, t[, D)\right) \otimes \Gamma\left(L^2([t, +\infty), D)\right)$$

and the fact that $\mathbf{\Omega} \cong \mathbf{\Omega} \otimes \mathbf{\Omega}$ holds for the vacuum vector with respect to this factorization, one can show that the increments of $(j_{st})_{0 \leq s \leq t}$ are boson independent. The family

$$\left(j_{st}: \mathfrak{g} \longrightarrow \mathcal{L}_{\mathrm{AH}}\left(\Gamma\left(L^2(\mathbb{R}_+, D)\right)\right)\right)_{0 \leq s \leq t}$$

extends to a unique family of unital $*$-representations of $\mathcal{U}(\mathfrak{g})$, since the elements of $\mathfrak{g}$ generate $\mathcal{U}(\mathfrak{g})$. We denote this family again by $(j_{st})_{0 \leq s \leq t}$.

The following theorem can be traced back to the works of Araki and Streater. In the form given here it is a special case of Schürmann's representation theorem for Lévy processes on involutive bialgebras, cf. [106].

Theorem 8.2.2 *Let $\mathfrak{g}$ be a real Lie algebra. Then there is a one-to-one correspondence (modulo equivalence) between Lévy processes on $\mathfrak{g}$ and Schürmann triples on $\mathfrak{g}$. Precisely, given (ρ, η, L) a Schürmann triple on $\mathfrak{g}$ over D,*

$$j_{st}(X) := a_{st}^{\circ}\big(\rho(X)\big) + a_{st}^{+}\big(\eta(X)\big) - a_{st}^{-}\big(\eta(X)\big) + (t-s)L(X)I_d, \qquad (8.3)$$

$0 \leq s \leq t$, $X \in \mathfrak{g}$, *defines Lévy process on $\mathfrak{g}$ over a dense subspace $H \subseteq \Gamma\big(L^2(\mathbb{R}_+, D)\big)$, with respect to the vacuum vector Ω.*

The correspondence between (equivalence classes of) Lévy processes and Schürmann triples is one-to-one and the representation (8.2) is universal.

Theorem 8.2.3 *[106]*

i) *Two Lévy processes on $\mathfrak{g}$ are equivalent if and only if their Schürmann triples are unitarily equivalent on the subspace $\rho\big(\mathcal{U}(\mathfrak{g})\big)\eta(\mathfrak{g})$.*

ii) *A Lévy process $(k_{st})_{0 \leq s \leq t}$ with generating functional L and Schürmann triple (ρ, η, ψ) is equivalent to the Lévy process $(j_{st})_{0 \leq s \leq t}$ associated to (ρ, η, L) defined in Equation (8.2).*

Due to Theorem 8.2.3, the problem of characterising and constructing all Lévy processes on a given real Lie algebra can be decomposed into the following steps.

a) First, classify all representations of $\mathfrak{g}$ by anti-Hermitian operators (modulo unitary equivalence). This gives the possible choices for the representation ρ in the Schürmann triple.

b) Next, determine all ρ-1-cocycles. We distinguish between trivial cocycles, *i.e.*, cocycles which are of the form

$$\eta(X) = \rho(X)\omega, \qquad X \in \mathfrak{g}$$

for some vector $\omega \in D$ in the representation space of ρ, and non-trivial cocycles, *i.e.*, cocycles, which cannot be written in this form.

c) Finally, determine all functionals L that turn a pair (ρ, η) into a Schürmann triple (ρ, η, L).

The last step can also be viewed as a cohomological problem. If η is a ρ-1-cocycle then the bilinear map

$$(X, Y) \longmapsto \langle \eta(X), \eta(Y) \rangle - \langle \eta(X), \eta(Y) \rangle$$

is a 2-cocycle for the trivial representation, *i.e.*, it is antisymmetric and it satisfies

$$\langle \eta([X,Y]), \eta(Z) \rangle - \langle \eta(Z), \eta([X,Y]) \rangle$$
$$- \langle \eta(X), \eta([Y,Z]) \rangle + \langle \eta([Y,Z]), \eta(X) \rangle = 0,$$

for all $X, Y, Z \in \mathfrak{g}$. For L we can take any functional that has the map

$$(X,Y) \longmapsto \langle \eta(X), \eta(Y) \rangle - \langle \eta(Y), \eta(Z) \rangle$$

as coboundary, *i.e.*, $L(X)$ is imaginary and

$$L([X,Y]) = \langle \eta(Y), \eta(X) \rangle - \langle \eta(X), \eta(Y) \rangle$$

for all $X, Y \in \mathfrak{g}$. If η is trivial then such a functional always exists as we can take $L(X) = \langle \omega, \rho(X)\omega \rangle$. But for a nontrivial cocycle such a functional may not exist. For a given pair (ρ, η), L is determined only up to a Hermitian 0–1-cocycle, *i.e.*, a Hermitian functional ℓ that satisfies $\ell([X,Y]) = 0$ for all $X, Y \in \mathfrak{g}$.

Trivial changes to the cocycle are equivalent to a conjugation of the corresponding Lévy process by a unitary cocycle of the time shift on the Fock space. More generally, we have the following proposition.

Proposition 8.2.4 *Let* $\mathfrak{g}$ *be a real Lie algebra,* $(j_{st})_{0 \le s \le t}$ *a Lévy process on* $\mathfrak{g}$ *with Schürmann triple* (ρ, η, ψ) *over the pre-Hilbert space D, B a unitary operator on D, and $\omega \in D$. Then $(\tilde{\rho}, \tilde{\eta}, \tilde{\psi})$ with*

$$
\begin{cases}
\tilde{\rho}(X) := B^* \rho(X) B, \\[2mm]
\tilde{\eta}(X) := B^* \eta(X) - B^* \rho(X) B \omega, \\[2mm]
\tilde{\psi}(X) := \psi(X) - \langle B\omega, \eta(X) \rangle + \langle \eta(X), B\omega \rangle + \langle B\omega, \rho(X)B\omega \rangle \\[2mm]
\qquad\quad = \psi(X) - \langle \omega, \tilde{\eta}(X) \rangle + \langle \tilde{\eta}(X), \omega \rangle - \langle \omega, \tilde{\rho}(X)\omega \rangle, \quad X \in \mathfrak{g},
\end{cases}
$$

is also a Schürmann triple on $\mathfrak{g}$.

Proof: The operator process $(U_t)_{t \in \mathbb{R}_+}$ is given by

$$U_t = e^{t(ih - \|\omega\|^2/2)} \exp\left(-a_t^+(B\omega)\right) a_t^\circ(B) \exp\left(a_t^-(\omega)\right),$$

where $\Gamma_t(B)$ denotes the second quantisation of B. It is unitary and adapted to the Fock space. Furthermore, for $0 \le s \le t$, we can write $U_s^* U_t$ as

$$U_s^* U_t = \mathrm{Id} \otimes u_{st} \otimes \mathrm{Id}$$

on the Fock space with respect to the factorisation

$$\Gamma\big(L^2([0,s[,D)\big) \otimes \Gamma\big(L^2([s,t[,D)\big) \otimes \Gamma\big(L^2([t,+\infty),D)\big).$$

This allows us to show that $(\tilde{j}_{st})_{0 \leq s \leq t}$ is again a Lévy process on $\mathfrak{g}$. $\qquad\square$

Corollary 8.2.5 *If the generating functional*

$$L: \mathcal{U}_0(\mathfrak{g}) \longrightarrow \mathbb{C}$$

of (j_{st}) can be extended to a positive functional $\tilde{L}$ on $\mathcal{U}(\mathfrak{g})$, then (j_{st}) is equivalent to the cocycle conjugate

$$U_t^* \Gamma_{st}\big(\rho(\cdot)\big) U_t$$

of the second quantisation of ρ.

Proof: Applying the standard GNS construction to $\tilde{L}$ on $\mathcal{U}(\mathfrak{g})$, we obtain a pre-Hilbert space D, a projection $\eta: \mathcal{U}(\mathfrak{g}) \longrightarrow D$, and a representation ρ of $\mathcal{U}(\mathfrak{g})$ on D such that

$$L(u) = \langle \omega, \rho(u)\omega \rangle,$$

for all $u \in \mathcal{U}(\mathfrak{g})$, where $\omega = \eta(\mathbf{1})/\tilde{L}(\mathbf{1})$. It is not difficult to check that $(\rho|_{\mathfrak{g}}, \eta|_{\mathfrak{g}}, \tilde{L}|_{\mathfrak{g}})$ is a Schürmann triple for (j_{st}). Taking

$$U_t = \exp\big(A_t(\omega) - A_t^*(\omega)\big)$$

and applying Proposition 8.2.4, we obtain the desired result. $\qquad\square$

Remark 8.2.6 *If the generating functional of Lévy process (j_{st}) can be written in the form $L(u) = \langle \omega, \rho(u)\omega \rangle$ for all $u \in \mathcal{U}_0(\mathfrak{g})$, then we call (j_{st}) a compound Poisson process.*

In the next examples we will work with the complexification $\mathfrak{g}_{\mathbb{C}}$ of a real Lie algebra $\mathfrak{g}$ and the involution

$$(X + iY)^* = -X + iY, \qquad X, Y \in \mathfrak{g}.$$

One can always recover $\mathfrak{g}$ as the real subspace

$$\mathfrak{g} = \big\{ X \in \mathfrak{g}_{\mathbb{C}} : X^* = -X \big\}$$

of antisymmetric elements in $\mathfrak{g}_{\mathbb{C}}$.

8.2.1 Quantum stochastic differentials

Since we can define quantum stochastic integrals and know the *Itô table* for the four Hudson–Parthasarathy integrators,

•	$da^+(u)$	$da^\circ(F)$	$da^-(u)$	dt
$da^+(v)$	0	0	0	0
$da^\circ(G)$	$da^+(Gu)$	$da^\circ(GF)$	0	0
$da^-(v)$	$\langle v, u \rangle dt$	$da^-(F^*v)$	0	0
dt	0	0	0	0

for all $F, G \in \mathcal{L}(D)$, $u, v \in D$, we get a quantum stochastic calculus for Lévy processes on $\mathfrak{g}$, too. The map d_L associating elements u of the universal enveloping algebra to the corresponding quantum stochastic differentials $d_L u$ defined by

$$d_L u = da^\circ\big(\rho(u)\big) + da^+\big(\eta(u)\big) + da^-\big(\eta(u^*)\big) + \psi(u)dt, \qquad (8.4)$$

is a $*$-homomorphism from $\mathcal{U}_0(\mathfrak{g})$ to the Itô algebra over D, see [45, Proposition 4.4.2]. It follows that the dimension of the Itô algebra generated by $\{d_L X : X \in \mathfrak{g}\}$ is at least the dimension of D (since η is supposed surjective) and not bigger than $(\dim D + 1)^2$. If D is infinite-dimensional, then its dimension is also infinite. Note that it depends on the choice of the Lévy process.

Proposition 8.2.7 *The Lévy process of $(\tilde{\rho}, \tilde{\eta}, \tilde{\psi})$ in Proposition 8.2.4 is equivalent to the Lévy process defined by*

$$\tilde{j}_{st}(X) = U_t^* j_{st}(X) U_t, \qquad 0 \leq s \leq t, \quad X \in \mathfrak{X},$$

where $(U_t)_{t \in \mathbb{R}_+}$ is the solution of

$$U_t = \mathrm{Id} + \int_0^t U_s \left(da_s^-(\omega) - da_s^+(B\omega) + da_s^\circ(B - \mathrm{Id}) + \left(ih - \frac{\|\omega\|^2}{2} ds \right) \right).$$

Proof: Using the quantum Itô table, one can show that $(\tilde{j}_{st})_{0 \leq s \leq t}$ is of the form

$$\begin{aligned}
\tilde{j}_{st}(X) = {} & a_{st}^\circ\big(B^*\rho(X)B\big) + a_{st}^+\big(B^*\eta(X) - B^*\rho(X)B\omega\big) \\
& - a_{st}^-\big(B^*\eta(X) - B^*\rho(X)B\omega\big) \\
& + (t - s)\big(\psi(X) - \langle B\omega, \eta(X) \rangle + \langle \eta(X), B\omega \rangle + \langle B\omega, \rho(X)B\omega \rangle\big)I_d.
\end{aligned}$$

$\square$

8.3 Lévy processes on $\mathfrak{hw}$ and $\mathfrak{osc}$

A Lévy process $\left(j_{st}\colon \mathfrak{g} \longrightarrow \mathcal{L}_{\mathrm{AH}}(D)\right)_{0 \le s \le t}$ on $\mathfrak{g}$ with generating functional $L\colon \mathcal{U}_0(\mathfrak{g}) \longrightarrow \mathbb{C}$ and Schürmann triple (ρ, η, ψ) is called *Gaussian* if ρ is the trivial representation of $\mathfrak{g}$, *i.e.*, if

$$\rho(X) = 0, \qquad X \in \mathfrak{g}.$$

Recall that the Heisenberg–Weyl Lie algebra $\mathfrak{hw}$ is the three-dimensional Lie algebra with basis $\{A^+, A^-, E\}$, commutation relations

$$[A^-, A^+] = E, \quad [A^\pm, E] = 0,$$

and the involution $(A^-)^* = A^+$, $E^* = E$. We begin with the classification of all Gaussian generating functionals on the $\mathfrak{hw}$ algebra.

Proposition 8.3.1 *Let $v_1, v_2 \in \mathbb{C}^2$ and $z \in \mathbb{C}$. Then*

$$
\begin{cases}
\rho(A^+) = \rho(A^-) = \rho(E) = 0, \\[2mm]
\eta(A^+) = v_1, \quad \eta(A^-) = v_2, \quad \eta(E) = 0, \\[2mm]
L(A^+) = z, \quad L(A^-) = \bar{z}, \quad L(E) = ||v_1||^2 - ||v_2||^2,
\end{cases}
$$

defines the Schürmann triple on $D = \operatorname{span}\{v_1, v_2\}$ of a Gaussian generating functional on $\mathcal{U}_0(\mathfrak{hw})$.

Proof: One checks that for all these cocycles there do indeed exist generating functionals and computes their general form.   □

Therefore from (8.4) we get, for an arbitrary Gaussian Lévy process on $\mathfrak{hw}$:

$$
\begin{cases}
d_L A^+ = da^+(v_1) + da^-(v_2) + z\,dt, \\[2mm]
d_L A^- = da^+(v_2) + da^-(v_1) + \bar{z}\,dt, \\[2mm]
d_L E = \left(||v_1||^2 - ||v_2||^2\right)dt,
\end{cases}
$$

and the Itô table

$\bullet$	$d_L A^+$	$d_L A^-$	$d_L E$
$d_L A^+$	$\langle v_2, v_1\rangle dt$	$\langle v_2, v_2\rangle dt$	0
$d_L A^-$	$\langle v_1, v_1\rangle dt$	$\langle v_1, v_2\rangle dt$	0
$d_L E$	0	0	0

For $||v_1||^2 = 1$ and $v_2 = 0$, this is the usual Itô table for the creation and annihilation process in Hudson–Parthasarathy calculus.

We now consider the oscillator Lie algebra $\mathfrak{osc}$ which is obtained by addition of a Hermitian element N with commutation relations

$$[N, A^\pm] = \pm A^\pm, \quad [N, E] = 0.$$

The elements E and $NE - A^+ A^-$ generate the center of $\mathcal{U}_0(\mathfrak{osc})$. If we want an irreducible representation of $\mathcal{U}(\mathfrak{osc})$, which has nontrivial cocycles, they have to be represented by zero. Since we are only interested in $*$-representations, this also implies $\rho(A^+) = \rho(A^-) = 0$ as in the next proposition whose proof is similar to that of Proposition 8.3.1.

Proposition 8.3.2 *The Schürmann triples of Gaussian generating functionals on* $\mathcal{U}_0(\mathfrak{osc})$ *are all of the form*

$$\begin{cases} \rho(N) = \rho(A^+) = \rho(A^-) = \rho(E) = 0, \\[2mm] \eta(N) = v, \quad \eta(A^+) = \eta(A^-) = \eta(E) = 0, \\[2mm] L(N) = b, \quad L(A^+) = L(A^-) = L(E) = 0, \end{cases}$$

with $v \in \mathbb{C}, b \in \mathbb{R}$.

In particular, letting $v_1, v_2 \in \mathbb{C}^2$ and $b \in \mathbb{R}$, $\rho = \rho_1$, and

$$\begin{cases} \eta(N) = v_1, \quad \eta(A^+) = v_2, \quad \eta(A^-) = \eta(E) = 0, \\[2mm] L(N) = b, \quad L(E) = ||v_2||^2, \quad L(A^+) = \overline{L(A^-)} = \langle v_1, v_2 \rangle, \end{cases}$$

also defines a Schürmann triple on $\mathfrak{osc}$ acting on $D = \operatorname{span}\{v_1, v_2\}$. The corresponding quantum stochastic differentials are

$$\begin{cases} d_L N = da^\circ(\mathrm{Id}) + da^+(v_1) + da^-(v_1) + b\, dt, \\[2mm] d_L A^+ = da^+(v_2) + \langle v_1, v_2 \rangle\, dt, \\[2mm] d_L A^- = da^-(v_2) + \langle v_2, v_1 \rangle\, dt, \\[2mm] d_L E = ||v_2||^2\, dt, \end{cases}$$

and they satisfy the Itô table

•	$d_L A^+$	$d_L N$	$d_L A^-$	$d_L E$
$d_L A^+$	0	0	0	0
$d_L N$	$d_L A^+$	$d_L N + (\|v_1\|^2 - b)dt$	0	0
$d_L A^-$	$d_L E$	$d_L A^-$	0	0
$d_L E$	0	0	0	0

Note that for $\|v_1\|^2 = b$, this is the usual Itô table of the four fundamental noises of Hudson–Parthasarathy calculus.

For $n \geq 2$ we may also consider the real Lie algebra with basis $X_0, X_1, \ldots X_n$ and the commutation relations

$$[X_0, X_k] = \begin{cases} X_{k+1}, & \text{if } 1 \leq k < n, \\[2mm] 0, & \text{otherwise,} \end{cases} \tag{8.5}$$

and

$$[X_k, X_\ell] = 0, \qquad 1 \leq k, \ell \leq n.$$

For $n = 2$ this algebra coincides with the Heisenberg–Weyl Lie algebra $\mathfrak{hw}$, while for $n > 2$ it is a $n - 1$-step nilpotent Lie algebra. Its irreducible unitary representations can be described and constructed using the "orbit method" (*i.e.*, there exists exactly one irredicible unitary representation for each orbit of the coadjoint representation), see, *e.g.*, [101, 102].

8.4 Classical processes

Let $(j_{st})_{0 \leq s \leq t}$ be a Lévy process on a real Lie algebra $\mathfrak{g}_\mathbb{R}$ over $\Gamma = \Gamma\big(L^2(\mathbb{R}_+, D)\big)$. Denote by $\mathfrak{g}^{\mathbb{R}_+}$ the space of $\mathfrak{g}$-valued simple step functions

$$\mathfrak{g}^{\mathbb{R}_+} := \Big\{ \sum_{k=1}^n X_k \mathbf{1}_{[s_k, t_k)} : 0 \leq s_1 \leq t_1 \leq s_2 \leq \cdots \leq t_n < \infty, X_1, \ldots, X_n \in \mathfrak{g} \Big\}.$$

Then $\mathfrak{g}^{\mathbb{R}_+}$ is a real Lie algebra with the pointwise Lie bracket and the Lévy process $(j_{st})_{0 \leq s \leq t}$ on $\mathfrak{g}$ defines a representation π of $\mathfrak{g}^{\mathbb{R}_+}$ via

$$\pi(X) = \sum_{k=1}^n j_{s_k t_k}(X_k), \quad \text{for } X = \sum_{k=1}^n X_k \mathbf{1}_{[s_k, t_k)} \in \mathfrak{g}^{\mathbb{R}_+}. \tag{8.6}$$

Denote also by $\Sigma(\mathbb{R}_+)$ the space of real-valued simple step functions

$$\Sigma(\mathbb{R}_+) := \Big\{ \sum_{k=1}^n \phi_k \mathbf{1}_{[s_k,t_k)} : 0 \le s_1 \le t_1 \le s_2 \le \cdots \le t_n < \infty, \phi_1,\ldots,\phi_n \in \mathbb{R} \Big\}.$$

Given an Hermitian element Y of $\mathfrak{g}_\mathbb{R}$ such that $Y^* = Y$, we define a map

$$y \colon \Sigma(\mathbb{R}_+) \longrightarrow \mathcal{L}(\Gamma)$$

where $\mathcal{L}(\Gamma)$ is defined in (8.1), by

$$y_\phi := \sum_{k=1}^n \phi_k j_{s_k t_k}(Y), \quad \text{for } \phi = \sum_{k=1}^n \phi_k \mathbf{1}_{[s_k,t_k)} \in \Sigma(\mathbb{R}_+).$$

Clearly, the operators $\{y_\phi : \phi \in \Sigma(\mathbb{R}_+)\}$ commute, since y is the restriction of

$$\pi \colon \mathfrak{g}^{\mathbb{R}_+} \ni \phi = \sum_{k=1}^n \phi_k \mathbf{1}_{[s_k,t_k)} \longmapsto \sum_{k=1}^n j_{s_k t_k}(\phi_k) \in \mathcal{L}(\Gamma)$$

to the abelian current algebra $\mathbb{C}Y^{\mathbb{R}_+}$ over $\mathbb{C}Y$. Furthermore, if ϕ is real-valued, then y_ϕ is Hermitian, since Y is Hermitian. Therefore, there exists a classical stochastic process $(\tilde{Y}_t)_{t \in \mathbb{R}_+}$ whose moments are given by

$$\mathbf{E}\big[\tilde{Y}_{t_1} \cdots \tilde{Y}_{t_n}\big] = \langle \boldsymbol{\Omega}, y\mathbf{1}_{[0,t_1)} \cdots y\mathbf{1}_{[0,t_n)} \boldsymbol{\Omega} \rangle, \qquad t_1,\ldots,t_n \in \mathbb{R}_+.$$

Since the expectations of $(j_{st})_{0 \le s \le t}$ factorise, we can choose $(\tilde{Y}_t)_{t \in \mathbb{R}_+}$ to be a Lévy process, and if $j_{st}(Y)$ is even essentially self-adjoint then the marginal distributions of $(\tilde{Y}_t)_{t \in \mathbb{R}_+}$ are uniquely determined.

In order to characterize the process $(\tilde{Y}_t)_{t \in \mathbb{R}_+}$ in Theorem 8.4.3 below, we will need the following analogues of the splitting Lemma 6.1.2 in the framework of quantum Lévy processes.

Lemma 8.4.1 *Let $X \in \mathcal{L}(D)$, $u, v \in D$, and suppose further that the series*

$$\sum_{n=0}^\infty \frac{t^n}{n!} X^n w \quad \text{and} \quad \sum_{n=0}^\infty \frac{t^n}{n!} (X^*)^n w \tag{8.7}$$

converge in D for all $w \in D$. Then we have

$$\begin{cases} e^{a^\circ(X)} a^-(v) = a^-\left(e^{-X^*} v\right) e^{a^\circ(X)}, \\[2mm] e^{a^+(u)} a^-(v) = \left(a^-(v) - \langle v, u \rangle\right) e^{a^+(u)}, \\[2mm] e^{a^+(u)} a^\circ(X) = \left(a^\circ(X) - a^+(Xu)\right) e^{a^+(u)}, \end{cases}$$

on the algebraic boson Fock space over D.

Proof: This can be deduced from the formula

$$\text{Ad}e^X Y = e^X Y e^{-X} = Y + [X, Y] + \frac{1}{2}[X, [X, Y]] + \cdots = e^{\text{ad}X} Y$$

for the adjoint actions. □

The following lemma, which is the Lévy process analogue of Lemma 6.1.3, provides the normally ordered form of the generalised Weyl operators, and it is a key tool to calculate the characteristic functions of classical subprocesses of Lévy processes on real Lie algebras.

Lemma 8.4.2 *Let $X \in \mathcal{L}(D)$ and $u, v \in D$ and suppose further that the series (8.7) converge in D for all $w \in D$. Then we have*

$$\exp\left(\alpha + a^\circ(X) + a^+(u) + a^-(v)\right) = e^{\tilde{\alpha}} \exp\left(a^+(\tilde{u})\right) \exp\left(a^\circ(X)\right) \exp\left(a^-(\tilde{v})\right)$$

on the algebraic boson Fock space over D, where $\alpha \in \mathbb{C}$ and

$$\tilde{u} = \sum_{n=1}^{\infty} \frac{X^{n-1}}{n!} u, \quad \tilde{v} = \sum_{n=1}^{\infty} \frac{(X^*)^{n-1}}{n!} v, \quad \tilde{\alpha} = \alpha + \sum_{n=2}^{\infty} \frac{1}{n!} \langle v, X^{n-2} u \rangle.$$

Proof: Let $\omega \in D$ and set $\omega_1(t) = \exp t\left(\alpha + a^\circ(X) + a^+(u) + a^-(v)\right)\omega$ and

$$\omega_2(t) = e^{\tilde{\alpha}(t)} \exp\left(a^+(\tilde{u}(t))\right) \exp\left(ta^\circ(X)\right) \exp\left(a^-(\tilde{v}(t))\right)\omega$$

for $t \in [0, 1]$, where

$$\begin{cases} \tilde{u}(t) = \sum_{n=1}^{\infty} \frac{t^n}{n!} X^{n-1} u, \\ \tilde{v}(t) = \sum_{n=1}^{\infty} \frac{t^n}{n!} (X^*)^{n-1} v, \\ \tilde{\alpha}(t) = t\alpha + \sum_{n=2}^{\infty} \frac{t^n}{n!} \langle v, X^{n-2} u \rangle, \end{cases}$$

with $\omega_1(0) = \omega = \omega_2(0)$. Using Lemma 6.1.2 we also check that

$$\omega_1'(t) = \left(\alpha + a^\circ(X) + a^+(u) + a^-(v)\right)\omega \exp t\left(\alpha + a^\circ(X) + a^+(u) + a^-(v)\right)\omega,$$

and

$$\omega_2'(t) = e^{\tilde{\alpha}(t)} a^+ \left(\frac{d\tilde{u}}{dt}(t) \right) \exp\left(a^+ \left(\tilde{u}(t)\right)\right) \exp\left(ta^\circ(X)\right) \exp\left(a^- \left(\tilde{v}(t)\right)\right)\omega$$

$$+ e^{\tilde{\alpha}(t)} \exp\left(a^+\left(\tilde{u}(t)\right)\right) a^\circ(X) \exp\left(ta^\circ(X)\right) \exp\left(a^-\left(\tilde{v}(t)\right)\right)\omega$$

$$+ e^{\tilde{\alpha}(t)} \exp\left(a^+\left(\tilde{u}(t)\right)\right) \exp\left(ta^\circ(X)\right) a^- \left(\frac{d\tilde{v}}{dt}(t)\right) \exp\left(a^-\left(\tilde{v}(t)\right)\right)\omega$$

$$+ e^{\tilde{\alpha}(t)} \exp\left(a^+\left(\tilde{u}(t)\right)\right) \exp\left(ta^\circ(X)\right) \exp\left(a^-\left(\tilde{v}(t)\right)\right) \frac{d\tilde{\alpha}}{dt}(t)\omega$$

coincide for all $t \in [0, 1]$. Therefore we have $\omega_1(1) = \omega_2(1)$. $\qquad\square$

In the next theorem we compute the characteristic exponent $(\tilde{Y}_t)_{t\in\mathbb{R}_+}$ by application of the splitting Lemma 8.4.2.

Theorem 8.4.3 *Let $(j_{st})_{0\leq s\leq t}$ be a Lévy process on a real Lie algebra $\mathfrak{g}_\mathbb{R}$ with Schürmann triple (ρ, η, L). Then for any Hermitian element Y of $\mathfrak{g}_\mathbb{R}$ such that $\eta(Y)$ is analytic for $\rho(Y)$, the associated classical Lévy process $(\tilde{Y}_t)_{t\in\mathbb{R}_+}$ has characteristic exponent*

$$\Psi(\lambda) = i\lambda L(Y) + \sum_{n=2}^\infty i^n \frac{\lambda^n}{n!} \langle \eta(Y^*), \rho(Y)^{n-2}\eta(Y) \rangle,$$

for λ in a neighborhood of zero, with $\rho(Y)^0 = \mathrm{Id}$.

Proof: From Lemma 6.1.3 we have

$$\mathbf{E}\left[e^{i\lambda\tilde{Y}_t}\right] = \exp\left(it\lambda L(Y) + t\sum_{n=2}^\infty i^n \frac{\lambda^n}{n!} \langle \eta(Y^*), \rho(Y)^{n-2}\eta(Y) \rangle \right),$$

which yields the characteristic exponent

$$\Psi(\lambda) = \frac{1}{t} \log \mathbf{E}\left[e^{i\lambda\tilde{Y}_t}\right] = \frac{1}{t}\log\langle\boldsymbol{\Omega}, e^{i\lambda j_{0t}(Y)}\boldsymbol{\Omega}\rangle, \quad \lambda \in \mathbb{R},$$

for

$$j_{0t}(Y) = a_{0t}^\circ\left(\rho(Y)\right) + a_{0t}^+\left(\eta(Y)\right) + a_{0t}^-\left(\eta(Y)\right) + tL(Y).$$

$\qquad\square$

A more direct proof of the theorem is also possible using the convolution of functionals on $\mathcal{U}(\mathfrak{g})$ instead of the boson Fock space realisation of $(j_{st})_{0\leq s\leq t}$. We note $\Psi(\lambda)$ also coincides with

$$\Psi(\lambda) = \sum_{n=1}^\infty i^n \frac{\lambda^n}{n!} L(Y^n).$$

Next, we give two corollaries of Theorem 8.4.3; the first of them justifies our definition of Gaussian generating functionals.

Corollary 8.4.4 *Let L be a Gaussian generating functional on $\mathfrak{g}_{\mathbb{R}}$ with corresponding Lévy process $(j_{st})_{0 \leq s \leq t}$. For any Hermitian element Y the associated classical Lévy process $(\tilde{Y}_t)_{t \in \mathbb{R}_+}$ is Gaussian with mean and variance*

$$\mathbf{E}[\tilde{Y}_t] = tL(Y), \qquad \mathbf{E}[\tilde{Y}_t^2] = ||\eta(Y)||^2 t, \qquad t \in \mathbb{R}_+.$$

We see that in this case we can take

$$(\tilde{Y}_t)_{t \in \mathbb{R}_+} = \left(||\eta(Y)||B_t + L(Y)t \right)_{t \in \mathbb{R}_+},$$

where $(B_t)_{t \in \mathbb{R}_+}$ is a standard Brownian motion. The next corollary deals with the case where L is the restriction to $\mathcal{U}_0(\mathfrak{g})$ of a positive functional on $\mathcal{U}(\mathfrak{g})$.

Corollary 8.4.5 *Let (ρ, η, L) be a Schürmann triple on $\mathfrak{g}_{\mathbb{R}}$ whose cocycle is trivial, i.e., there exists a vector $\omega \in D$ such that*

$$\eta(u) = \rho(u)\omega, \qquad u \in \mathcal{U}_0(\mathfrak{g}),$$

with generating functional of the form

$$L(u) = \langle \omega, \rho(u)\omega \rangle, \qquad u \in \mathcal{U}_0(\mathfrak{g}).$$

Suppose further that the vector ω is analytic for $\rho(Y)$, i.e.,

$$\exp(u\rho(Y))\,\omega := \sum_{n=1}^{\infty} \frac{u^n}{n!} \rho(Y)^n \omega,$$

converges for sufficiently small u. Then the classical stochastic process $(\tilde{Y}_t)_{t \in \mathbb{R}_+}$ associated to $(j_{st})_{0 \leq s \leq t}$ and to Y is a compound Poisson process with characteristic exponent given by

$$\Psi(u) = \left\langle \omega, \left(e^{iu\rho(Y)} - 1 \right) \omega \right\rangle.$$

The above corollary suggests to call a Lévy process on $\mathfrak{g}$ with trivial cocycle $\eta(u) = \rho(u)\omega$ and generating functional $L(u) = \langle \omega, \rho(u)\omega \rangle$ for $u \in \mathcal{U}_0(\mathfrak{g})$ a *Poisson process* on $\mathfrak{g}$. Note that in case the operator $\rho(Y)$ is (essentially) self-adjoint, the Lévy measure of $(\tilde{Y}_t)_{t \in \mathbb{R}_+}$ can be obtained by evaluating its spectral measure

$$\mu(d\lambda) = \langle \omega, d\mathbb{P}_\lambda \omega \rangle$$

in the state ω, where $\rho(Y) = \int \lambda d\mathbb{P}_\lambda$ is the spectral resolution of (the closure of) $\rho(Y)$.

By choosing a commutative subalgebra of $\pi(\mathfrak{g}^{\mathbb{R}_+})$ we can also obtain a classical process, using the fact that the product fX of an element $X \in \mathfrak{g}^{\mathbb{R}_+}$ with a function $f \in \Sigma(\mathbb{R}_+)$ is again in $\mathfrak{g}^{\mathbb{R}_+}$.

Theorem 8.4.6 *Let $(j_{st})_{0 \leq s \leq t}$ be a Lévy process on a real Lie algebra $\mathfrak{g}$ and let π be as in Equation (8.6). Choose $X \in \mathfrak{g}^{\mathbb{R}_+}$, and define*

$$X(f) := i\pi(fX), \qquad f \in \Sigma(\mathbb{R}_+).$$

Then there exists a classical stochastic process $(\hat{X}_t)_{t \in \mathbb{R}_+}$ with independent increments that has the same finite distributions as X, i.e.,

$$\langle \Omega, g_1(X(f_1)) \cdots g_n(X(f_n))\Omega \rangle = \mathbb{E}\left[g_1(\hat{X}(f_1)) \cdots g_n(\hat{X}(f_n))\right]$$

for all $n \in \mathbb{N}, f_1, \ldots, f_n \in \Sigma(\mathbb{R}_+), g_1, \ldots, g_n \in C_0(\mathbb{R})$, where

$$\hat{X}(f) = \int_{\mathbb{R}_+} f(t)d\hat{X}_t = \sum_{k=1}^{n} \alpha_k(\hat{X}_{t_k} - \hat{X}_{s_k}),$$

for $f = \sum_{k=1}^{n} \alpha_k \mathbf{1}_{[s_k, t_k)} \in \Sigma(\mathbb{R}_+)$.

The existence of $(\hat{X}_t)_{t \in \mathbb{R}_+}$ follows as in [1, Section 4], and

$$g_1(X(f_1)), \ldots, g_n(X(f_n))$$

can be defined by the usual functional calculus for the (essentially) self-adjoint operators $X(f_1), \ldots, X(f_n)$.

Notes

Lévy processes on real Lie algebras form a special case of Lévy processes on involutive bialgebras, see [106], [79, chapter VII], [45]. They have already been studied under the name factorisable representations of current algebras in the sixties and seventies, see [109] for a historical survey and for references. They are at the origin of the theory of quantum stochastic differential calculus. See Section 5 of [109] for more references and a historical survey on the theory of factorisable representations of current groups and algebra and its relation to quantum stochastic calculus. Among future problems we can mention the study of the cohomology of representations and the classification of all Lévy processes on Lie algebras. We refer to [51] for the cohomology of Lie algebras and Lie groups. It is known that the cohomology groups of all simple nontrivial representations of the Lie algebra defined in (8.5) are trivial, see [51, Proposition II.6.2].

Exercises

Exercise 8.1 Example of classical Lévy process. Let $Y = B^+ + B^- + \beta M$ with $\beta \in \mathbb{R}$ and $Me_0 = m_0 e_0$. This exercise aims at characterising the *classical* Lévy process $(\tilde{Y}_t)_{t \in \mathbb{R}_+}$ associated to Y and $(j_{st})_{0 \leq s \leq t}$ in the manner described earlier. Corollary 8.4.5 tells us that $(\tilde{Y}_t)_{t \in \mathbb{R}_+}$ is a compound Poisson process with characteristic exponent

$$\Psi(u) = \langle e_0, \left(e^{iuX} - 1\right) e_0 \rangle.$$

We want to determine the Lévy measure of $(\tilde{Y}_t)_{t \in \mathbb{R}_+}$, *i.e.*, we want to determine the measure μ on $\mathbb{R}$, for which

$$\Psi(u) = \int_{-\infty}^{\infty} \left(e^{iux} - 1\right) \mu(dx).$$

This is the spectral measure of X evaluated in the state $\langle e_0, \cdot \, e_0 \rangle$. Consider the polynomials $p_n(x) \in \mathbb{R}[x]$ defined by the condition

$$e_n = p_n(X)e_0, \qquad n \in \mathbb{N}.$$

1. Show that the polynomials $p_n(x)$ are orthogonal with respect to μ, *i.e.*,

$$\int_{-\infty}^{\infty} p_n(x)p_m(x)\mu(dx) = \delta_{nm}, \qquad n, m \in \mathbb{N}.$$

2. Find the three-term recurrence relation satisfied by the polynomials $p_n(x)$.
3. Determine the polynomials $p_n(x)$ according to the value of β.
4. Determine the density μ with respect to which the polynomials $p_n(x)$ are orthogonal.

9

A guide to the Malliavin calculus

I do not think that 150 years from now, people will photocopy pages
from Bourbaki to rhapsodize on them. Some lines in this memoir by
Poisson, on the other hand, are beaming with life . . .

(P. Malliavin, in Dialogues Autour de la Création
Mathématique, *1997.)*

This chapter is an introduction to the Malliavin calculus, as a preparation for
the noncommutative setting of Chapters 11 and 12. We adopt the point of view
of normal martingales in a general framework that encompasses Brownian
motion and the Poisson process as particular cases, as in [98]. The Malliavin
calculus originally requires a heavy functional analysis apparatus, here we
assume a basic knowledge of stochastic calculus; proofs are only outlined and
the reader is referred to the literature for details.

9.1 Creation and annihilation operators

Let $(\Omega, \mathcal{F}, \mathbb{P})$ be a probability space equipped with a right-continuous filtration
$(\mathcal{F}_t)_{t \in \mathbb{R}_+}$, *i.e.*, an increasing family of sub σ-algebras of $\mathcal{F}$ such that

$$\mathcal{F}_t = \bigcap_{s>t} \mathcal{F}_s, \qquad t \in \mathbb{R}_+.$$

In our presentation of stochastic integration we work in the framework of
normal martingales, which are square-integrable martingales $(M_t)_{t \in \mathbb{R}_+}$ such
that

$$\mathbb{E}\left[(M_t - M_s)^2 \mid \mathcal{F}_s \right] = t - s, \qquad 0 \le s < t. \tag{9.1}$$

As will be seen in the next sections, the family of normal martingales contains
Brownian motion and the compensated standard Poisson process as particular
cases.

Every square-integrable process $(M_t)_{t \in \mathbb{R}_+}$ with centered independent increments and generating the filtration $(\mathcal{F}_t)_{t \in \mathbb{R}_+}$ satisfies

$$\mathbb{E}\left[(M_t - M_s)^2 \,\middle|\, \mathcal{F}_s\right] = \mathbb{E}\left[(M_t - M_s)^2\right], \qquad 0 \le s \le t.$$

In particular, a square-integrable process $(M_t)_{t \in \mathbb{R}_+}$ with centered independent increments is a normal martingale if and only if

$$\mathbb{E}\left[(M_t - M_s)^2\right] = t - s, \qquad 0 \le s \le t.$$

Note that a martingale $(M_t)_{t \in \mathbb{R}_+}$ is normal if and only if $(M_t^2 - t)_{t \in \mathbb{R}_+}$ is a martingale, *i.e.*,

$$\mathbb{E}\left[M_t^2 - t \mid \mathcal{F}_s\right] = M_s^2 - s, \qquad 0 \le s < t.$$

9.1.1 Multiple stochastic integrals

Let $L^2(\mathbb{R}_+)^{\circ n}$ denote the subspace of $L^2(\mathbb{R}_+)^{\otimes n} = L^2(\mathbb{R}_+^n)$, made of symmetric functions f_n in n variables (see the Appendix A.8. for a review of tensor products). The multiple stochastic integral of a symmetric function $f_n \in L^2(\mathbb{R}_+)^{\circ n}$ is defined as an iterated integral. First we let

$$I_1(f) = \int_0^\infty f(t) dM_t, \qquad f \in L^2(\mathbb{R}_+).$$

As a convention we identify $L^2(\mathbb{R}_+)^{\circ 0}$ to $\mathbb{R}$ and let

$$I_0(f_0) = f_0, \qquad f_0 \in L^2(\mathbb{R}_+)^{\circ 0} \simeq \mathbb{R}.$$

As a consequence of (9.1) we can prove the following.

Proposition 9.1.1 *The multiple stochastic integral*

$$I_n(f_n) := n! \int_0^\infty \int_0^{t_n} \cdots \int_0^{t_2} f_n(t_1, \ldots, t_n) dM_{t_1} \cdots dM_{t_n} \qquad (9.2)$$

of $f_n \in L^2(\mathbb{R}_+)^{\circ n}$ satisfies the isometry formula

$$\mathbb{E}[I_n(f_n) I_m(g_m)] = n! 1_{\{n=m\}} \langle f_m, g_m \rangle_{L^2(\mathbb{R}_+^n)},$$

$f_n \in L^2(\mathbb{R}_+)^{\circ n}, f_m \in L^2(\mathbb{R}_+)^{\circ m}, n, m \in \mathbb{N}.$

In particular we have $\mathbb{E}[I_n(f_n)] = 0$ for all $n \ge 1$. As a consequence of Proposition 9.1.1 the multiple stochastic integral operator I_n induces an isometric isomorphism between $L^2(\Omega)$ and the Fock space over $L^2(\mathbb{R}_+)$.

Lemma 9.1.2 *For all $f_n \in L^2(\mathbb{R}_+)^{\circ n}$, $n \geq 1$, we have*

$$\mathbb{E}[I_n(f_n) \mid \mathcal{F}_t] = I_n\left(f_n \mathbf{1}_{[0,t]^n}\right), \qquad t \in \mathbb{R}_+.$$

Proof: Since the indefinite Itô integral is a martingale from (9.2) we have

$$\mathbb{E}[I_n(f_n) \mid \mathcal{F}_t] = n!\mathbb{E}\left[\int_0^\infty \int_0^{t_n} \cdots \int_0^{t_2} f_n(t_1, \ldots, t_n) dM_{t_1} \cdots dM_{t_n} \Big| \mathcal{F}_t\right]$$

$$= n! \int_0^t \int_0^{t_n} \cdots \int_0^{t_2} f_n(t_1, \ldots, t_n) dM_{t_1} \cdots dM_{t_n}$$

$$= I_n\left(f_n \mathbf{1}_{[0,t]^n}\right). \qquad \square$$

9.1.2 Annihilation operator

Consider the spaces $\mathcal{S}$ and $\mathcal{U}$ defined by

$$\mathcal{S} = \left\{ \sum_{k=0}^n I_k(f_k) : f_k \in L^4(\mathbb{R}_+)^{\circ k}, \; k = 0, \ldots, n, \; n \in \mathbb{N} \right\}, \qquad (9.3)$$

and

$$\mathcal{U} = \left\{ \sum_{i=1}^n \mathbf{1}_{[t_i, t_{i-1})} F_i \; : \; F_i \in \mathcal{S}, \; 0 = t_0 \leq t_1 < \cdots < t_n, \; n \geq 1 \right\},$$

which is contained in

$$\tilde{\mathcal{U}} := \left\{ \sum_{k=0}^n I_k(g_k(*, \cdot)) \; : \; g_k \in L^2(\mathbb{R}_+)^{\circ k} \otimes L^2(\mathbb{R}_+), \; k = 0, \ldots, n, \; n \in \mathbb{N} \right\},$$

where the symmetric tensor product $\circ$ is defined in the Appendix A.8. Next we state the definition of the operators D and δ on multiple stochastic integrals (random variables and processes), whose linear combinations span $\mathcal{S}$ and $\mathcal{U}$.

Definition 9.1.3 *Let $D: \mathcal{S} \longrightarrow L^2(\Omega \times \mathbb{R}_+)$ be the linear operator defined by*

$$D_t I_n(f_n) = n I_{n-1}(f_n(*, t)), \qquad d\mathbb{P} \times dt - a.e.,$$

$f_n \in L^2(\mathbb{R}_+)^{\circ n}$.

Due to its role as a lowering operator on the degree of multiple stochastic integrals, the operator D identifies to an *annihilation* operator on the boson Fock space over $L^2(\mathbb{R}_+)$.

Proposition 9.1.4   *The domain* $\mathrm{Dom}(D) = D([0,\infty))$ *of* D *consists in the space of square-integrable random variables with chaos expansion*

$$F = \sum_{n=0}^{\infty} I_n(f_n), \qquad (9.4)$$

such that the series

$$\sum_{k=1}^{n} k I_{k-1}(f_k(*, \cdot))$$

converges in $L^2(\Omega \times \mathbb{R}_+)$ *as* n *goes to infinity.*

Given $F \in \mathrm{Dom}(D)$ with the expansion (9.4) we have

$$\mathbf{E}\left[\|DF\|_{L^2(\mathbb{R}_+)}^2\right] = \sum_{k=1}^{\infty} k k! \|f_k\|_{L^2(\mathbb{R}_+^k)}^2 < \infty,$$

and

$$D_t F = f_1(t) + \sum_{k=1}^{\infty} k I_{k-1}(f_k(*, t)), \qquad dt d\mathbb{P} - a.e.$$

In particular, the exponential vector

$$\xi_t(u) := \sum_{n=0}^{\infty} \frac{1}{n!} I_n\left(\left(u \mathbf{1}_{[0,t]}\right)^{\otimes n}\right), \qquad t \in \mathbb{R}_+,$$

belongs to $\mathrm{Dom}(D)$ for all $u \in L^2(\mathbb{R}_+)$ and we have

$$D_s \xi_t(u) = \mathbf{1}_{[0,t]}(s) u(s) \xi_t(u), \qquad s, t \in \mathbb{R}.$$

Since S defined by (9.3) is assumed to be dense in $L^2(\Omega)$, $(M_t)_{t \in \mathbb{R}_+}$ has the chaos representation property.

Definition 9.1.5   *Let*

$$\delta : \tilde{\mathcal{U}} \longrightarrow L^2(\Omega)$$

be the linear operator defined by

$$\delta(I_n(f_{n+1}(*, \cdot))) = I_{n+1}(\tilde{f}_{n+1}), \qquad f_{n+1} \in L^2(\mathbb{R}_+)^{\circ n} \otimes L^2(\mathbb{R}_+),$$

where $\tilde{f}_{n+1}$ *is the symmetrisation of* f_{n+1} *in* $n+1$ *variables defined as*

$$\tilde{f}_{n+1}(t_1, \ldots, t_{n+1}) = \frac{1}{n+1} \sum_{k=1}^{n+1} f_{n+1}(t_1, \ldots, t_{k-1}, t_{k+1}, \ldots, t_{n+1}, t_k).$$

In particular we have

$$f \circ g_n(t_1, \ldots, t_{n+1}) = \frac{1}{n+1} \sum_{k=1}^{n+1} f(t_k) g_n(t_1, \ldots, t_{k-1}, t_{k+1}, \ldots, t_{n+1}),$$

i.e., $f \circ g_n$ is the symmetrisation of $f \otimes g_n$ in $n+1$ variables. Similarly, the operator δ is usually referred to as a *creation* operator, due to the fact that it raises the degree of multiple stochastic integrals. The operator δ is also called the Skorohod integral. Note that we have

$$\delta(f) = I_1(f) = \int_0^\infty f(t) dM_t, \quad f \in L^2(\mathbb{R}_+),$$

and, in particular,

$$\delta(u I_n(f_n))$$
$$= n \int_0^\infty I_n(f_n(*, s) \circ u . \mathbf{1}_{[0,s]^n}(*, \cdot)) dM_s + \int_0^\infty u_s I_n(f_n \mathbf{1}_{[0,s]^n}) dM_s,$$

$u \in L^2(\mathbb{R}_+)$, $g_n \in L^2(\mathbb{R}_+)^{\circ n}$, where as a convention "$*$" denotes the $n-1$ first variables and "$\cdot$" denotes the last integration variable in I_n.

By the isomorphism between $L^2(\Omega)$ and $\Gamma(L^2(\mathbb{R}_+))$ we can deduce that canonical commutation relation satisfied by the operators D and δ, *i.e.*, for any $u \in \tilde{\mathcal{U}}$ we have

$$D_t \delta(u) = u_t + \delta(D_t u), \quad t \in \mathbb{R}_+.$$

9.1.3 Duality relation

The next proposition states the duality relation satisfied by D and δ.

Proposition 9.1.6 *The operators D and δ satisfy the duality relation*

$$\mathbf{E}[F\delta(u)] = \mathbf{E}[\langle DF, u \rangle_{L^2(\mathbb{R}_+)}], \quad F \in \mathcal{S}, \quad u \in \mathcal{U}.$$

Proof: We consider $F = I_n(f_n)$ and $u_t = I_m(g_{m+1}(*, t))$, $t \in \mathbb{R}_+$, $f_n \in L^2(\mathbb{R}_+)^{\circ n}$, $g_{m+1} \in L^2(\mathbb{R}_+)^{\circ m} \otimes L^2(\mathbb{R}_+)$. We have

$$\mathbf{E}[F\delta(u)] = \mathbf{E}[I_{m+1}(\tilde{g}_{m+1}) I_n(f_n)]$$
$$= n! \mathbf{1}_{\{n=m+1\}} \langle f_n, \tilde{g}_n \rangle_{L^2(\mathbb{R}_+^n)}$$
$$= n! \mathbf{1}_{\{n=m+1\}} \langle f_n, g_n \rangle_{L^2(\mathbb{R}_+^n)}$$

$$= n! \mathbf{1}_{\{n-1=m\}}$$

$$\times \int_0^\infty \cdots \int_0^\infty f_n(s_1, \ldots, s_{n-1}, t) g_n(s_1, \ldots, s_{n-1}, t) ds_1 \cdots ds_{n-1} dt$$

$$= n \mathbf{1}_{\{n-1=m\}} \int_0^\infty \mathbf{E}[I_{n-1}(f_n(*, t)) I_{n-1}(g_n(*, t))] dt$$

$$= \mathbf{E}[\langle D.I_n(f_n), I_m(g_{m+1}(*, \cdot)) \rangle_{L^2(\mathbb{R}_+)}]$$

$$= \mathbf{E}[\langle DF, u \rangle_{L^2(\mathbb{R}_+)}]. \qquad \qquad \square$$

Remark 9.1.7 *By construction, the operator D satisfies the stability assumption thus we have $D_s F = 0$, $s > t$, for any $\mathcal{F}_t$-measurable $F \in \mathcal{S}$, $t \in \mathbb{R}_+$.*

From now on we will assume that $\mathcal{S}$ is dense in $L^2(\Omega)$, which is equivalent to saying that $(M_t)_{t \in \mathbb{R}_+}$ has the chaos representation property. As a consequence of Proposition 9.1.6 we have the following.

Proposition 9.1.8 *The operators D and δ are closable on $L^2(\Omega)$ and $L^2(\Omega \times \mathbb{R}_+)$ respectively.*

It also follows from the density of $\mathcal{S}$ in $L^2(\Omega)$ that $\mathcal{U}$ is dense in $L^2(\Omega \times \mathbb{R}_+)$. More generally, the following proposition follows from the fact that the denseness of $\mathcal{S}$ is equivalent to the chaos representation property.

Proposition 9.1.9 *If $(M_t)_{t \in \mathbb{R}_+}$ has the chaos representation property then it has the predictable representation property.*

The domain $\text{Dom}(\delta)$ of δ is the space of processes $(u_t)_{t \in \mathbb{R}_+} \in L^2(\Omega \times \mathbb{R}_+)$ with

$$u_t = \sum_{n=0}^\infty I_n(f_{n+1}(*, t)),$$

and such that

$$\mathbf{E}[|\delta(u)|^2] = \sum_{n=1}^\infty (n+1)! \|\tilde{f}_n\|_{L^2(\mathbb{R}_+^{n+1})}^2 < \infty.$$

The creation operator δ satisfies the following Itô–Skorohod type isometry.

Proposition 9.1.10 *Let $u \in \text{Dom}(\delta)$ such that $u_t \in \text{Dom}(D)$, dt-a.e., and $(D_s u_t)_{s,t \in \mathbb{R}_+} \in L^2(\Omega \times \mathbb{R}_+^2)$. We have*

$$\mathbf{E}[|\delta(u)|^2] = \mathbf{E}\left[\|u\|_{L^2(\mathbb{R}_+)}^2 \right] + \mathbf{E}\left[\int_0^\infty \int_0^\infty D_s u_t D_t u_s ds dt \right]. \qquad (9.5)$$

By bilinearity, we also have

$$\langle \delta(u), \delta(v) \rangle_{L^2(\Omega)} = \langle u, v \rangle_{L^2(\Omega \times \mathbb{R}_+)} + \int_0^\infty \int_0^\infty \langle D_s u_t, D_t v_s \rangle_{L^2(\Omega)} ds dt,$$

for u and v satisfying the conditions of Proposition 9.1.10.

Definition 9.1.11 *Let $L_{p,1}$ denote the space of stochastic processes $(u_t)_{t \in \mathbb{R}_+}$ such that $u_t \in \mathrm{Dom}(D)$, dt-a.e., and*

$$\|u\|_{p,1}^p := \mathbf{E}\left[\|u\|_{L^2(\mathbb{R}_+)}^p \right] + \mathbf{E}\left[\int_0^\infty \int_0^\infty |D_s u_t|^p ds dt \right] < \infty.$$

The next result is a direct consequence of Proposition 9.1.10 and Definition 9.1.11 for $p = 2$.

Proposition 9.1.12 *We have $L_{2,1} \subset \mathrm{Dom}(\delta)$.*

As a consequence of Proposition 9.1.6, the operator δ coincides with the Itô integral with respect to $(M_t)_{t \in \mathbb{R}_+}$ on the square-integrable adapted processes, as stated in the next proposition.

Proposition 9.1.13 *Let $(u_t)_{t \in \mathbb{R}_+} \in L_{ad}^2(\Omega \times \mathbb{R}_+)$ be a square-integrable adapted process. We have*

$$\delta(u) = \int_0^\infty u_t dM_t.$$

Note that when $(u_t)_{t \in \mathbb{R}_+} \in L_{ad}^2(\Omega \times \mathbb{R}_+)$ is a square-integrable adapted process, then Relation (9.5) becomes the Itô isometry as a consequence of Proposition 9.1.13, *i.e.*, we have

$$\|\delta(u)\|_{L^2(\Omega)} = \left\| \int_0^\infty u_t dM_t \right\|_{L^2(\Omega)}$$

$$= \|u\|_{L^2(\Omega \times \mathbb{R}_+)}, \qquad u \in L_{ad}^2(\Omega \times \mathbb{R}_+),$$

as follows from Remark 9.1.7 since $D_t u_s = 0$, $0 \le s \le t$.

9.2 Wiener space

In this section we focus on the case where $(M_t)_{t \in \mathbb{R}_+}$ is a standard Brownian motion, *i.e.*, $(M_t)_{t \in \mathbb{R}_+}$ is a *normal martingale* that solves the structure equation

$$[M, M]_t = t, \qquad t \in \mathbb{R}_+.$$

The reader is referred to [58, 76, 83, 84, 98, 113] for more details. Brownian motion can also be defined from a linear map $W \colon \mathfrak{h} \longrightarrow L^2(\Omega)$ on a real separable Hilbert space $\mathfrak{h}$, such that the $W(h)$ are centered Gaussian random variables with covariances given by

$$\mathbb{E}[W(h)W(k)] = \langle h, k \rangle, \qquad h, k \in \mathfrak{h}$$

on a probability space $(\Omega, \mathcal{F}, \mathbb{P})$.

Setting $\mathcal{H}_1 = W(\mathfrak{h})$ yields a closed Gaussian subspace of $L^2(\Omega)$, and $W : \mathfrak{h} \longrightarrow \mathcal{H}_1 \subseteq L^2(\Omega)$ is an isometry, and we will assume that the σ-algebra $\mathcal{F}$ is generated by the elements of $\mathcal{H}_1$.

Let

$$\phi_d^\sigma(s_1, \ldots, s_d) = \frac{1}{(2\pi)^{d/2}} e^{-(s_1^2 + \cdots + s_d^2)/2}, \qquad (s_1, \ldots, s_d) \in \mathbb{R}^d,$$

denote the standard Gaussian density function with covariance $\sigma^2 \mathrm{Id}$ on $\mathbb{R}^n$. The multiple stochastic integrals $I_n(f_n)$ of $f_n \in L^2(\mathbb{R}_+)^{\circ n}$ with respect to $(B_t)_{t \in \mathbb{R}_+}$ satisfy the multiplication formula

$$I_n(f_n)I_m(g_m) = \sum_{k=0}^{n \wedge m} k! \binom{n}{k}\binom{m}{k} I_{n+m-2k}(f_n \otimes_k g_m),$$

where $f_n \otimes_k g_m$ is the contraction

$$(t_{k+1}, \ldots, t_n, s_{k+1}, \ldots, s_m) \longmapsto$$

$$\int_0^\infty \cdots \int_0^\infty f_n(t_1, \ldots t_n) g_m(t_1, \ldots, t_k, s_{k+1}, \ldots, s_m) dt_1 \ldots dt_k,$$

$t_{k+1}, \ldots, t_n, s_{k+1}, \ldots, s_m \in \mathbb{R}_+$. In particular, we have

$$I_1(u)I_n(v^{\otimes n}) = I_{n+1}(v^{\otimes n} \circ u) + n\langle u, v \rangle_{L^2(\mathbb{R}_+)} I_{n-1}(v^{\otimes(n-1)}) \qquad (9.6)$$

for $n \geq 1$, and

$$I_1(u)I_1(v) = I_2(v \circ u) + \langle u, v \rangle_{L^2(\mathbb{R}_+)}$$

for $n = 1$. The Hermite polynomials, cf. Section A.1 in the appendix, will be used to represent the multiple Wiener integrals.

Proposition 9.2.1 *For any orthogonal family $\{u_1, \ldots, u_d\}$ in $L^2(\mathbb{R}_+)$ we have*

$$I_n(u_1^{\otimes n_1} \circ \cdots \circ u_d^{\otimes n_d}) = \prod_{k=1}^d H_{n_k}(I_1(u_k); \|u_k\|_2^2),$$

where $n = n_1 + \cdots + n_d$.

Proof: We have

$$H_0(I_1(u); \|u\|_2^2) = I_0(u^{\otimes 0}) = 1 \quad \text{and} \quad H_1(I_1(u); \|u\|_2^2) = I_1(u),$$

hence the proof follows by induction on $n \geq 1$, by comparison of the recurrence formula (A.1) with the multiplication formula (9.6). □

In particular we have

$$I_n\left(1_{[0,t]}^{\otimes n}\right) = n! \int_0^t \int_0^{s_n} \cdots \int_0^{s_2} dB_{s_1} \cdots dB_{s_n} = H_n(B_t; t),$$

and

$$I_n\left(1_{[t_0,t_1]}^{\otimes n_1} \circ \cdots \circ 1_{[t_{d-1},t_d]}^{\otimes n_d}\right) = \prod_{k=1}^d I_{n_k}\left(1_{[t_{k-1},t_k]}^{\otimes n_k}\right)$$

$$= \prod_{k=1}^d H_{n_k}(B_{t_k} - B_{t_{k-1}}; t_k - t_{k-1}).$$

From this we recover the orthonormality properties of the Hermite polynomials with respect to the Gaussian density:

$$\int_{-\infty}^{\infty} H_n(x; t)\check{H}_m(x; t)e^{-x^2/(2t)} \frac{dx}{\sqrt{2\pi t}} = \mathbf{E}[H_n(B_t; t)H_m(B_t; t)]$$

$$= \mathbf{E}\left[I_n\left(1_{[0,t]}^{\otimes n}\right)I_m\left(1_{[0,t]}^{\otimes m}\right)\right]$$

$$= 1_{\{n=m\}}n!t^n.$$

In addition, by Lemma 9.1.2 we have that

$$H_n(B_t; t) = I_n\left(1_{[0,t]}^{\otimes n}\right) = \mathbf{E}\left[I_n\left(1_{[0,T]}^{\otimes n}\right) \Big| \mathcal{F}_t\right], \qquad t \in \mathbb{R}_+,$$

is a martingale which, from Itô's formula, can be written as

$$H_n(B_t; t) = I_n\left(1_{[0,t]}^{\otimes n}\right)$$

$$= H_n(0; 0) + \int_0^t \frac{\partial H_n}{\partial x}(B_s; s)dB_s + \frac{1}{2}\int_0^t \frac{\partial^2 H_n}{\partial x^2}(B_s; s)ds$$

$$+ \int_0^t \frac{\partial H_n}{\partial s}(B_s; s)ds$$

$$= n\int_0^t I_{n-1}\left(1_{[0,s]}^{\otimes(n-1)}\right)dB_s$$

$$= n\int_0^t H_{n-1}(B_s; s)dB_s.$$

Given $f_n \in L^2(\mathbb{R}_+)^{\otimes n}$ with orthogonal expansion

$$f_n = \sum_{\substack{n_1 + \cdots + n_d = n \\ k_1, \ldots, k_d \geq 0}} a_{k_1, \ldots, k_d}^{n_1, \ldots, n_d} e_{k_1}^{\otimes n_1} \circ \cdots \circ e_{k_d}^{\otimes n_d},$$

in an orthonormal basis $(e_n)_{n \in \mathbb{N}}$ of $L^2(\mathbb{R}_+)$, we have

$$I_n(f_n) = \sum_{\substack{n_1 + \cdots + n_d = n \\ k_1, \ldots, k_d \geq 0}} a_{k_1, \ldots, k_d}^{n_1, \ldots, n_d} H_{n_1}(I_1(e_{k_1}); 1) \cdots H_{n_d}(I_1(e_{k_d}); 1),$$

where the coefficients $a_{k_1, \ldots, k_d}^{n_1, \ldots, n_d}$ are given by

$$a_{k_1, \ldots, k_d}^{n_1, \ldots, n_d} = \frac{1}{n_1! \cdots n_d!} \langle I_n(f_n), I_k(e_{k_1}^{\otimes n_1} \circ \cdots \circ u_{k_d}^{\otimes n_d}) \rangle_{L^2(\Omega)}$$

$$= \langle f_n, e_{k_1}^{\otimes n_1} \circ \cdots \circ e_{k_d}^{\otimes n_d} \rangle_{L^2(\mathbb{R}_+^n)}.$$

The following relation for exponential vectors, can be recovered independently using the Hermite polynomials.

Proposition 9.2.2 *We have*

$$\xi(u) = \sum_{k=0}^{\infty} \frac{1}{n!} I_n(u^{\otimes n}) = \exp\left(I_1(u) - \frac{1}{2} \|u\|_{L^2(\mathbb{R}_+)}^2\right). \tag{9.7}$$

Proof: Relation (9.7) follows from Proposition A.1.3-*i*) and Proposition 9.2.1 which reads $I_n(u^{\otimes n}) = H_n(I_1(u); \|u\|_{L^2(\mathbb{R}_+)}^2)$, $n \geq 1$. $\qquad\square$

The following property can be proved by a Fourier transform argument and the density property in $L^2(\Omega)$ of the linear space spanned by the exponential vectors

$$\left\{ \exp\left(I_1(u) - \frac{1}{2} \|u\|_{L^2(\mathbb{R}_+)}^2\right) : u \in L^2(\mathbb{R}_+) \right\},$$

cf. *e.g.*, Theorem 4.1, p. 134 of [52].

Proposition 9.2.3 *The Brownian motion* $(B_t)_{t \in \mathbb{R}_+}$ *has the chaos representation property, i.e., any* $F \in L^2(\Omega)$ *admits a chaos decomposition*

$$F = \sum_{k=0}^{\infty} I_k(g_k).$$

Assume that F has the form $F = g(I_1(e_1), \ldots, I_1(e_k))$ for some

$$g \in L^2\left(\mathbb{R}^k, \frac{1}{(2\pi)^{k/2}} e^{-|x|^2/2} dx\right),$$

and admits the chaos expansion $F = \sum_{n=0}^{\infty} I_n(f_n)$. Then for all $n \geq 1$ there exists a (multivariate) Hermite polynomial P_n of degree n such that

$$I_n(f_n) = P_n(I_1(e_1), \dots, I_1(e_k)).$$

9.2.1 Gradient and divergence operators

In the Brownian case, the operator D has the derivation property, *i.e.*,

$$D_t(FG) = FD_tG + GD_tF, \qquad F, G \in \mathcal{S}.$$

More precisely, introducing the algebra of bounded smooth functionals

$$\mathcal{S} = \left\{ F = f\big(W(h_1), \dots, W(h_n)\big) \ : \ n \in \mathbb{N}, \, f \in C_b^{\infty}(\mathbb{R}^n), \, h_1, \dots, h_n \in \mathfrak{h} \right\},$$

the derivation operator

$$D \colon \mathcal{S} \longrightarrow L^2(\Omega) \otimes \mathfrak{h} \cong L^2(\Omega; \mathfrak{h})$$

is given by

$$DF = \sum_{i=1}^{n} \frac{\partial f}{\partial x_i}\big(W(h_1), \dots, W(h_n)\big) \otimes h_i$$

for $F = f\big(W(h_1), \dots, W(h_n)\big) \in \mathcal{S}$. In particular, D is a derivation with respect to the natural $L^{\infty}(\Omega)$-bimodule structure of $L^2(\Omega; \mathfrak{h})$, *i.e.*,

$$D(FG) = F(DG) + (DG)F, \qquad F, G \in \mathcal{S}.$$

We can also define the gradient $D_u F = \langle u, DF \rangle$ with respect to $\mathfrak{h}$-valued random variables $u \in L^2(\Omega; \mathfrak{h})$, this is $L^{\infty}(\Omega)$-linear in the first argument and a derivation in the second, *i.e.*,

$$D_{Fu}G = FD_uG \quad \text{and} \quad D_u(FG) = F(D_uG) + (D_uF)G.$$

The derivation operator D is a closable operator from $L^p(\Omega)$ to $L^p(\Omega; \mathfrak{h})$ for $1 \leq p \leq \infty$. We will denote its closure again by D. Given that $L^2(\Omega)$ and $L^2(\Omega; \mathfrak{h})$ are Hilbert spaces (with the obvious inner products), the closability of D implies that it has an adjoint. We will call the adjoint of

$$D \colon L^2(\Omega) \longrightarrow L^2(\Omega; \mathfrak{h})$$

the divergence operator and denote it by

$$\delta \colon L^2(\Omega; \mathfrak{h}) \longrightarrow L^2(\Omega).$$

Denoting by

$$S_{\mathfrak{h}} = \left\{ u = \sum_{j=1}^{n} F_j \otimes h_j \; : \; F_1, \ldots, F_n \in S, h_1, \ldots, h_n \in \mathfrak{h}, n \in \mathbb{N} \right\}$$

the smooth elementary $\mathfrak{h}$-valued random variables, $\delta(u)$ is then given by

$$\delta(u) = \sum_{j=1}^{n} F_j W(h_j) - \sum_{j=1}^{n} \langle h_j, DF_j \rangle_{L^2(\mathbb{R}_+)}$$

for $u = \sum_{j=1}^{n} F_j \otimes h_j \in S_{\mathfrak{h}}$. If we take, *e.g.*, $\mathfrak{h} = L^2(\mathbb{R}_+)$, then $B_t = W(\mathbf{1}_{[0,t]})$ is a standard Brownian motion, and the $\mathfrak{h}$-valued random variables can also be interpreted as stochastic processes indexed by $\mathbb{R}_+$.

Proposition 9.2.4 *Let $u_1, \ldots, u_n \in L^2(\mathbb{R}_+)$ and*

$$F = f(I_1(u_1), \ldots, I_1(u_n)),$$

where f is a polynomial or $f \in C_b^1(\mathbb{R}^n)$. We have

$$D_t F = \sum_{i=1}^{n} u_i(t) \frac{\partial f}{\partial x_i}(I_1(u_1), \ldots, I_1(u_n)), \qquad t \in \mathbb{R}_+. \tag{9.8}$$

In particular for f polynomial and for $f \in C_b^1(\mathbb{R}^n)$ we have

$$D_t f(B_{t_1}, \ldots B_{t_n}) = \sum_{i=1}^{n} \mathbf{1}_{[0,t_i]}(t) \frac{\partial f}{\partial x_i}(B_{t_1}, \ldots B_{t_n}),$$

$0 \le t_1 < \cdots < t_n$, and (9.8) can also be written as

$$\langle DF, h \rangle_{L^2(\mathbb{R}_+)}$$
$$= \frac{d}{d\varepsilon} f\left(\int_0^\infty u_1(t)(dB(t) + \varepsilon h(t)dt), \ldots, \int_0^\infty u_n(t)(dB(t) + \varepsilon h(t)dt) \right)_{|\varepsilon=0}$$
$$= \frac{d}{d\varepsilon} F(\omega + \epsilon h)_{|\varepsilon=0},$$

$h \in L^2(\mathbb{R}_+)$, where the limit exists in $L^2(\Omega)$. We refer to the above identity as the *probabilistic interpretation* of the gradient operator D on the Wiener space. In other words the scalar product $\langle h, DF \rangle_{L^2(\mathbb{R}_+)}$ coincides with the Fréchet derivative

$$D_h F = \frac{d}{d\varepsilon}\bigg|_{\varepsilon=0} f\left(W(h_1) + \varepsilon \langle h, h_1 \rangle, \ldots, W(h_n) + \varepsilon \langle h, h_n \rangle \right)$$

for all $F = f(W(h_1), \ldots, W(h_n)) \in \mathcal{S}$ and all $h \in \mathfrak{h}$. We also have the integration by parts formulas

$$\mathbf{E}[FW(h)] = \mathbf{E}[\langle h, DF \rangle_{L^2(\mathbb{R}_+)}], \tag{9.9}$$

and

$$\mathbf{E}[FGW(h)] = \mathbf{E}[\langle h, DF \rangle_{L^2(\mathbb{R}_+)} G + F \langle h, DG \rangle_{L^2(\mathbb{R}_+)}], \tag{9.10}$$

for all $F, G \in \mathcal{S}, h \in \mathfrak{h}$. The derivation operator D and the divergence operator δ satisfy the commutation relation

$$D_h(\delta(u)) = \langle h, u \rangle_{L^2(\mathbb{R}_+)} + \delta(D_h u), \tag{9.11}$$

and the Skorohod isometry

$$\mathbf{E}[\delta(u)\delta(v)] = \mathbf{E}[\langle u, v \rangle_{L^2(\mathbb{R}_+)}] + \mathbf{E}[\mathrm{Tr}(Du \circ Dv)],$$

for $h \in \mathfrak{h}, u, v \in \mathcal{S}_{\mathfrak{h}}, F \in \mathcal{S}$. In addition, we have the divergence formula of the next proposition.

Proposition 9.2.5 *For all $u \in \mathcal{U}$ and $F \in \mathcal{S}$ we have*

$$\delta(u)F = \delta(uF) + \langle DF, u \rangle_{L^2(\mathbb{R}_+)}. \tag{9.12}$$

From Proposition 9.1.13, the Skorohod integral $\delta(u)$ coincides with the Itô integral of $u \in L^2(W; H)$ with respect to Brownian motion, *i.e.*,

$$\delta(u) = \int_0^\infty u_t dB_t,$$

when u is square-integrable and adapted with respect to the Brownian filtration $(\mathcal{F}_t)_{t \in \mathbb{R}_+}$. In this case the divergence operator is also called the Hitsuda–Skorohod integral.

The operator D can be extended in the obvious way to $\mathfrak{h}$-valued random variables, *i.e.*, as $D \otimes \mathrm{Id}_{\mathfrak{h}}$. Thus Du is an $\mathfrak{h} \otimes \mathfrak{h}$-valued random variable and can also be interpreted as a random variable whose values are (Hilbert–Schmidt) operators on $\mathfrak{h}$. If $\{e_j : j \in \mathbb{N}\}$ is a complete orthonormal system on $\mathfrak{h}$, then $\mathrm{Tr}(Du \circ Dv)$ can be computed as

$$\mathrm{Tr}(Du \circ Dv) = \sum_{i,j=1}^\infty D_{e_i}\langle u, e_j \rangle_{L^2(\mathbb{R}_+)} D_{e_j}\langle v, e_i \rangle.$$

9.3 Poisson space

Let X be a σ-compact metric space with a diffuse Radon measure σ. The space of configurations of X is the set of Radon measures

$$\Omega^X := \left\{ \omega = \sum_{k=0}^{n} \epsilon_{x_k} \; : \; (x_k)_{k=0}^{k=n} \subset X, \; n \in \mathbb{N} \cup \{\infty\} \right\}, \tag{9.13}$$

where ϵ_x denotes the Dirac measure at $x \in X$, *i.e.*,

$$\epsilon_x(A) = \mathbf{1}_A(x), \qquad A \in \mathcal{B}(X),$$

and Ω defined in (9.13) is restricted to locally finite configurations.

The configuration space Ω^X is endowed with the vague topology and its associated σ-algebra denoted by $\mathcal{F}^X$, cf. [6]. Under the measure π_σ^X on $(\Omega^X, \mathcal{F}^X)$, the $\mathbb{N}^n$-valued vector

$$\omega \longmapsto (\omega(A_1), \ldots, \omega(A_n))$$

has independent components with Poisson distributions of respective parameters $\sigma(A_1), \ldots, \sigma(A_n)$, whenever $A_1, \ldots, A_n$ are compact disjoint subsets of X. When X is compact we will consider Poisson functionals of the form

$$F(\omega) = f_0 \mathbf{1}_{\{\omega(X)=0\}} + \sum_{n=1}^{\infty} \mathbf{1}_{\{\omega(X)=n\}} f_n(x_1, \ldots, x_n),$$

where $f_n \in L^1(X^n, \sigma^{\otimes n})$ is symmetric in n variables, $n \geq 1$.

Recall that the Fourier transform of π_σ^X via the Poisson stochastic integral

$$\int_X f(x)\omega(dx) = \sum_{x \in \omega} f(x), \quad f \in L^1(X, \sigma)$$

is given by

$$\mathbf{E}_{\pi_\sigma} \left[\exp\left(i \int_X f(x)\omega(dx) \right) \right] = \exp\left(\int_X (e^{if(x)} - 1)\sigma(dx) \right), \tag{9.14}$$

$f \in L^1(X, \sigma)$, which shows that

$$\mathbf{E}\left[\int_X f(x)\omega(dx) \right] = \int_X f(x)\sigma(dx), \quad f \in L^1(X, \sigma),$$

and

$$\mathbf{E}\left[\left(\int_X f(x)(\omega(dx) - \sigma(dx)) \right)^2 \right] = \int_X |f(x)|^2 \sigma(dx), \quad f \in L^2(X, \sigma).$$

When $f \in L^2(X, \sigma)$, Relation (9.14) extends as

$$\mathbf{E}_{\pi_\sigma}\left[\exp\left(i\int_X f(x)(\omega(dx) - \sigma(dx))\right)\right] = \exp\left(\int_X (e^{if(x)} - if(x) - 1)\sigma(dx)\right).$$

The standard Poisson process $(N_t)_{t \in \mathbb{R}_+}$ with intensity $\lambda > 0$ can be constructed as

$$N_t(\omega) = \omega([0, t]), \qquad t \in \mathbb{R}_+,$$

on the Poisson space

$$\Omega = \left\{ \omega = \sum_{k=1}^{n} \epsilon_{t_k} \ : \ 0 \le t_1 < \cdots < t_n, \ n \in \mathbb{N} \cup \{\infty\} \right\}$$

over $X = \mathbb{R}_+$, with the intensity measure

$$\nu(dx) = \lambda dx, \qquad \lambda > 0.$$

In this setting, every configuration $\omega \in \Omega$ can be viewed as the ordered sequence $\omega = (T_k)_{k \ge 1}$ of jump times of $(N_t)_{t \in \mathbb{R}_+}$ on $\mathbb{R}_+$.

Proposition 9.3.1 *Let $f_n \colon \mathbb{R}_+^n \longmapsto \mathbb{R}$ be continuous with compact support in $\mathbb{R}_+^n$. Then we have the $\mathbb{P}(d\omega)$-almost sure equality*

$$I_n(f_n)(\omega) = n! \int_0^\infty \int_0^{t_n^-} \cdots \int_0^{t_2^-} f_n(t_1, \ldots, t_n)(\omega(dt_1) - dt_1) \cdots (\omega(dt_n) - dt_n).$$

The above formula can also be written as

$$I_n(f_n) = n! \int_0^\infty \int_0^{t_n^-} \cdots \int_0^{t_2^-} f_n(t_1, \ldots, t_n) d(N_{t_1} - t_1) \cdots d(N_{t_n} - t_n),$$

and by symmetry of f_n in n variables we have

$$I_n(f_n) = \int_{\Delta_n} f_n(t_1, \ldots, t_n)(\omega(dt_1) - dt_1) \cdots (\omega(dt_n) - dt_n),$$

with

$$\Delta_n = \{(t_1, \ldots, t_n) \in \mathbb{R}_+^n \ : \ t_i \ne t_j, \ \forall i \ne j\}.$$

Letting

$$\Delta_n^X = \{(x_1, \ldots, x_n) \in X^n \ : \ x_i \ne x_j, \ \forall i \ne j\},$$

we have

$$I_n^X(f_n)(\omega) = \int_{\Delta_n^X} f_n(x_1, \ldots, x_n)(\omega(dx_1) - \sigma(dx_1)) \cdots (\omega(dx_n) - \sigma(dx_n)).$$

The integral $I_n^X(f_n)$ extends to symmetric functions in $f_n \in L^2(X)^{\circ n}$ via the isometry formula

$$\mathbf{E}_{\pi_\sigma}\left[I_n^X(f_n)I_m^X(g_m)\right] = n!\mathbf{1}_{\{n=m\}}\langle f_n, g_m\rangle_{L^2(X,\sigma)^{\circ n}},$$

for all symmetric functions $f_n \in L^2(X,\sigma)^{\circ n}$, $g_m \in L^2(X,\sigma)^{\circ m}$.

Proposition 9.3.2 *For $u, v \in L^2(X,\sigma)$ such that $uv \in L^2(X,\sigma)$ we have*

$$I_1^X(u)I_n^X(v^{\otimes n})$$

$$= I_{n+1}^X(v^{\otimes n} \circ u) + nI_n^X((uv) \circ v^{\otimes(n-1)}) + n\langle u, v\rangle_{L^2(X,\sigma)}I_{n-1}^X(v^{\otimes(n-1)}).$$

We have the multiplication formula

$$I_n^X(f_n)I_m^X(g_m) = \sum_{s=0}^{2(n\wedge m)} I_{n+m-s}^X(h_{n,m,s}),$$

$f_n \in L^2(X,\sigma)^{\circ n}$, $g_m \in L^2(X,\sigma)^{\circ m}$, where

$$h_{n,m,s} = \sum_{s\leq 2i\leq 2(s\wedge n\wedge m)} i!\binom{n}{i}\binom{m}{i}\binom{i}{s-i}f_n \circ_i^{s-i} g_m,$$

and $f_n \circ_k^l g_m$, $0 \leq l \leq k$, is the symmetrisation of

$$(x_{l+1},\ldots,x_n,y_{k+1},\ldots,y_m) \longmapsto$$

$$\int_{X^l} f_n(x_1,\ldots,x_n)g_m(x_1,\ldots,x_k,y_{k+1},\ldots,y_m)\sigma(dx_1)\cdots\sigma(dx_l)$$

in $n + m - k - l$ variables. The multiple Poisson stochastic integral of the function

$$\mathbf{1}_{A_1}^{\otimes k_1} \circ \cdots \circ \mathbf{1}_{A_d}^{\otimes k_d}$$

is linked to the Charlier polynomials by the relation

$$I_n^X(\mathbf{1}_{A_1}^{\otimes k_1} \circ \cdots \circ \mathbf{1}_{A_d}^{\otimes k_d})(\omega) = \prod_{i=1}^{d} C_{k_i}(\omega(A_i),\sigma(A_i)),$$

provided $A_1,\ldots,A_d$ are mutually disjoint compact subsets of X and $n = k_1 + \cdots + k_d$. The following expression of the exponential vector

$$\xi(u) = \sum_{k=0}^{\infty} \frac{1}{n!}I_n^X(u^{\otimes n})$$

is referred to as the Doléans exponential and satisfies

$$\xi(u) = \exp\left(\int_X u(x)(\omega(dx) - \sigma(dx))\right) \prod_{x \in \omega}((1 + u(x))e^{-u(x)}),$$

$u \in L^2(X)$. We note that the Poisson measure has the chaos representation property, *i.e.*, every square-integrable functional $F \in L^2(\Omega^X, \pi_\sigma)$ admits the orthogonal Wiener–Poisson decomposition

$$F = \sum_{n=0}^{\infty} I_n^X(f_n)$$

in series of multiple stochastic integrals.

9.3.1 Finite difference gradient

In this section we study the probabilistic interpretation and the extension to the Poisson space on X of the operators D and δ defined in Definitions 9.1.3 and 9.1.5. Consider the spaces $\mathcal{S}$ and $\mathcal{U}$ of random variables and processes given by

$$\mathcal{S} = \left\{ \sum_{k=0}^{n} I_k^X(f_k) : f_k \in L^4(X)^{\circ k}, \ k = 0, \dots, n, \ n \in \mathbb{N} \right\},$$

and

$$\mathcal{U} = \left\{ \sum_{k=0}^{n} I_k^X(g_k(*, \cdot)) : g_k \in L^2(X)^{\circ k} \otimes L^2(X), \ k = 0, \dots, n, \ n \in \mathbb{N} \right\}.$$

Definition 9.3.3 *Let the linear, unbounded, closable operators*

$$D^X : L^2(\Omega^X, \pi_\sigma) \to L^2(\Omega^X \times X, \mathbb{P} \otimes \sigma)$$

and

$$\delta^X : L^2(\Omega^X \times X, \mathbb{P} \otimes \sigma) \to L^2(\Omega^X, \mathbb{P})$$

be defined on $\mathcal{S}$ and $\mathcal{U}$ respectively by

$$D_x^X I_n^X(f_n) := n I_{n-1}^X(f_n(*, x)), \quad \pi_\sigma(d\omega) \otimes \sigma(dx) - a.e.,$$

$n \in \mathbb{N}, f_n \in L^2(X, \sigma)^{\circ n}$, *and*

$$\delta^X(I_n^X(f_{n+1}(*, \cdot))) := I_{n+1}^X(\tilde{f}_{n+1}), \quad \pi_\sigma(d\omega) - a.s.,$$

$n \in \mathbb{N}, f_{n+1} \in L^2(X, \sigma)^{\circ n} \otimes L^2(X, \sigma).$

In particular we have

$$\delta^X(f) = I_1^X(f) = \int_X f(x)(\omega(dx) - \sigma(dx)), \quad f \in L^2(X, \sigma),$$

and

$$\delta^X(\mathbf{1}_A) = \omega(A) - \sigma(A), \quad A \in \mathcal{B}(X),$$

and the Skorohod integral has zero expectation:

$$\mathbb{E}[\delta^X(u)] = 0, \quad u \in \text{Dom}(\delta^X).$$

In case $X = \mathbb{R}_+$ we simply write D and δ instead of $D^{\mathbb{R}_+}$ and $\delta^{\mathbb{R}_+}$. The commutation relation between D^X and δ^X is given by

$$D_x^X \delta^X(u) = u(x) + \delta^X(D_x^X u), \quad u \in \mathcal{U}.$$

Let $\text{Dom}(D^X)$ denote the set of functionals $F \colon \Omega^X \longrightarrow \mathbb{R}$ with the expansion

$$F = \sum_{n=0}^{\infty} I_n^X(f_n), \quad \text{such that} \quad \sum_{n=1}^{\infty} n! n \|f_n\|_{L^2(X^n, \sigma^{\otimes n})}^2 < \infty,$$

and let $\text{Dom}(\delta^X)$ denote the set of processes $u \colon \Omega^X \times X \longrightarrow \mathbb{R}$ with the expansion

$$u(x) = \sum_{n=0}^{\infty} I_n^X(f_{n+1}(*, x)), \ x \in X, \text{ such that } \sum_{n=1}^{\infty} n! \|\tilde{f}_n\|_{L^2(X^n, \sigma^{\otimes n})}^2 < \infty.$$

The following duality relation can be obtained by transfer from Proposition 9.1.6.

Proposition 9.3.4 *The operators D^X and δ^X satisfy the duality relation*

$$\mathbb{E}[\langle D^X F, u \rangle_{L^2(X, \sigma)}] = \mathbb{E}[F \delta^X(u)],$$

$F \in \text{Dom}(D^X)$, $u \in \text{Dom}(\delta^X)$.

The next lemma gives the probabilistic interpretation of the gradient D^X, as an extension of the finite difference operator (3.3) to spaces of random configurations.

Lemma 9.3.5 *For any F of the form*

$$F = f(I_1^X(u_1), \ldots, I_1^X(u_n)), \tag{9.15}$$

with $u_1, \ldots, u_n \in \mathcal{C}_c(X)$, and f is a bounded and continuous function, or a polynomial on $\mathbb{R}^n$, we have $F \in \mathrm{Dom}(D^X)$ and

$$D_x^X F(\omega) = F(\omega \cup \{x\}) - F(\omega), \qquad \mathbb{P} \otimes \sigma(d\omega, dx) - a.e., \qquad (9.16)$$

where as a convention we identify $\omega \in \Omega^X$ with its support.

Definition 9.3.6 *Given a mapping $F: \Omega^X \longrightarrow \mathbb{R}$, let*

$$\varepsilon_x^+ F: \Omega^X \longrightarrow \mathbb{R} \quad \text{and} \quad \varepsilon_x^- F: \Omega^X \longrightarrow \mathbb{R},$$

$x \in X$, be defined by

$$(\varepsilon_x^- F)(\omega) = F(\omega \backslash x), \quad \text{and} \quad (\varepsilon_x^+ F)(\omega) = F(\omega \cup x), \qquad \omega \in \Omega^X.$$

Note that Relation (9.16) can be written as

$$D_x^X F = \varepsilon_x^+ F - F, \qquad x \in X.$$

On the other hand, the result of Lemma 9.3.5 is clearly verified on simple functionals. For instance when $F = I_1^X(u)$ is a single Poisson stochastic integral, we have

$$D_x^X I_1^X(u)(\omega) = I_1^X(u)(\omega \cup \{x\}) - I_1^X(u)(\omega)$$

$$= \int_X u(y)(\omega(dy) + \epsilon_x(dy) - \sigma(dy)) - \int_X u(y)(\omega(dy) - \sigma(dy))$$

$$= \int_X u(y)\epsilon_x(dy)$$

$$= u(x), \qquad x \in X.$$

As in [126], the law of the mapping $(x, \omega) \longmapsto \omega \cup \{x\}$ under $\mathbf{1}_A(x)\sigma(dx)\pi_\sigma(d\omega)$ is absolutely continuous with respect to π_σ. In particular, $(\omega, x) \longmapsto F(\omega \cup \{x\})$ is well-defined, $\pi_\sigma \otimes \sigma$, and this justifies the extension of Lemma 9.3.5 in the next proposition.

Proposition 9.3.7 *For any $F \in \mathrm{Dom}(D^X)$ we have*

$$D_x^X F(\omega) = F(\omega \cup \{x\}) - F(\omega), \qquad \pi_\sigma(d\omega)\sigma(dx) - a.e.$$

Proof: There exists a sequence $(F_n)_{n\in\mathbb{N}}$ of functionals of the form (9.15), such that $(D^X F_n)_{n\in\mathbb{N}}$ converges everywhere to $D^X F$ on a set A_F such that $(\pi_\sigma \otimes \sigma)(A_F^c) = 0$. For each $n \in \mathbb{N}$, there exists a measurable set $B_n \subset \Omega^X \times X$ such that $(\pi_\sigma \otimes \sigma)(B_n^c) = 0$ and

$$D_x^X F_n(\omega) = F_n(\omega \cup \{x\}) - F_n(\omega), \qquad (\omega, x) \in B_n.$$

Taking the limit as n goes to infinity on $(\omega, x) \in A_F \cap \bigcap_{n=0}^{\infty} B_n$, we get

$$D_x^X F(\omega) = F(\omega \cup \{x\}) - F(\omega), \qquad \pi_\sigma(d\omega)\sigma(dx) - a.e.$$

$\square$

Proposition 9.3.7 implies that D^X satisfies the following finite difference product rule.

Proposition 9.3.8 *We have for* $F, G \in \mathcal{S}$:

$$D_x^X(FG) = F D_x^X G + G D_x^X F + D_x^X F D_x^X G, \qquad \pi_\sigma(d\omega)\sigma(dx) - a.e.$$

9.3.2 Divergence operator

The adjoint δ^X of D^X satisfies the following divergence formula.

Proposition 9.3.9 *Let* $u\colon X \times \Omega^X \longrightarrow \mathbb{R}$ *and* $F\colon \Omega^X \longrightarrow \mathbb{R}$ *such that* $u(\cdot, \omega)$, $D_. ^X F(\omega)$, *and* $u(\cdot, \omega) D_. ^X F(\omega) \in L^1(X, \sigma)$, $\omega \in \Omega^X$. *We have*

$$F\delta^X(u) = \delta^X(uF) + \langle u, D^X F \rangle_{L^2(X, \sigma)} + \delta^X(u D^X F).$$

The relation also holds if the series and integrals converge, or if $F \in \mathrm{Dom}(D^X)$ *and* $u \in \mathrm{Dom}(\delta^X)$ *is such that* $u D^X F \in \mathrm{Dom}(\delta^X)$.

In the next proposition, Relation (9.17) can be seen as a generalisation of (A.3c) in Proposition A.1.5:

$$C_{n+1}(k, t) = k C_n(k - 1, t) - t C_n(k, t),$$

which is recovered by taking $u = \mathbf{1}_A$ and $t = \sigma(A)$. The following statement provides a connection between the Skorohod integral and the Poisson stochastic integral.

Proposition 9.3.10 *For all* $u \in \mathrm{Dom}(\delta^X)$ *we have*

$$\delta^X(u) = \int_X u_x(\omega \setminus \{x\})(\omega(dx) - \sigma(dx)). \qquad (9.17)$$

9.4 Sequence models

In this section we describe a construction of differential operators and integration by parts formulas based on *sequence models*, in which

$$\Omega = \left\{ \omega = (\omega_k)_{k \in \mathbb{N}} \; : \; \omega_k \in \mathbb{R}^{d+2} \right\},$$

is a linear space of sequences, where $\omega_k = (\omega_k^0, \ldots, \omega_k^{d+1}) \in \mathbb{R}^{d+2}$, $k \in \mathbb{N}$, $d \geq 1$, with the norm

$$\|\omega\|_\Omega = \sup_{k \in \mathbb{N}} \frac{\|\omega_k\|_{\mathbb{R}^{d+2}}}{k+1},$$

and associated Borel σ-algebra $\mathcal{F}$. This is in connection with the notion of *numerical model* in § I-4.3 of [77] in the case of Brownian motion be built from a sequence of independent standard Gaussian random variables. In that spirit, the Malliavin calculus on real sequences is also developed in § I-6.2 of [86].

Consider the finite measure λ on $\mathbb{R}^{d+2}$ with density

$$d\lambda(t_0, t_1, \ldots, t_{d+1})$$

$$= \frac{1}{2^{d-1}\sqrt{2\pi}} e^{-t_0^2/2} e^{-t_1} \mathbf{1}_{[0,\infty)}(t_1) \mathbf{1}_{[-1,1]^d}(t_2, \ldots, t_{d+1}) dt_0 \cdots dt_{d+1}.$$

Definition 9.4.1 *We denote by $\mathbb{P}$ the probability defined on $(\Omega, \mathcal{F})$ via its expression*

$$\mathbb{P}(\{\omega = (\omega_k)_{k \in \mathbb{N}} \in \Omega : (\omega_0, \ldots, \omega_n) \in A\}) = \lambda^{\otimes n+1}(A),$$

on cylinder sets of the form

$$\{\omega = (\omega_k)_{k \in \mathbb{N}} \in \Omega : (\omega_0, \ldots, \omega_n) \in A\}$$

where A is a Borel set in $(\mathbb{R}^{d+2})^{n+1}$ and $n \in \mathbb{N}$.

We denote by

$$\tau_k = (\tau_k^0, \ldots, \tau_k^{d+1}) : \Omega \longrightarrow \mathbb{R}^{d+2} \qquad k \in \mathbb{N},$$

the coordinate functionals defined as

$$\tau_k(\omega) = (\tau_k^0(\omega), \ldots, \tau_k^{d+1}(\omega)) = (\omega_k^0, \ldots, \omega_k^{d+1}) = \omega_k.$$

The sequences $(\tau_k^0)_{k \in \mathbb{N}}$, $(\tau_k^1)_{k \in \mathbb{N}}$, $(\tau_k^i)_{k \in \mathbb{N}}$, $i = 2, \ldots, d+2$, are independent and respectively Gaussian, exponential, and uniform on $[-1, 1]$. Letting

$$E = \mathbb{R} \times (0, \infty) \times (-1, 1)^{d-1}, \qquad \bar{E} = \mathbb{R} \times [0, \infty) \times [-1, 1]^d,$$

we construct a random point process γ as the sequence

$$\gamma = \{T_k : k \geq 1\} \subset \mathbb{R}_+ \times [-1, 1]^d,$$

of random points defined by

$$T_k(\omega) = \left(\sum_{i=0}^{k-1} \tau_i^1(\omega), \tau_k^2(\omega), \ldots, \tau_k^{d+1}(\omega) \right), \qquad \omega \in \Omega, \quad k \geq 1.$$

On the other hand the standard Brownian motion indexed by $t \in [0, 1]$ can be constructed as the Paley–Wiener series

$$W(t) = t\tau_0^0 + \frac{1}{\pi\sqrt{2}} \sum_{n=1}^{\infty} \frac{\tau_n^0}{n} \sin(2n\pi t), \qquad t \in [0, 1],$$

with

$$\tau_n^0 = \sqrt{2} \int_0^1 \sin(2\pi nt) dW(t), \, n \geq 1, \qquad \tau_0^0 = \int_0^1 dW(t) = W(1),$$

and if $(z(t))_{t \in [0,1]}$ is an adapted process given as

$$z(t) = F(0, 0) + \sqrt{2} \sum_{n=1}^{\infty} F(n, 0) \cos(2n\pi t), \qquad t \in [0, 1],$$

then the stochastic integral of $(z(t))_{t \in [0,1]}$ with respect to $(W(t))_{t \in [0,1]}$ is written as

$$\int_0^1 z(t) dW(t) = \sum_{n=0}^{\infty} F(n, 0)\tau_n^0.$$

In this framework, the shift of Brownian motion by a process $(\psi(s))_{s \in [0,1]}$ and of the point process γ by a random diffeomorphism

$$\phi \colon \mathbb{R}_+ \times [-1, 1]^d \longrightarrow \mathbb{R}_+ \times [-1, 1]^d$$

will be replaced by a random variable $F \colon \Omega \longrightarrow H$ whose components are denoted by $(F(k, i))_{k \in \mathbb{N}, i=0,1,\ldots,d+1}$. The link between F and ψ, ϕ is the following:

$$\begin{cases} F(k, 0) = \begin{cases} \sqrt{2} \displaystyle\int_0^1 \sin(2\pi kt)\psi(t) dt, & k \geq 1, \\[2mm] \displaystyle\int_0^1 \psi(t) dt, & k = 0, \end{cases} \\[6mm] \tau_k^1 + F(k, 1) = \phi^1(T_{k+1}) - \phi^1(T_k), & k \geq 0, \\[4mm] \tau_k^i + F(k, i) = \phi^i(T_k), & k \geq 0, \, i = 2, \ldots, d+1. \end{cases}$$

We now introduce a gradient and a divergence operator in the sequence model. Given X a real separable Hilbert space with orthonormal basis $(h_i)_{i \in \mathbb{N}}$, let $H \otimes X$ denote the completed Hilbert–Schmidt tensor product of H with X. Let S be the set of functionals on Ω of the form $f(\tau_{k_1}, \ldots, \tau_{k_n})$, where $n \in \mathbb{N}$, $k_1, \ldots, k_n \in \mathbb{N}$, and f is a polynomial or $f \in C_c^\infty(E^n)$. We define a set of smooth vector-valued functionals as

$$\mathcal{S}(X) = \left\{ \sum_{i=0}^{n} F_i h_i \; : \; F_0, \ldots, F_n \in \mathcal{S}, \; h_0, \ldots, h_n \in X, \; n \in \mathbb{N} \right\},$$

which is dense in $L^2(\Omega, \mathbb{P}; X)$.

Definition 9.4.2 *Let $D: \mathcal{S}(X) \to L^2(\Omega, H \otimes X)$ be defined via*

$$\langle DF(x), h \rangle_{H \otimes X} = \lim_{\varepsilon \to 0} \frac{F(\omega + \varepsilon h) - F(\omega)}{\varepsilon}, \quad \omega \in \Omega, \quad h \in H.$$

Let $(e_k)_{k \geq 0}$ the canonical basis of

$$H = \ell^2(\mathbb{N}, \mathbb{R}^{d+2}) = \ell^2(\mathbb{N}) \otimes \mathbb{R}^{d+2}, \quad \text{with} \quad e_k = (e_k^0, \ldots, e_k^{d+1}), \quad k \in \mathbb{N}.$$

We denote $DF = (D_k^i F)_{(k,i) \in \mathbb{N} \times \{0,1,\ldots,d+1\}} \in L^2(\Omega; H \otimes X)$, and for $u \in \mathcal{S}(H \otimes X)$, we write

$$u = \sum_{k=0}^{\infty} \sum_{i=0}^{d-1} u_k^i e_k^i, \quad u_k \in \mathcal{S}(X), \quad k \in \mathbb{N},$$

Let also

$$E_i = \begin{cases} \mathbb{R}^{d+2}, & i = 0, \\[2mm] \left\{ (y^0, \ldots, y^{d+1}) \in \mathbb{R}^{d+2} \; : \; y^1 = 0 \right\}, & i = 1, \\[2mm] \left\{ (y^0, \ldots, y^{d+1}) \in \mathbb{R}^{d+2} \; : \; y^i \in \{-1, 1\} \right\}, & i = 2, \ldots, d+1, \end{cases}$$

and

$$\Omega_k^i = \{ \omega \in \Omega \; : \; \omega_k \in E_i \}, \quad k \in \mathbb{N}, \; i = 1, \ldots, d+1,$$

and let

$$\mathcal{U}(X) := \left\{ u \in \mathcal{S}(H \otimes X) \; : \; u_k^i = 0 \text{ on } \Omega_k^i, \; k \in \mathbb{N}, \; i = 0, 1, \ldots, d+1 \right\},$$

which is dense in $L^2(\Omega; H \otimes X)$.

Proposition 9.4.3 *The operator $D: L^2(\Omega; X) \to L^2(\Omega; H \otimes X)$ is closable and has an adjoint operator $\delta: \mathcal{U}(X) \to L^2(\Omega; X)$, with*

$$\mathbf{E}_{\mathbb{P}}\left[\langle DF, u \rangle_{H \otimes X} \right] = \mathbf{E}\left[\langle \delta(u), F \rangle_X \right], \quad u \in \mathcal{U}(X), F \in \mathcal{S}(X),$$

where δ is defined as

$$\delta(u) = \sum_{k \in \mathbb{N}} \tau_k^0 u_k^0 + u_k^1 - \text{trace} D_k u_k, \qquad u \in \mathcal{U}(X),$$

with

$$\text{trace} D_k u_k := D_k^0 u_k^0 + \cdots + D_k^{d+1} u_k^{d+1}, \qquad u \in \mathcal{U}(X).$$

Proof: This result is proved by finite dimensional integration by parts with respect to λ, under the boundary conditions imposed on elements of $\mathcal{U}(X)$. □

Definition 9.4.4 *For $p \geq 1$, we call $D_{p,1}(X)$ the completion of $\mathcal{S}(X)$ with respect to the norm*

$$\|F\|_{D_{p,1}(X)} = \| \|F\|_X \|_{L^p(\Omega)} + \| \|DF\|_{H \otimes X} \|_{L^p(\Omega)}.$$

In particular, $D_{p,1}^{\mathcal{U}}(H)$ is the completion of $\mathcal{U}(\mathbb{R})$ with respect to the norm $\| \cdot \|_{D_{p,1}(H)}$. For $p = 2$, let $\text{Dom}(\delta; X)$ denote the domain of the closed extension of δ. As shown in the following proposition, $D_{2,1}^{\mathcal{U}}(H)$ is a Hilbert space contained in $\text{Dom}(\delta; X)$.

Proposition 9.4.5 *The operator δ is continuous from $D_{2,1}^{\mathcal{U}}(H)$ into $L^2(\Omega)$ with*

$$\|\delta(F)\|_{L^2(\Omega)}^2 \leq (d+2)\|F\|_{D_{2,1}^{\mathcal{U}}(H)}^2, \qquad F \in D_{2,1}^{\mathcal{U}}(H).$$

Proof: Let $F \in \mathcal{U}(\mathbb{R})$. We have

$$\delta(F) = \sum_{k=0}^{\infty} \left(\tau_k^0 F(k,0) + F(k,1) - \sum_{i=0}^{d+1} D_k^i F(k,i) \right),$$

and

$$(\delta(F))^2 \leq (d+2) \left(\sum_{k=0}^{\infty} \tau_k^0 F(k,0) - D_k^0 F(k,0) \right)^2$$

$$+ (d+2) \left(\sum_{k=0}^{\infty} F(k,1) - D_k^1 F(k,1) \right)^2 + (d+2) \sum_{i=2}^{d+1} \left(\sum_{k=0}^{\infty} D_k^i F(k,i) \right)^2,$$

hence from the Gaussian, exponential and uniform cases, cf. [103], [94], [96], we have

$$\|\delta(F)\|^2_{L^2(\Omega)} \le (d+2)\, \mathbf{E}_{\mathbb{P}} \left[\sum_{k=0}^{\infty} (F(k,0))^2 \right]$$

$$+ (d+2)\, \mathbf{E}_{\mathbb{P}} \left[\sum_{k,l=0}^{\infty} (D_k^0 F(l,0))^2 + (D_k^1 F(l,1))^2 + \sum_{i=2}^{d+1} (D_k^i F(l,i))^2 \right]$$

$$\le (d+2)\, \|\pi^0 F\|^2_{D^{\mathcal{U}}_{2,1}(H)}.$$

$\square$

Based on the duality relation between D and δ and on the density of $\mathcal{U}(X)$ in $L^2(\Omega; H \otimes X)$, it can be shown that the operators D and δ are local, *i.e.*, for $F \in D_{2,1}(X)$, resp. $F \in \mathrm{Dom}\,(\delta; X)$, we have $DF = 0$ almost surely on $\{F = 0\}$, resp. $\delta(F) = 0$ almost surely on $\{F = 0\}$.

Notes

Infinite-dimensional analysis has a long history: it began in the sixties (work of Gross [49], Hida, Elworthy, Krée, . . .), but it is Malliavin [75] who has applied it to diffusions in order to give a probabilistic proof of Hörmander's theorem. Proposition 9.2.4 is usually taken as a definition of the Malliavin derivative D, see, *e.g.*, [84]. The relation between multiple Wiener integrals and Hermite polynomials originates in [107]. Finding the probabilistic interpretation of D for normal martingales other than the Brownian motion or the Poisson process, *e.g.*, for the Azéma martingales, is still an open problem.

Exercises

Exercise 9.1 Consider $(B_t)_{t \in \mathbb{R}_+}$ and $(N_t)_{t \in \mathbb{R}_+}$ as two *independent* standard Brownian motion and Poisson process. Compute the *mean* and *variance* of the following stochastic integrals:

$$\int_0^T B_{e^t}\,dB_t, \ \int_0^T B_t\,dB_t, \ \int_0^T (N_t - t)\,d(N_t - t), \ \int_0^T B_t\,d(N_t - t), \ \int_0^T (N_t - t)\,dB_t.$$

Exercise 9.2 Let $(B_t)_{t \in [0,T]}$ denote a standard Brownian motion. Compute the expectation

$$\mathbf{E}\left[\exp\left(\beta \int_0^T B_t dB_t\right)\right]$$

for all $\beta < 1/T$. *Hint:* expand $(B_T)^2$ by Itô's calculus.

Exercise 9.3 Let $(B_t)_{t \in [0,T]}$ denote a standard Brownian motion generating the filtration $(\mathcal{F}_t)_{t \in [0,T]}$ and let $f \in L^2([0,T])$. Compute the conditional expectation

$$\mathbf{E}\left[e^{\int_0^T f(s) dB_s} \mid \mathcal{F}_t\right], \qquad 0 \le t \le T.$$

Exercise 9.4 Let $(B_t)_{t \in [0,T]}$ denote a standard Brownian motion and let $\alpha \in \mathbb{R}$. Solve the stochastic differential equation

$$dX_t = \alpha X_t dt + dB_t, \qquad 0 \le t \le T.$$

Exercise 9.5 Consider $(B_t)_{t \in \mathbb{R}_+}$ a standard Brownian motion generating the filtration $(\mathcal{F}_t)_{t \in \mathbb{R}_+}$, and let $(S_t)_{t \in \mathbb{R}_+}$ denote the solution of the stochastic differential equation

$$dS_t = r S_t dt + \sigma S_t dB_t. \qquad (9.18)$$

1. Solve the stochastic differential equation (9.18).
2. Find the function $f(t,x)$ such that
$$f(t, S_t) = \mathbf{E}[(S_T)^2 \mid \mathcal{F}_t], \qquad 0 \le t \le T.$$
3. Show that the process $t \longmapsto f(t, S_t)$ is a martingale.
4. Using the Itô formula, compute the process $(\zeta_t)_{t \in [0,T]}$ in the predictable representation
$$f(t, S_t) = \mathbf{E}[\phi(S_T)] + \int_0^t \zeta_s dB_s.$$

Exercise 9.6 Consider the stochastic integral representation

$$F = \mathbf{E}[F] + \int_0^\infty u_t dM_t \qquad (9.19)$$

with respect to the normal martingale $(M_t)_{t \in \mathbb{R}_+}$.

1. Show that the process u in (9.19) is unique in $L^2(\Omega \times \mathbb{R}_+)$.
2. Using the Clark–Ocone formula (cf. *e.g.*, § 5.5 of [98]), find the process $(u_t)_{t \in \mathbb{R}_+}$ in the following cases:

a) $M_t = B_t$ is a standard Brownian motion, and

(*i*) $F = (B_T)^3$, (*ii*) $F = e^{aB_T}$ for some $a \in \mathbb{R}$.

b) $M_t = N_t - t$ is a standard (compensated) Poisson process, and

(*i*) $F = (N_T)^3$, (*ii*) $F = (1 + a)^{N_T}$ for some $a > -1$.

Exercise 9.7 Consider the multiple stochastic integral expansion

$$F = \mathbf{E}[F] + \sum_{n=1}^{\infty} I_n(f_n) \tag{9.20}$$

with respect to the normal martingale $(M_t)_{t \in \mathbb{R}_+}$.

1. Show that the decomposition (9.20) is unique.
2. Find the sequence $(f_n)_{n \geq 1}$ in the following cases:

   a) $M_t = B_t$ is a standard Brownian motion, and

   (*i*) $F = (B_T)^3$, (*ii*) $F = e^{aB_T}$ for some $a \in \mathbb{R}$.

   b) $M_t = N_t - t$ is a standard (compensated) Poisson process, and

   (*i*) $F = (N_T)^3$, (*ii*) $F = (1 + a)^{N_T}$ for some $a > -1$.

Exercise 9.8 Consider $(B_t)_{t \in \mathbb{R}_+}$ a standard Brownian motion generating the filtration $(\mathcal{F}_t)_{t \in \mathbb{R}_+}$, and let $(S_t)_{t \in \mathbb{R}_+}$ denote the solution of the stochastic differential equation

$$dS_t = rS_t dt + \sigma S_t dB_t. \tag{9.21}$$

1. Solve the stochastic differential equation (9.21).
2. Given ϕ a $\mathcal{C}^1$ bounded function on $\mathbb{R}$, show that there exists a function $f(t, x)$ such that

$$f(t, S_t) = \mathbf{E}[\phi(S_T) \mid \mathcal{F}_t], \qquad 0 \leq t \leq T,$$

with $f(T, x) = \phi(x)$, $x \in \mathbb{R}$.
3. Show that the process $t \longmapsto f(t, S_t)$ is a martingale.
4. Using the Itô formula, compute the process $(\zeta_t)_{t \in [0,T]}$ in the predictable representation

$$f(t, S_t) = \mathbf{E}[\phi(S_T)] + \int_0^t \zeta_s dB_s.$$

5. What is the partial differential equation satisfied by $f(t, x)$?

Exercise 9.9 Consider $(N_t)_{t \in \mathbb{R}_+}$ a standard Poisson process generating the filtration $(\mathcal{F}_t)_{t \in \mathbb{R}_+}$, and let $(S_t)_{t \in \mathbb{R}_+}$ denote the solution of the stochastic differential equation

$$dS_t = rS_t dt + \sigma S_{t^-} d(N_t - t). \tag{9.22}$$

1. Solve the stochastic differential equation (9.22).
2. Given ϕ a C^1 bounded function on $\mathbb{R}$, show that there exists a function $f(t,x)$ such that

$$f(t, S_t) = \mathbf{E}[\phi(S_T) \mid \mathcal{F}_t], \qquad 0 \le t \le T.$$

3. Show that the process $t \longmapsto f(t, S_t)$ is a martingale.
4. Using the Itô formula, compute the process $(\zeta_t)_{t \in [0,T]}$ in the predictable representation

$$f(T, S_T) = \mathbf{E}[\phi(S_T)] + \int_0^T \zeta_{t^-} (dN_t - dt).$$

5. What is the difference-differential equation satisfied by $f(t,x)$?

Exercise 9.10 Let $(N_t)_{t \in \mathbb{R}_+}$ denote a standard Poisson process on $\mathbb{R}_+$. Given $f \in L^1(\mathbb{R}_+)$ and bounded we let

$$\int_0^\infty f(y)(dN_y - dy)$$

denote the compensated Poisson stochastic integral of f, and

$$L(s) := \mathbf{E}\left[\exp\left(s \int_0^\infty f(y)(dN_y - dy) \right) \right], \quad s \in \mathbb{R}_+.$$

1. Show that we have

$$L'(s) = \int_0^\infty f(y)(e^{sf(y)} - 1)dy\, \mathbf{E}\left[\exp\left(s \int_0^\infty f(y)(dN_y - dy) \right) \right].$$

2. Show that we have

$$\frac{L'(s)}{L(s)} \le h(s) := \alpha^2 \frac{e^{sK} - 1}{K}, \qquad s \in \mathbb{R}_+,$$

provided $f(t) \le K$, dt-a.e., for some $K > 0$.
3. Show that

$$L(t) \le \exp\left(\int_0^t h(s)ds \right) = \exp\left(\alpha^2 \int_0^t \frac{e^{sK} - 1}{K}ds \right), \qquad t \in \mathbb{R}_+,$$

provided in addition that $\int_0^\infty |f(y)|^2 dy \le \alpha^2$, for some $\alpha > 0$.
4. Show, using Chebyshev's inequality, that

$$\mathbb{P}\left(\int_0^\infty f(y)(dN_y - dy) \ge x \right) \le e^{-tx}\, \mathbf{E}\left[\exp\left(t \int_0^\infty f(y)dN_y \right) \right],$$

and that

$$\mathbb{P}\left(\int_0^\infty f(y)(dN_y - dy) \ge x \right) \le \exp\left(-tx + \alpha^2 \int_0^t \frac{e^{sK} - 1}{K}ds \right).$$

5. By minimisation in t, show that

$$\mathbb{P}\left(\int_0^\infty f(y)dN_y - \int_0^\infty f(y)dx \geq x\right) \leq e^{x/K}\left(1 + \frac{xK}{\alpha^2}\right)^{-x/K-\alpha^2/K^2},$$

for all $x > 0$, and that

$$\mathbb{P}\left(\int_0^\infty f(x)dN_x - \int_0^\infty f(x)dx \geq x\right) \leq \left(1 + \frac{xK}{\alpha^2}\right)^{-x/2K},$$

for all $x > 0$.

10

Noncommutative Girsanov theorem

Be not astonished at new ideas; for it is well-known to you that a thing
does not therefore cease to be true because it is not accepted by many.
(B. Spinoza.)

In this chapter we derive quasi-invariance results and Girsanov density formulas for classical stochastic processes with independent increments, which are obtained as components of Lévy processes on real Lie algebras. The examples include Brownian motion as well as the Poisson process, the gamma process, and the Meixner process. By restricting ourselves to commutative subalgebras of the current algebra that have dimension one at every point, we can use techniques from the representation theory of Lie algebras in order to get explicit expressions on both sides of our quasi-invariance formulas.

10.1 General method

We will use results from Chapter 8 on Lévy processes on real Lie algebras and their associated classical increment processes. Let $(j_{st})_{0\leq s\leq t}$ be a Lévy process on a real Lie algebra $\mathfrak{g}$, defined as in (8.3) and fix $X \in \mathfrak{g}^{\mathbb{R}_+}$ with classical version $(\hat{X}_t)_{t\in\mathbb{R}_+}$. In addition to the conditions of Definition 8.1.1, we assume that the representation ρ in the Schürmann triple can be exponentiated to a continuous unitary representation of the Lie group associated to $\mathfrak{g}$. These assumptions guarantee that j_{st} can also be exponentiated to a continuous unitary group representation. By Nelson's theorem, this implies that D contains a dense subspace whose elements are analytic vectors for all $\rho(X)$, $X \in \mathfrak{g}$, and any finite set of operators of the form $j_{st}(X)$, $0 \leq s \leq t$, $X \in \mathfrak{g}$, is essentially selfadjoint on some common domain. Furthermore, the vacuum vector Ω is an analytic vector for all $j_{st}(X)$, $0 \leq s \leq t$, $X \in \mathfrak{g}$, and we will assume that $\eta(\mathfrak{g})$ consists of analytic vectors.

Denote by $g = e^X$ an element of the simply connected Lie group G associated to $\mathfrak{g}$. Our assumptions guarantee that $\eta(g)$ and $L(g)$ can be defined for X in a sufficiently small neighborhood of 0. For an explicit expression for the action of $U_{st}(g)$ on exponential vectors, see also [106, Proposition 4.1.2].

In order to get a quasi-invariance formula for $(\hat{X}_t)_{t\in\mathbb{R}_+}$ we choose an element $Y \in \mathfrak{g}^{\mathbb{R}_+}$ that does not commute with X and let the unitary operator $U = e^{\pi(Y)}$ act on the algebra

$$\mathcal{A}_X = \mathrm{alg}\,\{X(f) : f \in \mathcal{S}(\mathbb{R}_+)\}$$

generated by X. By letting U^* act on the vacuum state $\mathbf{\Omega}$, we obtain a new state vector $\mathbf{\Omega}' = U^*\mathbf{\Omega}$. If $\mathbf{\Omega}$ is cyclic for $\mathcal{A}_X$, then $\mathbf{\Omega}'$ can be approximated by elements of the form $G\mathbf{\Omega}$ with $G \in \mathcal{A}_X$. It is actually possible to find an element G which is affiliated to the von Neumann algebra generated by $\mathcal{A}_X$ such that $G\mathbf{\Omega} = \mathbf{\Omega}'$, as follows from the BT theorem, see [104, Theorem 2.7.14].

The following calculation then shows that the finite marginal distributions of $(\hat{X}'_t)_{t\in\mathbb{R}_+}$ are absolutely continuous with respect to those of $(\hat{X}_t)$,

$$
\begin{aligned}
\mathbf{E}\left[g(\hat{X}'(f))\right] &= \langle \mathbf{\Omega}, g(X'(f))\mathbf{\Omega} \rangle \\
&= \langle \mathbf{\Omega}, g(UX(f)U^*)\mathbf{\Omega} \rangle \\
&= \langle \mathbf{\Omega}, Ug(X(f))U^*\mathbf{\Omega} \rangle \\
&= \langle U^*\mathbf{\Omega}, g(X(f))U^*\mathbf{\Omega} \rangle \\
&= \langle \mathbf{\Omega}', g(X(f))\mathbf{\Omega}' \rangle \\
&= \langle G\mathbf{\Omega}, g(X(f))G\mathbf{\Omega} \rangle \\
&= \mathbf{E}\left[g(X(f))|\hat{G}|^2\right].
\end{aligned}
$$

Here, G is a "function" of X and $\hat{G}$ is obtained from G by replacing X by $\hat{X}$. This is possible, because $\mathcal{A}_X$ is commutative, and requires only standard functional calculus.

The density relating the law of $(\hat{X}'_t)_{t\in\mathbb{R}_+}$ to that of $(\hat{X}_t)_{t\in\mathbb{R}_+}$ is therefore given by $|\hat{G}|^2$. The same calculation also applies to finite joint distributions, *i.e.*, we also have

$$\mathbf{E}\left[g_1(\hat{X}'(f_1)) \cdots g_n(\hat{X}'(f_n))\right] = \mathbf{E}\left[g_1(\hat{X}(f_1)) \cdots g_n(\hat{X}(f_n))|\hat{G}|^2\right],$$

for all $n \in \mathbb{N}$, $f_1, \ldots, f_n \in \mathcal{S}(\mathbb{R}_+)$, $g_1, \ldots, g_n \in \mathcal{C}_0(\mathbb{R})$.

In the following section we will show several examples how quasi-invariance formulas for Brownian motion, the Poisson process, the gamma process [111, 112], and the Meixner process can be obtained in a noncommutative framework. We present explicit calculations for several classical increment processes related to the oscillator algebra and the Lie algebra $\mathfrak{sl}_2(\mathbb{R})$ of real 2×2 matrices with trace zero.

Note that by letting U act on $\mathcal{A}_X$ directly we obtain a different algebra

$$\mathcal{A}_{X'} = \text{alg}\,\{UX(f)U^* : f \in \mathcal{S}(\mathbb{R}_+)\},$$

generated by $X'(f) = UX(f)U^*$, $f \in \mathcal{S}(X)$. Since this algebra is again commutative, there exists a classical process $(\hat{X}'_t)_{t \in \mathbb{R}_+}$ that has the same expectation values as X' with respect to $\mathbf{\Omega}$, i.e.,

$$\langle \mathbf{\Omega}, g_1\big(X'(f_1)\big)\cdots g_n\big(X'(f_n)\big)\mathbf{\Omega}\rangle = \mathbf{E}\left[g_1\big(\hat{X}'(f_1)\big)\cdots g_n\big(\hat{X}'(f_n)\big)\right]$$

for all $n \in \mathbb{N}$, $g_1,\ldots,g_n \in \mathcal{C}_0(\mathbb{R})$, $f_1,\ldots,f_n \in \mathcal{S}(\mathbb{R}_+)$, where

$$\hat{X}'(f) = \int_0^\infty f(t)d\hat{X}'_t, \quad \text{for} \quad f = \sum_{k=1}^n f_k \mathbf{1}_{[s_k,t_k)} \in \mathcal{S}(\mathbb{R}_+).$$

If $X'(f)$ is a function of $X(f)$, then $\mathcal{A}_X$ is invariant under the action of U. In this case the classical process $(\hat{X}'_t)_{t \in \mathbb{R}_+}$ can be obtained from $(\hat{X}_t)_{t \in \mathbb{R}_+}$ by a pathwise transformation, see (10.5) and (10.6). But even if this is not the case, we can still get a quasi-invariance formula that states that the law of $(\hat{X}'_t)_{t \in \mathbb{R}_+}$ is absolutely continuous with respect to the law of $(\hat{X}_t)_{t \in \mathbb{R}_+}$.

10.2 Quasi-invariance on $\mathfrak{osc}$

In this section we explicitly compute the density $|G|^2$ on several examples based on Gaussian and Poisson random variables.

The oscillator Lie algebra is the four dimensional Lie algebra $\mathfrak{osc}$ with basis $\{N, A^+, A^-, E\}$ and the Lie bracket given by

$$[N, A^\pm] = \pm A^\pm, \quad [A^-, A^+] = E, \quad [E, N] = [E, A^\pm] = 0,$$

with the involution $N^* = N$, $(A^+)^* = A^-$, and $E^* = E$.

Letting $Y = i(wA^+ + \overline{w}A^-)$, by Lemma 3.2.2 we can compute the adjoint action of $g_t = e^{tY}$ on a general Hermitian element $X_{\alpha,\zeta,\beta}$ of $\mathfrak{osc}$ written as

$$X_{\alpha,\zeta,\beta} = \alpha N + \zeta A^+ + \overline{\zeta}A^- + \beta E$$

with $\alpha, \beta \in \mathbb{R}$, $\zeta \in \mathbb{C}$, as

$$X(t) = \alpha N + (\zeta - i\alpha wt)A^+ + (\overline{\zeta} + i\alpha\overline{w}t)A^- + \big(\beta + 2t\Im(w\overline{\zeta}) + \alpha|w|^2 t^2\big)E,$$

where $\Im(z)$ denotes the imaginary part of z. Recall that by Proposition 6.1.1, the distribution of $\rho(X_{\alpha,\zeta,\beta})$ in the vacuum vector e_0 is either a Gaussian random variable with variance $|\zeta|^2$ and mean β or Poisson random variable with "jump size" α, intensity $|\zeta|^2/\alpha^2$, and drift $\beta - |\zeta|^2/\alpha$. We interpret the result of the next proposition as

$$\mathbf{E}\left[g\big(X(t)\big)\right] = \mathbf{E}\left[g(X_{\alpha,\zeta,\beta})\big|G(X_{\alpha,\zeta,\beta},t)\big|^2\right], \qquad g \in \mathcal{C}_0(\mathbb{R}).$$

Proposition 10.2.1 *Letting*

$$X_{\alpha,\zeta,\beta} = \alpha N + \zeta a^+ + \overline{\zeta}a^- + \beta E,$$

$Y = i(wa^+ + \overline{w}a^-)$ *and*

$$X_t := e^{tY}X_{\alpha,\zeta,\beta}e^{-tY},$$

we have

$$\langle e_0, g(X_t)e_0\rangle = \langle e_0, g(X_{\alpha,\zeta,\beta})\big|G(X_{\alpha,\zeta,\beta},t)\big|^2 e_0\rangle$$

for all $g \in \mathcal{C}_0(\mathbb{R})$, with

$$|G(x,t)|^2 = \left(1 + 2t\alpha\Im\left(\frac{w}{\zeta}\right) + t^2\alpha^2\frac{|w|^2}{|\zeta|^2}\right)^{(x-\beta)/\alpha+|\zeta|^2/\alpha^2}$$
$$\times \exp\left(\frac{|\zeta|^2}{\alpha^2}\left(2t\alpha\Im\left(\frac{w}{\zeta}\right) + t^2\alpha^2\frac{|w|^2}{|\zeta|^2}\right)\right). \qquad (10.1)$$

Proof: We have $Y^* = -Y$ and

$$\begin{aligned}
\langle e_0, g(X_t)e_0\rangle &= \langle e_0, g\big(e^{tY}X_{\alpha,\zeta,\beta}e^{-tY}\big)e_0\rangle \\
&= \langle e_0, e^{tY}g(X_{\alpha,\zeta,\beta})e^{-tY}e_0\rangle \\
&= \langle v(t), g(X_{\alpha,\zeta,\beta})v(t)\rangle \\
&= \langle G(X_{\alpha,\zeta,\beta},t)e_0, g(X_{\alpha,\zeta,\beta})G(X_{\alpha,\zeta,\beta},t)e_0\rangle \\
&= \langle e_0, g(X_{\alpha,\zeta,\beta})\big|G(X_{\alpha,\zeta,\beta},t)\big|^2 e_0\rangle.
\end{aligned}$$

As a consequence of Lemma 6.1.4, the function

$$v(t) := e^{-tY}e_0$$

can be written in the form

$$v(t) = \sum_{k=0}^{\infty} c_k(t)X_{\alpha,\zeta,\beta}^k e_0 = G(X_{\alpha,\zeta,\beta},t)e_0.$$

In order to compute the function G we consider

$$v'(t) = -\exp\big(-t\rho(Y)\big)\rho(Y)e_0 = -i\exp\big(-t\rho(Y)\big)we_1,$$

with

$$
\begin{aligned}
v'(t) &= -e^{-tY} Y e_0 \\
&= -iw e^{-tY} e_1 \\
&= -\frac{iw}{\tilde{\zeta}(t)} e^{-tY} \left(X_{\alpha,\tilde{\zeta}(t),\tilde{\beta}(t)} e_0 - \tilde{\beta}(t) e_0 \right) \\
&= -\frac{iw}{\tilde{\zeta}(t)} \left(X_{\alpha,\zeta,\beta} - \tilde{\beta}(t) \right) e^{-tY} e_0 \\
&= -\frac{iw}{\tilde{\zeta}(t)} \left(X_{\alpha,\zeta,\beta} - \tilde{\beta}(t) \right) v(t),
\end{aligned}
$$

where

$$
\tilde{\zeta}(t) = \zeta - i\alpha w t, \quad \text{and} \quad \tilde{\beta}(t) = \beta + 2t\Im(w\bar{\zeta}) + \alpha|w|^2 t^2.
$$

This is satisfied provided $G(x, t)$ satisfies the differential equation

$$
\frac{\partial G}{\partial t}(x, t) = -\frac{iw}{\tilde{\zeta}(t)} \left(x - \tilde{\beta}(t) \right) G(x, t)
$$

with initial condition $G(x, 0) = 1$, which shows that

$$
\begin{aligned}
G(x, t) &= \exp\left(-iw \int_0^t \left(\frac{x - \tilde{\beta}(s)}{\tilde{\zeta}(s)} \right) ds \right) \\
&= \left(1 - i\frac{\alpha w t}{\zeta} \right)^{(x-\beta)/\alpha + |\zeta|^2/\alpha^2} \exp\left(i\frac{w\bar{\zeta}}{\alpha} - \frac{t^2}{2}|w|^2 \right),
\end{aligned}
$$

and yields (10.1). □

After letting α go to 0 we get

$$
|G(x, t)|^2 = \exp\left(2t(x - \beta)i\frac{w}{\zeta} - \frac{t^2}{2}|\zeta|^2 \left(2i\frac{w}{\zeta} \right)^2 \right).
$$

When $\alpha = 0$, this identity gives the relative density of two Gaussian random variables with the same variance, but different means. For $\alpha \neq 0$, it gives the relative density of two Poisson random variables with different intensities. Note that the classical analogue of this limiting procedure is

$$
\lim_{\alpha \to 0} (1 + \alpha)^{\alpha\lambda(N_\alpha - \lambda/\alpha^2) + \lambda^2/\alpha} = e^{\lambda X - \lambda^2/2},
$$

where N_α is a Poisson random variable with intensity $\lambda > 0$ and $\lambda(N_\alpha - \lambda/\alpha^2)$ converges in distribution to a standard Gaussian variable X. No such normalisation is needed in the quantum case.

10.3 Quasi-invariance on $\mathfrak{sl}_2(\mathbb{R})$

Let us now consider the three-dimensional Lie algebra $\mathfrak{sl}_2(\mathbb{R})$, with basis B^+, B^-, M, Lie bracket

$$[M, B^\pm] = \pm 2B^\pm, \qquad [B^-, B^+] = M,$$

and the involution $(B^+)^* = B^-, M^* = M$. Letting $Y = B^- - B^+$, $\beta \in \mathbb{R}$, and $X_\beta = B^+ + B^- + \beta M$ in $\mathfrak{sl}_2(\mathbb{R})$, we can compute

$$[Y, X_\beta] = 2\beta B^+ + 2\beta B^- + 2M = 2\beta X_{1/\beta}$$

and by Lemma 3.3.1 the adjoint action of $g_t = e^{tY}$ on $X_\beta = B^+ + B^- + \beta M$ is given by

$$e^{tY/2} X_\beta e^{-tY/2} = e^{\frac{t}{2} \operatorname{ad} Y} X_\beta = \big(\cosh(t) + \beta \sinh(t)\big) X_{\gamma(\beta, t)},$$

where $\gamma(\beta, t) = \dfrac{\beta \cosh(t) + \sinh(t)}{\cosh(t) + \beta \sinh(t)}$. By Proposition 6.2.2 the distribution of $\rho_c(X_\beta)$ in the state vector e_0 is given by its Fourier–Laplace transform

$$\langle e_0, e^{\lambda \rho_c(X_\beta)} e_0 \rangle = \left(\frac{\sqrt{\beta^2 - 1}}{\sqrt{\beta^2 - 1} \cosh\left(\lambda \sqrt{\beta^2 - 1}\right) - \beta \sinh\left(\lambda \sqrt{\beta^2 - 1}\right)} \right)^c.$$

Proposition 10.3.1   *Letting* $Y = B^- - B^+$, *we have*

$$\langle e_0, g(e^{tY} X_\beta e^{-tY}) e_0 \rangle = \langle e_0, g(X_\beta) |G(X_\beta, t)|^2 e_0 \rangle$$

for all $g \in \mathcal{C}_0(\mathbb{R})$, *with*

$$G(x, t) = \exp\left(\frac{1}{2} \int_0^t \frac{x - c\big(\beta \cosh(s) + \sinh(s)\big)}{\cosh(s) + \beta \sinh(s)} ds \right). \tag{10.2}$$

Proof: We have $Y^* = -Y$ and

$$\begin{aligned}
\langle e_0, g(e^{t\rho_c(Y)/2} X_\beta e^{-t\rho_c(Y)/2}) e_0 \rangle &= \langle e_0, g\big(e^{-t\rho_c(Y)/2} X_\beta e^{-t\rho_c(Y)/2}\big) e_0 \rangle \\
&= \langle e_0, e^{-t\rho_c(Y)/2} g(X_\beta) e^{-t\rho_c(Y)/2} e_0 \rangle \\
&= \langle v(t), g(X_\beta) v(t) \rangle \\
&= \langle G(X_\beta, t) e_0, g(X_\beta) G(X_\beta, t) e_0 \rangle \\
&= \langle e_0, g(X_\beta) |G(X_\beta, t)|^2 e_0 \rangle.
\end{aligned}$$

By Lemma 6.2.3 the lowest weight vector e_0 is cyclic for $\rho_c(X_\beta)$ for all $\beta \in \mathbb{R}$, $c > 0$, therefore the function

$$v(t) := e^{-t\rho_c(Y)/2} e_0$$

can be written in the form

$$v(t) = \sum_{k=0}^{\infty} c_k(t)\rho_c(X_\beta)^k e_0 = G(X_\beta, t)e_0.$$

In order to compute the function G, we consider

$$v'(t) = -\frac{1}{2}e^{-t\rho_c(Y)/2}\rho_c(Y)e_0 = \frac{1}{2}e^{-t\rho_c(Y)}\sqrt{c}e_1.$$

As shown earlier we introduce X_β into this equation to get

$$v'(t) = \frac{1}{2}e^{-t\rho_c(Y)/2}\big(\rho_c(X_{\gamma(\beta,t)}) - c\gamma(\beta,t)\big)e_0$$

$$= \frac{\rho_c(X_\beta) - c\big(\beta\cosh(t) + \sinh(t)\big)}{2\cosh(t) + 2\beta\sinh(t)}e^{-t\rho_c(Y)/2}e_0,$$

which is satisfied under the ordinary differential equation

$$\frac{\partial G}{\partial t}(x,t) = \frac{x - c\big(\beta\cosh(t) + \sinh(t)\big)}{2\cosh(t) + 2\beta\sinh(t)}G(x,t),$$

for $G(x,t)$ with initial condition $G(x,0) = 1$. We check that the solution of this ODE is given by (10.2). $\qquad\Box$

If $|\beta| < 1$, then we can write G in the form

$$G(x,t) = \exp\big(\Phi(\beta,t)x - c\Psi(\beta,t)\big),$$

where

$$\Phi(\beta,t) = \frac{1}{\sqrt{1-\beta^2}}\left(\arctan\left(e^t\sqrt{\frac{1+\beta}{1-\beta}}\right) - \arctan\left(\sqrt{\frac{1+\beta}{1-\beta}}\right)\right), \quad (10.3)$$

and

$$\Psi(\beta,t) = \frac{t}{2} + \frac{1}{2}\log\frac{1 + \beta + e^{-2t}(1-\beta)}{2}. \quad (10.4)$$

10.4 Quasi-invariance on $\mathfrak{h}\,\mathfrak{w}$

In this section we use the Weyl operators and notation of Section 1.3. Recall that in Chapter 7, a continuous map O from $L^p(\mathbb{R}^2)$, $1 \le p \le 2$, into the space of bounded operators on $\mathfrak{h}$ has been defined via

$$O(f) = \int_{\mathbb{R}^2}(\mathcal{F}^{-1}f)(x,y)e^{ixP+iyQ}dxdy,$$

where $\mathcal{F}$ denotes the Fourier transform, with the bound

$$\|O(f)\| \leq C_p \|f\|_{L^p(\mathbb{R}^2)},$$

and the relation

$$O(e^{iux+ivy}) = \exp(iuP + ivQ), \quad u, v \in \mathbb{R}.$$

In the next proposition we state the Girsanov theorem on the noncommutative Wiener space, in which we conjugate $O_h(\varphi)$ with $U(-k_2/2, k_1/2)$, $k \in \mathfrak{h} \otimes \mathbb{R}^2$, which amounts to a translation of the argument of φ by $(\langle k_1, h_1\rangle, \langle k_2, h_2\rangle)$.

Proposition 10.4.1 *Let* $h, k \in \mathfrak{h} \otimes \mathbb{R}^2$ *and* $\varphi \in \text{Dom } O_h$. *Then we have*

$$U(-k_2/2, k_1/2)O_h(\varphi)U(-k_2/2, k_1/2)^* = O_h\big(T_{(\langle k_1, h_1\rangle, \langle k_2, h_2\rangle)}\varphi\big)$$

where $T_{(x_0, y_0)}\varphi(x, y) = \varphi(x + x_0, y + y_0)$.

Proof: For $(u, v) \in \mathbb{R}^2$, we have

$$U(-k_2/2, k_1/2)e^{i(uP(h_1)+vQ(h_2))}U(-k_2/2, k_1/2)^*$$
$$= U(-k_2/2, k_1/2)U(uh_1, vh_2)U(-k_2/2, k_1/2)^*$$
$$= \exp\big(-i\big(u\langle k_1, h_1\rangle + v\langle k_2, h_2\rangle\big)\big) U(uh_1, vh_2)$$

and therefore

$$U(-k_2/2, k_1/2)O_h(\varphi)U(-k_2/2, k_1/2)^*$$
$$= \int_0^\infty \int_0^\infty \mathcal{F}^{-1}\varphi(u, v)e^{-i(u\langle k_1, h_1\rangle + v\langle k_2, h_2\rangle)}e^{i(uP(h_1)+vQ(h_2))}dudv$$
$$= \int_{\mathbb{C}^d} \mathcal{F}^{-1}T_{(\langle k_1, h_1\rangle, \langle k_2, h_2\rangle)}\varphi(u, v)e^{i(uP(h_1)+vQ(h_2))}dudv$$
$$= O_h\big(T_{(\langle k_1, h_1\rangle, \langle k_2, h_2\rangle)}\varphi\big).$$

$\square$

10.5 Quasi-invariance for Lévy processes

In this section we derive quasi-invariance or Girsanov formulas for Lévy processes on Lie algebras, such as Brownian motion, the Poisson process, the gamma process, and the Meixner process.

10.5.1 Brownian motion

Let now $(j_{st})_{0 \leq s \leq t}$ be the Lévy process on $\mathfrak{osc}$ with the Schürmann triple defined by $D = \mathbb{C}$,

$$\begin{cases} \rho(N) = 1, \quad \rho(A^\pm) = \rho(E) = 0, \\[2mm] \eta(A^+) = 1, \quad \eta(N) = \eta(A^-) = \eta(E) = 0, \\[2mm] L(N) = L(A^\pm) = 0, \quad L(E) = 1. \end{cases}$$

Taking $X = -i(A^+ + A^-)$ constant we get

$$X(f) = a^+(f) + a^-(f)$$

and the associated classical process $(\hat{X}_t)_{t\in\mathbb{R}_+}$ is Brownian motion. We choose for $Y = h(A^+ - A^-)$, with $h \in \mathcal{S}(\mathbb{R}_+)$. A similar calculation as in the previous subsection yields

$$X'(\mathbf{1}_{[0,t]}) = e^Y X(\mathbf{1}_{[0,t]})e^{-Y} = X(\mathbf{1}_{[0,t]}) - 2\int_0^t h(s)ds$$

i.e., $\mathcal{A}_X$ is invariant under e^Y and $(\hat{X}'_t)_{t\in\mathbb{R}_+}$ is obtained from $(\hat{X}_t)_{t\in\mathbb{R}_+}$ by adding a drift. Now $e^{\pi(Y)}$ is a Weyl operator and gives an exponential vector when it acts on the vacuum, *i.e.*, we have

$$e^{\pi(Y)}\mathbf{\Omega} = e^{-||h||^2/2}E(h)$$

see, *e.g.*, [79, 87]. But – up to the normalisation – we can create the same exponential vector also by acting on $\mathbf{\Omega}$ with $e^{X(h)}$,

$$e^{X(h)}\mathbf{\Omega} = e^{||h||^2/2}E(h).$$

Therefore, we get $G = \exp\left(X(h) - ||h||^2\right)$ and the well-known *Girsanov formula* for Brownian motion

$$\langle e_0, g(\hat{X}'(f))e_0\rangle = \left\langle e_0, \left(g(\hat{X}(f))\exp\left(2X(h) - 2\int_0^\infty h^2(s)ds\right)\right)e_0\right\rangle. \tag{10.5}$$

10.5.2 The Poisson process

Taking

$$X = -i(N + vA^+ + vA^- + v^2E)$$

constant we get

$$X(f) = a^\circ(f) + va^+(f) + va^-(f) + v^2\int_0^\infty f(s)ds$$

and the associated classical process $(\hat{X}_t)_{t\in\mathbb{R}_+}$ is a noncompensated Poisson process with intensity ν^2 and jump size 1. Given $h \in \mathcal{S}(\mathbb{R}_+)$ of the form

$$h(t) = \sum_{k=1}^{\tilde{n}} h_k \mathbf{1}_{[s_k,t_k)}(t),$$

with $h_k > -\nu^2$, let

$$w(t) = i(\sqrt{\nu^2 + h(t)} - \nu),$$

and $Y = w(A^+ - A^-)$. The aforementioned calculations show that

$$X'(\mathbf{1}_{[0,t]}) = e^Y X(\mathbf{1}_{[0,t]}) e^{-Y}$$

is a non-compensated Poisson process with intensity $\nu^2 + h(t)$. We have the Girsanov formula

$$\left\langle e_0, g\big(\hat{X}'(f)\big) e_0 \right\rangle$$

$$= \left\langle e_0, g\big(\hat{X}(f)\big) \prod_{k=1}^{n} \left(1 + \frac{h_k}{\nu^2}\right)^{\hat{X}(\mathbf{1}_{[s_k,t_k)})} e^{-\nu^2(t_k-s_k)h_k} e_0 \right\rangle$$

$$= \left\langle e_0, \left(g\big(\hat{X}(f)\big) e^{-\nu^2 \int_0^\infty h(s)ds} \prod_{k=1}^{n} \left(1 + \frac{h_k}{\nu^2}\right)^{\hat{X}(\mathbf{1}_{[s_k,t_k)})}\right) e_0 \right\rangle$$

$$= \left\langle e_0, \left(g\big(\hat{X}(f)\big) \exp\left(\hat{X}\left(\log\left(1 + \frac{h}{\nu^2}\right)\right) - \nu^2 \int_0^\infty h(s)ds\right)\right) e_0 \right\rangle.$$

10.5.3 The gamma process

Let now $(j_{st})_{0 \le s \le t}$ be the Lévy process on $\mathfrak{sl}_2(\mathbb{R})$ with Schürmann triple $D = \ell^2$, $\rho = \rho_2$, and

$$\eta(B^+) = e_0, \quad \eta(B^-) = \eta(M) = 0, \quad L(M) = 1, \quad L(B^\pm) = 0,$$

cf. [1, Example 3.1]. Taking $X = -i(B^+ + B^+ + M)$ constant, the random variables

$$X(\mathbf{1}_{[s,t]}) = a_{st}^\circ\big(\rho(X)\big) + a_{st}^+\big(e_0\big) + a_{st}^-(e_0) + (t-s)\mathrm{id}$$

are gamma distributed in the vacuum vector $\boldsymbol{\Omega}$. We also let $Y = h(B^- - B^+)$ where h is the simple function

$$h = \sum_{k=1}^{r} h_k \mathbf{1}_{[s_k,t_k)} \in \mathcal{S}(\mathbb{R}_+), \qquad 0 \le s_1 \le t_1 \le s_2 \le \cdots \le t_n,$$

and as in the previous subsection we get

$$X'(\mathbf{1}_{[s,t]}) = e^{\pi(Y)}X(\mathbf{1}_{[s,t]})e^{-\pi(Y)} = X(e^{2h}\mathbf{1}_{[s,t]}).$$

On the other hand, using the tensor product structure of the Fock space, we can calculate

$$
\begin{aligned}
e^{-\pi(Y)}\boldsymbol{\Omega} &= \exp\left(-\sum_{k=1}^{n} h_k j_{s_k t_k}(Y)\right)\boldsymbol{\Omega} \\
&= e^{-h_1 j_{s_1 t_1}(Y)}\boldsymbol{\Omega} \otimes \cdots \otimes e^{-h_n j_{s_n t_n}(Y)}\boldsymbol{\Omega} \\
&= \exp\left(\frac{X}{2}(1 - e^{-2h_1})\mathbf{1}_{[s_1,t_1)} - (t_1 - s_1)\int_0^\infty h(s)ds\right)\boldsymbol{\Omega} \otimes \cdots \\
&\quad \cdots \otimes \exp\left(\frac{X}{2}(1 - e^{-2h_n})\mathbf{1}_{[s_n,t_n)} - (t_n - s_n)\int_0^\infty h(s)ds\right)\boldsymbol{\Omega} \\
&= \exp\left(\frac{X}{2}(1 - e^{-2h}) - \int_0^\infty h(s)ds\right)\boldsymbol{\Omega},
\end{aligned}
$$

since j_{st} is equivalent to ρ_{t-s}.

Proposition 10.5.1 *Let $n \in \mathbb{N}, f_1, \ldots, f_n \in \mathcal{S}(\mathbb{R}_+), g_1, \ldots, g_n \in \mathcal{C}_0(\mathbb{R})$, then we have the Girsanov formula*

$$\langle e_0, \left(g_1(X'(f_1)) \cdots g_n(X'(f_n))\right)e_0\rangle \tag{10.6}$$
$$= \left\langle e_0, \left(g_1(X(f_1)) \cdots g_n(X(f_n)) \exp\left(X(1 - e^{-2h}) - 2\int_0^\infty h(s)ds\right)\right)e_0\right\rangle.$$

10.5.4 The Meixner process

We consider again the same Lévy process on $\mathfrak{sl}_2(\mathbb{R})$ as in the previous subsection. Let $\varphi, \beta \in \mathcal{S}(\mathbb{R}_+)$ with $|\beta(t)| < 1$ for all $t \in \mathbb{R}_+$, and set

$$X_{\varphi,\beta} = \varphi(B^+ + B^- + \beta M) \in \mathfrak{sl}_2(\mathbb{R})^{\mathbb{R}_+}.$$

Let Y again be given by $Y = h(B^- - B^+), h \in \mathcal{S}(\mathbb{R}_+)$. Then we get

$$
\begin{aligned}
X'(t) &= e^{Y(t)}X(t)e^{-Y(t)} \\
&= \varphi(t)\left(\cosh(2h) + \beta(t)\sinh(2h)\right)\left(B^+ + B^- + \gamma(\beta(t), 2h)M\right),
\end{aligned}
$$

i.e., $X' = X_{\varphi',\beta'}$ with

$$\varphi'(t) = \varphi(t)\left(\cosh(2h(t)) + \beta(t)\sinh(2h(t))\right), \text{ and } \beta'(t) = \gamma(\beta(t), 2h(t)).$$

As in the previous subsection, we can also calculate the function G,

$$e^{\pi(Y)}\Omega = \exp\left(\left(\frac{1}{2}X_{\Phi(\beta,2h),\beta} - \int_0^\infty \Psi\big(\beta(t),h(t)\big)\right)\right)\Omega,$$

where Φ, Ψ are defined as in Equations (10.3) and (10.4). As a consequence we get the following proposition.

Proposition 10.5.2 *The finite joint distributions of $X_{\varphi',\beta'}$ are absolutely continuous with respect to those of $X_{\varphi,\beta}$, and the mutual density is given by*

$$\exp\left(X_{\Phi(\beta,2h),\beta} - \int_0^\infty \Psi\big(\beta(t),2h(t)\big)\right).$$

Notes

The Girsanov formula for Brownian motion and gamma process appeared first in the context of factorisable representations of current groups [114], cf. [111, 112] for the gamma process. The quasi-invariance results of Section 10.2 for the Poisson, gamma, and Meixner processes have been proved for finite joint distributions. They can be extended to the distribution of the processes using continuity arguments for the states and endomorphisms on our operator algebras, or by the use of standard tightness arguments coming from classical probability. The general idea also applies to classical processes obtained by a different choice of the commutative subalgebra, cf. *e.g.*, [18]. The classical Girsanov theorem has been used by Bismut [22] in order to propose a simpler approach to the Malliavin calculus by the differentiation of related quasi-invariance formulas in order to obtain integration by parts formulas for diffusion processes, which where obtained by Malliavin in a different way.

Exercises

Exercise 10.1 Girsanov theorem for gamma random variables. Take $\beta = 1$ in the framework of Proposition 10.3.1. Show that we have

$$G(x,t) = \exp\left(-\frac{1}{2}(x(e^{-t} - 1) + ct)\right),$$

and that this recover the change of variable identity

$$\mathbf{E}\left[g(e^t Z)\right] = \mathbf{E}\left[g(Z)\exp\big(Z(1 - e^{-t}) - ct\big)\right]$$

for a gamma distributed random variable Z with parameter $c > 0$.

11

Noncommutative integration by parts

Mathematical thinking is logical and rational thinking. It's not like writing poetry.

(J. Nash, in Mathematicians: An Outer View of the Inner World.*)*

In this chapter we develop a calculus on noncommutative probability spaces. Our goal is to give a meaning to an integration by parts formula via a suitable gradient operator acting on noncommutative random variables. We focus in particular on the noncommutative Wiener space over a Hilbert space $\mathfrak{h}$. In this case, the integration by parts formula will show the closability of the gradient operator as in classical infinite-dimensional analysis. We also compute the matrix elements between exponential vectors for our divergence operator and use them to show that the divergence operator coincides with the noncausal creation and annihilation integrals defined by Belavkin [13, 14] and Lindsay [68] for integrable processes, and therefore with the Hudson–Parthasarathy [54] integral for adapted processes.

11.1 Noncommutative gradient operators

We use the notation of Section 7.4 in order to construct noncommutative gradient operators. Given a Lie group G with Lie algebra $\mathcal{G}$ whose dual is denoted by $\mathcal{G}^*$, we let $\mathrm{Ad}_g^\sharp \xi$, $\xi \in \mathcal{G}^*$, denote the co-adjoint action:

$$\langle \mathrm{Ad}_g^\sharp \xi, x \rangle_{\mathcal{G}^*, \mathcal{G}} = \langle \xi, \mathrm{Ad}_{g^{-1}} x \rangle_{\mathcal{G}^*, \mathcal{G}}, \qquad x \in \mathcal{G}.$$

We also let $\widetilde{\mathrm{Ad}}_g$, $g \in G$, be defined for $f : \mathcal{G}^* \to \mathbb{C}$ as

$$\widetilde{\mathrm{Ad}}_g f = f \circ \mathrm{Ad}_{g^{-1}}^\sharp,$$

and let $\widetilde{\mathrm{ad}}_x$ be the differential of $g \longmapsto \widetilde{\mathrm{Ad}}_g$. The following proposition, called covariance property, will provide an analogue of the integration by parts formula.

Proposition 11.1.1 *For any $x = (x_1, \ldots, x_n) \in \mathcal{G}$ and $f \in \mathrm{Dom}\, O$ we have*

$$[x_1 U(X_1) + \cdots + x_n U(X_n), O(f)] = O\left(\widetilde{\mathrm{ad}}(x)f\right).$$

Proof: Using the covariance condition

$$U(g)^* C^{-1} U(g) = \frac{C^{-1}}{\sqrt{\Delta(g^{-1})}}, \qquad g \in G,$$

cf. Relations (34), (44), (56) in [7], we have

$$W_{U(g)\rho U(g)^*}(\xi)$$

$$= \frac{\sqrt{\sigma(\xi)}}{(2\pi)^{n/2}} \int_{N_0} \sqrt{m(x)} e^{-i\langle \xi, x\rangle_{\mathcal{G}^*, \mathcal{G}}} \mathrm{Tr}[U(e^{-(x_1 X_1 + \cdots + x_n X_n)}) U(g)\rho U(g)^* C^{-1}] dx$$

$$= \frac{\sqrt{\sigma(\xi)}}{(2\pi)^{n/2}} \int_{N_0} e^{-i\langle \xi, x\rangle_{\mathcal{G}^*, \mathcal{G}}} \mathrm{Tr}\, U(g^{-1}) U(e^{-(x_1 X_1 + \cdots + x_n X_n)}) U(g)\rho C^{-1} \sqrt{\frac{m(x)}{\Delta(g^{-1})}} dx$$

$$= \frac{\sqrt{\sigma(\xi)}}{(2\pi)^{n/2}} \int_{N_0} e^{-i\langle \xi, x\rangle_{\mathcal{G}^*, \mathcal{G}}} \mathrm{Tr}\, e^{-\mathrm{Ad}_{g^{-1}}x} \rho C^{-1} \sqrt{m(x)\Delta(g)} dx$$

$$= \frac{\sqrt{\sigma(\xi)}}{(2\pi)^{n/2}} \int_{N_0} e^{-i\langle \xi, \mathrm{Ad}_g x\rangle_{\mathcal{G}^*, \mathcal{G}}} \mathrm{Tr}\, U(e^{-(x_1 X_1 + \cdots + x_n X_n)})$$

$$\times \rho C^{-1} \det(\mathrm{Ad}_g) \sqrt{m(\mathrm{Ad}_g x)\Delta(g)} dx$$

$$= \frac{\sqrt{\sigma(\mathrm{Ad}^{\sharp}_{g^{-1}}\xi)}}{(2\pi)^{n/2}} \int_{N_0} e^{-i\langle \mathrm{Ad}^{\sharp}_{g^{-1}}\xi, x\rangle_{\mathcal{G}^*, \mathcal{G}}} \mathrm{Tr}\, U(e^{-(x_1 X_1 + \cdots + x_n X_n)})\rho C^{-1} \sqrt{m(x)} dx$$

$$= W_\rho(\mathrm{Ad}^{\sharp}_{g^{-1}}\xi).$$

We proved the covariance property

$$W_{U(g)\rho U(g)^*}(\xi) = W_\rho(\mathrm{Ad}^{\sharp}_{g^{-1}}\xi).$$

By duality we have

$$\langle U(g)O(f)U(g)^* | \rho\rangle_{B_2(\mathfrak{h})} = \mathrm{Tr}[(U(g)O(f)U(g)^*)^* \rho]$$

$$= \mathrm{Tr}[U(g)O(f)^* U(g)^* \rho]$$

$$= \mathrm{Tr}[O(f)^* U(g)^* \rho U(g)]$$

$$= \langle O(f) | U(g)^* \rho U(g)\rangle_{B_2(\mathfrak{h})}$$

$$= \langle f | W_{U(g)^* \rho U(g)} \rangle_{B_2(\mathfrak{h})}$$

$$= \langle f | W_\rho \circ \mathrm{Ad}_g^\sharp \rangle_{L^2_{\mathbb{C}}(\mathcal{G}^*; d\xi/\sigma(\xi))}$$

$$= \langle f \circ \mathrm{Ad}_{g^{-1}}^\sharp | W_\rho \rangle_{L^2_{\mathbb{C}}(\mathcal{G}^*; d\xi/\sigma(\xi))}$$

$$= \langle O(f \circ \mathrm{Ad}_{g^{-1}}^\sharp) | \rho \rangle_{B_2(\mathfrak{h})},$$

which implies

$$U(g)O(f)U(g)^* = O\left(\widetilde{\mathrm{Ad}}_g f \right),$$

and the conclusion follows by differentiation. $\qquad\square$

11.2 Affine algebra

Here we turn to the case of the affine algebra generated by

$$X_1 = -i\frac{P}{2} \quad \text{and} \quad X_2 = i(Q + M),$$

which form a representation of the affine algebra with $[X_1, X_2] = X_2$. Next we define the gradient operator that will be used to show the smoothness of Wigner densities. For this we fix $\kappa \in \mathbb{R}$ and we let $\mathcal{S}_\mathfrak{h}$ denote the algebra of operators on $\mathfrak{h}$ that leave the Schwartz space $\mathcal{S}(\mathbb{R})$ invariant.

Definition 11.2.1 *For any $x = (x_1, x_2) \in \mathbb{R}^2$, the gradient operator D is defined as*

$$D_x F := -\frac{i}{2} x_1[P, F] + \frac{i}{2} x_2[Q + \kappa M, F], \qquad F \in \mathcal{S}_\mathfrak{h}.$$

Proposition 11.2.2 *For any $x = (x_1, x_2) \in \mathbb{R}^2$, the operator D_x is closable for the weak topology on the space $B(\mathfrak{h})$ of bounded operators on $\mathfrak{h}$.*

Proof: Let $\phi, \psi \in \mathcal{S}(\mathbb{R})$. Let $(B_\iota)_{\iota \in I}$ be a net of operators in $\mathcal{S}_\mathfrak{h} \cap B(\mathfrak{h})$ such that $B_\iota \xrightarrow{\iota \in I} 0$ and $D_x B_\iota \xrightarrow{\iota \in I} B \in B(\mathfrak{h})$ in the weak topology. We have

$$\langle \psi | B\phi \rangle_\mathfrak{h} = \lim_{\iota \in I} \langle \psi | D_x B_\iota \phi \rangle_\mathfrak{h}$$

$$= \lim_{\iota \in I} \langle \psi | -\frac{i}{2} x_1 (PB_\iota \phi - B_\iota P\phi) + \frac{i}{2} x_2 ((Q + \kappa M)B_\iota \phi - B_\iota(Q + \kappa M)\phi) \rangle_\mathfrak{h}$$

$$= \lim_{\iota \in I} -\frac{i}{2} x_1 (\langle P\psi | PB_\iota \phi \rangle_\mathfrak{h} - \langle \psi | B_\iota P\phi \rangle_\mathfrak{h})$$

$$+ \lim_{\iota \in I} -\frac{i}{2} x_2 (\langle (Q + \kappa M)\psi | B_\iota \phi \rangle_\mathfrak{h} - \langle \psi | B_\iota(Q + \kappa M)\phi \rangle_\mathfrak{h}) = 0,$$

hence $B = 0$. $\qquad\square$

The following is the affine algebra analogue of the integration by parts formula (2.1).

Proposition 11.2.3 *For any* $x = (x_1, x_2) \in \mathbb{R}^2$ *and* $f \in \mathrm{Dom}\, O$, *we have*

$$[x_1 U(X_1) + x_2 U(X_2), O(f)] = O(x_1 \xi_2 \partial_1 f(\xi_1, \xi_2) - x_2 \xi_2 \partial_2 f(\xi_1, \xi_2)).$$

Proof: This is a consequence of the covariance property since from (7.8), the co-adjoint action is represented by the matrix

$$\begin{bmatrix} 1 & ba^{-1} \\ 0 & a^{-1} \end{bmatrix},$$

i.e.,

$$\widetilde{\mathrm{Ad}}_g f(\xi_1, \xi_2) = f \circ \mathrm{Ad}^{\sharp}_{g^{-1}}(\xi_1, \xi_2) = f(\xi_1 + ba^{-1}\xi_2, a^{-1}\xi_2),$$

hence

$$\widetilde{\mathrm{ad}}_x f(\xi_1, \xi_2) = x_1 \xi_2 \partial_1 f(\xi_1, \xi_2) - x_2 \xi_2 \partial_2 f(\xi_1, \xi_2).$$

$\square$

For $\kappa = 1$, the integration by parts formula can also be written as

$$D_{(x_1, 2x_2)} O(f) = O(x_1 \xi_2 \partial_1 f - x_2 \xi_2 \partial_2 f).$$

The noncommutative integration by parts formulas on the affine algebra given in this section generalises the classical integration by parts formula (2.1) with respect to the gamma density on $\mathbb{R}$. We define the expectation of X as

$$\mathbf{E}[X] = \langle \mathbf{\Omega}, X\mathbf{\Omega} \rangle_{\mathfrak{h}},$$

where $\mathbf{\Omega} = \mathbf{1}_{\mathbb{R}_+}$ is the vacuum state in $\mathfrak{h}$. The results of this section are in fact valid for any representation $\{M, B^-, B^+\}$ of $\mathfrak{sl}_2(\mathbb{R})$ and any vector $\mathbf{\Omega} \in \mathfrak{h}$ such that $iP\mathbf{\Omega} = Q\mathbf{\Omega}$ and $M\mathbf{\Omega} = \beta\mathbf{\Omega}$.

Lemma 11.2.4 *Let* $x = (x_1, x_2) \in \mathbb{R}^2$. *We have*

$$\mathbf{E}[D_x F] = \frac{1}{2} \mathbf{E}[x_1 \{Q, F\} + x_2 \{P, F\}], \qquad F \in \mathcal{S}_{\mathfrak{h}}.$$

Proof: We use the relation $iP\mathbf{\Omega} = Q\mathbf{\Omega}$:

$$-\mathbf{E}[[iP, F]] = \langle \mathbf{\Omega}, -iPF\mathbf{\Omega} \rangle_{\mathfrak{h}} - \langle \mathbf{\Omega}, -iFP\mathbf{\Omega} \rangle_{\mathfrak{h}}$$
$$= \langle iP\mathbf{\Omega}, F\mathbf{\Omega} \rangle_{\mathfrak{h}} + \langle \mathbf{\Omega}, FQ\mathbf{\Omega} \rangle_{\mathfrak{h}}$$
$$= \langle Q\mathbf{\Omega}, F\mathbf{\Omega} \rangle_{\mathfrak{h}} + \langle \mathbf{\Omega}FQ\mathbf{\Omega} \rangle_{\mathfrak{h}}$$
$$= \langle Q\mathbf{\Omega}, F\mathbf{\Omega} \rangle_{\mathfrak{h}} + \langle \mathbf{\Omega}, FQ\mathbf{\Omega} \rangle_{\mathfrak{h}}$$
$$= \mathbf{E}[\{Q, F\}],$$

and

$$\mathbf{E}[[iQ, F]] = \langle \Omega, iQF\Omega \rangle_{\mathfrak{h}} - \langle \Omega, iFQ\Omega \rangle_{\mathfrak{h}}$$
$$= -\langle iQ\Omega, F\Omega \rangle_{\mathfrak{h}} + \langle \Omega, FP\Omega \rangle_{\mathfrak{h}}$$
$$= \langle P\Omega, F\Omega \rangle_{\mathfrak{h}} + \langle \Omega, FP\Omega \rangle_{\mathfrak{h}}$$
$$= \mathbf{E}[\{P, F\}],$$

and we note that $\mathbf{E}[[M, F]] = 0$. $\qquad \qquad \square$

In the sequel we fix a value of $\alpha \in \mathbb{R}$.

Definition 11.2.5 *For any* $x = (x_1, x_2) \in \mathbb{R}^2$ *and* $F \in \mathcal{S}_{\mathfrak{h}}$, *let*

$$\delta(F \otimes x) := \frac{x_1}{2} \{Q + \alpha(M - \beta), F\} + \frac{x_2}{2} \{P, F\} - D_x F.$$

Note also that

$$\delta(F \otimes x) = \frac{1}{2} (x_1(Q + iP + \alpha(M - \beta)) + x_2(P - i(Q + \kappa M))) F$$
$$+ \frac{F}{2} (x_1(Q - iP + \alpha(M - \beta)) + x_2(P + i(Q + \kappa M)))$$
$$= x_1(B^+ F + FB^-) - ix_2(B^+ F + FB^-)$$
$$+ \alpha \frac{x_1}{2} \{M - \beta, F\} - \frac{i}{2} x_2 \kappa [M, F]$$
$$= (x_1 - ix_2)(B^+ F + FB^-) + \alpha \frac{x_1}{2} \{M - \beta, F\} - \frac{i}{2} x_2 \kappa [M, F].$$

The following Lemma shows that the divergence operator has expectation zero.

Lemma 11.2.6 *For any* $x = (x_1, x_2) \in \mathbb{R}^2$ *we have*

$$\mathbf{E}[\delta(F \otimes x)] = 0, \qquad F \in \mathcal{S}_{\mathfrak{h}}.$$

Proof: It suffices to apply Lemma 11.2.4 and to note that $\langle \Omega, M\Omega \rangle_{\mathfrak{h}} = \beta$. $\qquad \square$

Compared to a classical commutative setup, the noncommutative case brings additional conceptual difficulties by requiring the definition of both a right-sided and a left-sided gradient, which can be combined to a two-sided symmetric gradient.

For $F, U, V \in \mathcal{S}_{\mathfrak{h}}$ and $x = (x_1, x_2) \in \mathbb{R}^2$ we let

$$U \overleftarrow{D}_x^F = (D_x U)F = -\frac{i}{2} x_1 [P, U]F + \frac{i}{2} x_2 [Q + \kappa M, U]F,$$

and

$$\overrightarrow{D}_x^F V = F D_x V = -\frac{i}{2} x_1 F[P, V] + \frac{i}{2} x_2 F[Q + \kappa M, V],$$

and we define a two-sided gradient by

$$
\begin{aligned}
U \overset{\leftrightarrow}{D}{}^F_x V &:= U \overset{\leftarrow}{D}{}^F_x V + U \vec{D}{}^F_x V \\
&= -\frac{i}{2} x_1 [P, U] FV - \frac{i}{2} x_1 UF[P, V] \\
&\quad + \frac{i}{2} x_2 [Q + \kappa M, U] FV + \frac{i}{2} x_2 UF[Q + \kappa M, V].
\end{aligned}
$$

Proposition 11.2.7 *Let* $x = (x_1, x_2) \in \mathbb{R}^2$ *and* $U, V \in \mathcal{S}_\mathfrak{h}$. *Assume that* $x_1 (Q + \alpha M) + x_2 P$ *commutes with* U *and with* V. *We have*

$$
\mathbf{E} \left[U \overset{\leftrightarrow}{D}{}^F_x V \right] = \mathbf{E} \left[U \delta(F \otimes x) V \right], \qquad F \in \mathcal{S}_\mathfrak{h}.
$$

Proof: By Lemma 11.2.6 we have

$$
\mathbf{E}[U \delta(F \otimes x) V]
$$

$$
= \frac{1}{2} \mathbf{E} \left[U \left(\{ x_1 (Q + \alpha(M - \beta)) + x_2 P, F \} + i x_1 [P, F] - i x_2 [Q + \kappa M, F] \right) V \right]
$$

$$
= \frac{1}{2} \mathbf{E}[\{ x_1 (Q + \alpha(M - \beta)) + x_2 P, UFV \} + i x_1 U[P, F]V - i x_2 U[Q + \kappa M, F]V]
$$

$$
= \frac{1}{2} \mathbf{E}[\{ x_1 (Q + \alpha(M - \beta)) + x_2 P, UFV \} + i x_1 [P, UFV]
$$

$$
\quad - i x_1 [P, U]FV + \mathbf{E}[-i x_1 UF[P, V] - i x_2 [Q + \kappa M, UFV]
$$

$$
\quad + i x_2 [Q + \kappa M, U]FV + i x_2 UF[Q + \kappa M, V]]
$$

$$
= \mathbf{E}[\delta(UFV \otimes x)] + \frac{1}{2} \mathbf{E}[-i x_1 [P, U]FV - i x_1 UF[P, V]
$$

$$
\quad + i x_2 [Q + \kappa M, U]FV + i x_2 UF[Q + \kappa M, V]]
$$

$$
= \mathbf{E}[U \overset{\leftrightarrow}{D}{}^F_x V].
$$

$\square$

The closability of δ can be proved using the same argument as in Proposition 11.2.2. Next is a commutation relation between D and δ.

Proposition 11.2.8 *For all* $\kappa = 0$ *and* $x = (x_1, x_2)$, $y = (y_1, y_2) \in \mathbb{R}^2$ *we have*

$$
D_x \delta(F \otimes y) - \delta(D_x F \otimes y)
$$

$$
= \frac{y_1 - i y_2}{2} (x_1 \{ M, F \} + i x_2 [M, F]) + \alpha \frac{y_1}{2} (x_1 \{ Q, F \} + x_2 \{ P, F \}),
$$

$F \in \mathcal{S}_\mathfrak{h}$.

Proof: We have

$$D_x \delta(F \otimes y)$$

$$= -\frac{i}{2}x_1[P, \delta(F \otimes y)] + \frac{i}{2}x_2[Q + \kappa M, \delta(F \otimes y)]$$

$$= -\frac{i}{2}x_1\left[P, y_1(B^+F + FB^-) - iy_2(B^+F + FB^-) + \frac{y_1}{2}\alpha\{M - \beta, F\}\right]$$

$$+ \frac{i}{2}x_2\left[Q + \kappa M, y_1(B^+F + FB^-) - iy_2(B^+F + FB^-) + \frac{y_1}{2}\alpha\{M - \beta, F\}\right]$$

$$= \delta(D_xF \otimes y) - \frac{i}{2}x_1(y_1[P, B^+]F + y_1F[P, B^-] - iy_2[P, B^+]F - iy_2F[P, B^-]$$

$$+ \frac{y_1}{2}\alpha[P, M]F + \frac{y_1}{2}\alpha F[P, M]) + \frac{i}{2}x_2(y_1[Q + \kappa M, B^+]F + y_1F[Q + \kappa M, B^-]$$

$$- iy_2[Q + \kappa M, B^+]F - iy_2F[Q + \kappa M, B^-] + \frac{y_1}{2}\alpha[Q, M]F + \frac{y_1}{2}\alpha F[Q, M])$$

$$= \delta(D_xF \otimes y) - \frac{i}{2}x_1(y_1\{iM, F\} - iy_2\{iM, F\} + \frac{y_1}{2}\alpha\{2iQ, F\})$$

$$+ \frac{i}{2}x_2(y_1[M, F] - iy_2[M, F] + iy_1\alpha\{P, F\})$$

$$= \delta(D_xF \otimes y) + \frac{1}{2}x_1y_1\{M + \alpha Q, F\} + x_2y_1\frac{i}{2}[M, F] + \frac{1}{2}x_2y_1\alpha\{P, F\}$$

$$- \frac{i}{2}x_1y_2\{M, F\} + \frac{1}{2}x_2y_2[M, F].$$

$\square$

Proposition 11.2.9   *For all $F, G \in \mathcal{S}_\mathfrak{h}$ we have*

$$\delta(GF \otimes x) = G\delta(F) - G\overleftarrow{D}_F - \frac{x_1}{2}[Q + \alpha M, G]F - \frac{x_2}{2}[P, G]F,$$

and

$$\delta(FG \otimes x) = \delta(F)G - \overrightarrow{D}_FG - \frac{x_1}{2}F[Q + \alpha M, G] - \frac{x_2}{2}F[P, G].$$

Proof: We have

$$\delta(GF \otimes x) = \frac{x_1}{2}(Q + iP + \alpha(M - \beta))GF + \frac{x_1}{2}GF(Q - iP + \alpha(M - \beta))$$

$$+ \frac{x_2}{2}(P - iQ)GF + \frac{x_2}{2}GF(P + iQ)$$

$$= \frac{x_1}{2}G(Q + iP + \alpha(M - \beta))F + \frac{x_1}{2}GF(Q - iP + \alpha M - \alpha/2)$$

$$+ \frac{x_2}{2}G(P - iQ)F + \frac{x_2}{2}GF(P + iQ)$$

$$+ \frac{i}{2}x_1[P, G]F - \frac{i}{2}x_2[Q, G]F - \frac{x_1}{2}[Q + \alpha M, G]F - \frac{x_2}{2}[P, G]F.$$

Similarly we have

$$
\begin{aligned}
\delta(FG \otimes x) &= \frac{x_1}{2}(Q + iP + \alpha(M - \beta))FG + \frac{x_1}{2}FG(Q - iP + \alpha(M - \beta)) \\
&+ \frac{x_2}{2}(P - iQ)FG + \frac{x_2}{2}FG(P + iQ) \\
&= \frac{x_1}{2}(Q + iP + \alpha(M - \beta))FG + \frac{x_1}{2}F(Q - iP + \alpha M - \alpha/2)G \\
&+ \frac{x_2}{2}(P - iQ)FG + \frac{x_2}{2}F(P + iQ)G \\
&+ \frac{i}{2}x_1 F[P, G] - \frac{i}{2}x_2 F[Q, G] - \frac{x_1}{2}F[Q + \alpha M, G] - \frac{x_2}{2}F[P, G].
\end{aligned}
$$

$\square$

11.3 Noncommutative Wiener space

In this section we define derivation and divergence operators which have properties similar to their commutative analogues defined in Chapter 9 in the classical Malliavin calculus on the Wiener space. The derivation operator will be used in Chapter 12 to provide sufficient conditions for the existence of smooth Wigner densities for pairs of operators satisfying the canonical commutation relations. However, since the vacuum state does not define a Hilbert space, the extension of the divergence to (noncommutative) vector fields will become more difficult.

11.3.1 Noncommutative gradient

On the Heisenberg–Weyl Lie algebra the statement of Proposition 11.1.1 reads

$$
\frac{i}{2}[uQ - vP, O(f)] = O(u\partial_1 f + v\partial_2 f).
$$

In this section we extend this relation to the noncommutative Wiener space of Section 1.3. To emphasise the analogy with the analysis on the Wiener space of Chapter 9, we call $(B(\mathfrak{h}), \mathbf{E})$ the noncommutative Wiener space over $\mathfrak{h}$, and denote by $\mathbf{E}$ the state defined by

$$
\mathbf{E}[X] = \langle \mathbf{\Omega}, X\mathbf{\Omega} \rangle, \qquad X \in B(\mathfrak{h}).
$$

We now define a derivation operator D on $B(\mathfrak{h})$ and a divergence operator δ on $B(\mathfrak{h}, \mathfrak{h} \otimes \mathfrak{h}_{\mathbb{C}} \otimes \mathbb{C}^2)$, as the adjoint of the two-sided gradient for cylindrical (noncommutative) vector fields on the algebra of bounded operators on the symmetric Fock space over the complexification of the real Hilbert space $\mathfrak{h}$.

Definition 11.3.1 *Let $k \in \mathfrak{h}_{\mathbb{C}} \otimes \mathbb{C}^2$. We set*

$$\operatorname{Dom} D_k = \Big\{ B \in B(\mathfrak{h}) :$$

$$\frac{i}{2}[Q(k_1) - P(k_2), B] \text{ defines a bounded operator on } \mathfrak{h} \Big\}$$

and for $B \in \operatorname{Dom} D_k$ we let

$$D_k B := \frac{i}{2}[Q(k_1) - P(k_2), B].$$

Note that $B \in \operatorname{Dom} D_k$ for some $k \in \mathfrak{h}_{\mathbb{C}} \otimes \mathbb{C}^2$ implies

$$B^* \in \operatorname{Dom} D_{\overline{k}} \quad \text{and} \quad D_{\overline{k}} B^* = (D_k B)^*.$$

Example 11.3.2

a) Let $k \in \mathfrak{h}_{\mathbb{C}} \otimes \mathbb{C}^2$ and consider a unit vector

$$\psi \in \operatorname{Dom} P(k_2) \cap \operatorname{Dom} Q(k_1) \cap \operatorname{Dom} P(\overline{k}_2) \cap \operatorname{Dom} Q(\overline{k}_1).$$

We denote by $\mathbb{P}_\psi$ the orthogonal projection onto the one-dimensional subspace spanned by ψ. Evaluating the commutator $[Q(k_1) - P(k_2), \mathbb{P}_\psi]$ on a vector $\phi \in \operatorname{Dom} P(k_2) \cap \operatorname{Dom} Q(k_1)$, we get

$$[Q(k_1) - P(k_2), \mathbb{P}_\psi]\phi$$
$$= \langle \psi, \phi \rangle \big(Q(k_1) - P(k_2) \big)(\psi) - \langle \psi, \big(Q(k_1) - P(k_2) \big)(\phi) \rangle \psi$$
$$= \langle \psi, \phi \rangle \big(Q(k_1) - P(k_2) \big)(\psi) - \langle \big(Q(\overline{k}_1) - P(\overline{k}_2) \big) \psi, \phi \rangle \psi.$$

We see that the range of $[Q(k_1) - P(k_2), \mathbb{P}_\psi]$ is two-dimensional, so it can be extended to a bounded operator on $\mathfrak{h}$. Therefore $\mathbb{P}_\psi \in \operatorname{Dom} D_k$, and we get

$$(D_k \mathbb{P}_\psi)\phi = \frac{i}{2} \Big(\langle \psi, \phi \rangle \big(Q(k_1) - P(k_2) \big)(\psi) - \langle \big(Q(\overline{k}_1) - P(\overline{k}_2) \big) \psi, \phi \rangle \psi \Big),$$

$\phi \in \mathfrak{h}$.

b) Let $h \in \mathfrak{h} \otimes \mathbb{R}^2$, $k \in \mathfrak{h}_{\mathbb{C}} \otimes \mathbb{C}^2$. Then

$$\frac{i}{2}[Q(k_1) - P(k_2), U(h_1, h_2)]$$

defines a bounded operator on $\mathfrak{h}$, and we get

$$D_k U(h_1, h_2) = i\big(\langle \overline{k}_1, h_1 \rangle + \langle \overline{k}_2, h_2 \rangle \big) U(h_1, h_2).$$

Proposition 11.3.3 *Let $k \in \mathfrak{h}_{\mathbb{C}} \otimes \mathbb{C}^2$. The operator D_k is a closable operator from $B(\mathfrak{h})$ to $B(\mathfrak{h})$ with respect to the weak topology.*

Proof: Let $(B_\iota)_{\iota \in I} \subseteq \text{Dom}\, D_k \subseteq B(\mathfrak{h})$ be any net such that $B_\iota \xrightarrow{\iota \in I} 0$ and $D_k B_\iota \xrightarrow{\iota \in I} \beta$ for some $\beta \in B(\mathfrak{h})$ in the weak topology. To show that D_k is closable, we have to show that this implies $\beta = 0$. Let us evaluate β between two exponential vectors $\mathcal{E}(h_1)$, $\mathcal{E}(h_2)$, $h_1, h_2 \in \mathfrak{h}_{\mathbb{C}}$, then we get

$$\langle \mathcal{E}(h_1), \beta \mathcal{E}(h_2) \rangle = \lim_{\iota \in I} \langle \mathcal{E}(h_1), D_k B_\iota \mathcal{E}(h_2) \rangle$$

$$= \frac{i}{2} \lim_{\iota \in I} \langle \big((Q(\bar{k}_1) - P(\bar{k}_2)) \mathcal{E}(h_1), B_\iota \mathcal{E}(h_2) \big\rangle$$

$$- \frac{i}{2} \lim_{\iota \in I} \langle \mathcal{E}(h_1), B_\iota \big(Q(k_1) - P(k_2) \big) \mathcal{E}(h_2) \rangle$$

$$= 0,$$

which implies $\beta = 0$ as desired. $\qquad \square$

11.3.2 Integration by parts

From the Girsanov theorem Proposition 10.4.1 we can derive an integration by parts formula that can be used to get the estimates that show the differentiability of the Wigner densities. In particular we interpret the expression in the next integration by parts formula as a directional or Fréchet derivative.

Proposition 11.3.4 *Let $h \in \mathfrak{h} \otimes \mathbb{R}^2$, $k \in \mathfrak{h}_{\mathbb{C}} \otimes \mathbb{C}^2$, and φ such that*

$$\varphi, \frac{\partial \varphi}{\partial x}, \frac{\partial \varphi}{\partial y} \in \text{Dom}\, O_h.$$

Then $[Q(\bar{k}_1) - P(\bar{k}_2), O_h(\varphi)]$ defines a bounded operator on $\mathfrak{h}$ and we have

$$\frac{i}{2} [Q(\bar{k}_1) - P(\bar{k}_2), O_h(\varphi)] = O_h \left(\langle k_1, h_1 \rangle \frac{\partial \varphi}{\partial x} + \langle k_2, h_2 \rangle \frac{\partial \varphi}{\partial y} \right).$$

Proof: For real k this is the infinitesimal version of the previous proposition, so we only need to differentiate

$$U \left(\varepsilon \frac{k_2}{2}, \varepsilon \frac{k_1}{2} \right) O_h(\varphi) U \left(\varepsilon \frac{k_2}{2}, \varepsilon \frac{k_1}{2} \right)^* = O_h \big(T_{(\varepsilon \langle k_1, h_1 \rangle, \varepsilon \langle k_2, h_2 \rangle)} \varphi \big)$$

with respect to ε and to set $\varepsilon = 0$. The conclusion for complex k follows by linearity. $\qquad \square$

We define a $\mathcal{S}$ of the smooth functionals as

$$\mathcal{S} = \text{alg} \left\{ O_h(\varphi) : h \in \mathfrak{h} \otimes \mathbb{R}; \ \varphi \in C^\infty(\mathbb{R}^2) \text{ satisfy} \right.$$

$$\left. \frac{\partial^{\kappa_1 + \kappa_2} \varphi}{\partial x^{\kappa_1} \partial y^{\kappa_2}} \in \text{Dom}\, O_h, \quad \kappa_1, \kappa_2 \geq 0 \right\}.$$

Note that $\mathcal{S}$ is weakly dense in $B(\mathfrak{h})$, *i.e.*, $\mathcal{S}'' = B(\mathfrak{h})$, since $\mathcal{S}$ contains the Weyl operators $U(h_1, h_2)$ with $h_1, h_2 \in \mathfrak{h}$. Next, we define

$$D: \mathcal{S} \longrightarrow B(\mathfrak{h}) \otimes \mathfrak{h}_{\mathbb{C}} \otimes \mathbb{C}^2$$

where the tensor product is the algebraic tensor product over $\mathbb{C}$, by setting $DO_h(\varphi)$ equal to

$$DO_h(\varphi) = \begin{pmatrix} O_h\left(\dfrac{\partial \varphi}{\partial x}\right) \otimes h_1 \\ O_h\left(\dfrac{\partial \varphi}{\partial y}\right) \otimes h_2 \end{pmatrix}$$

and extending it as a derivation with respect to the $B(\mathfrak{h})$-bimodule structure of $B(\mathfrak{h}) \otimes \mathfrak{h}_{\mathbb{C}} \otimes \mathbb{C}^2$ defined by

$$O \cdot \begin{pmatrix} O_1 \otimes k_1 \\ O_2 \otimes k_2 \end{pmatrix} = \begin{pmatrix} OO_1 \otimes k_1 \\ OO_2 \otimes k_2 \end{pmatrix}, \quad \begin{pmatrix} O_1 \otimes k_1 \\ O_2 \otimes k_2 \end{pmatrix} \cdot O = \begin{pmatrix} O_1 O \otimes k_1 \\ O_2 O \otimes k_2 \end{pmatrix}$$

for $O, O_1, O_2 \in B(\mathfrak{h})$ and $k \in \mathfrak{h}_{\mathbb{C}} \otimes \mathbb{C}^2$.

For example, when $h \in \mathfrak{h} \otimes \mathbb{R}^2$, we get

$$DU(h_1, h_2) = DO_h\Big(\exp i(x + y)\Big) = i \begin{pmatrix} U(h_1, h_2) \otimes h_1 \\ U(h_1, h_2) \otimes h_2 \end{pmatrix}$$

$$= iU(h_1, h_2) \otimes h.$$

Definition 11.3.5   *We define a $B(\mathfrak{h})$-valued inner product on $B(\mathfrak{h}) \otimes \mathfrak{h}_{\mathbb{C}} \otimes \mathbb{C}^2$ by $\langle \cdot, \cdot \rangle \colon B(\mathfrak{h}) \otimes \mathfrak{h}_{\mathbb{C}} \otimes \mathbb{C}^2 \times B(\mathfrak{h}) \otimes \mathfrak{h}_{\mathbb{C}} \otimes \mathbb{C}^2 \longrightarrow B(\mathfrak{h})$ by*

$$\left\langle \begin{pmatrix} O_1 \otimes h_1 \\ O_2 \otimes h_2 \end{pmatrix}, \begin{pmatrix} O_1' \otimes k_1 \\ O_2' \otimes k_2 \end{pmatrix} \right\rangle = O_1^* O_1' \langle h_1, k_1 \rangle + O_2^* O_2' \langle h_2, k_2 \rangle$$

For all $A, B \in B(\mathfrak{h}) \otimes \mathfrak{h}_{\mathbb{C}} \otimes \mathbb{C}^2$ and all $O \in B(\mathfrak{h})$ we have

$$\begin{cases} \langle B, A \rangle = \langle A, B \rangle^*, \\[2mm] O^* \langle A, B \rangle = \langle AO, B \rangle, \\[2mm] \langle A, B \rangle O = \langle A, BO \rangle, \\[2mm] \langle O^* A, B \rangle = \langle A, OB \rangle. \end{cases}$$

This turns $B(\mathfrak{h}) \otimes \mathfrak{h}_{\mathbb{C}} \otimes \mathbb{C}^2$ into a pre-Hilbert module over $B(\mathfrak{h})$, and by mapping $O \otimes k \in B(\mathfrak{h}) \otimes \mathfrak{h}_{\mathbb{C}} \otimes \mathbb{C}^2$ to the linear map

$$\mathfrak{h} \ni v \longmapsto Ov \otimes k \in \mathfrak{h} \otimes \mathfrak{h}_{\mathbb{C}} \otimes \mathbb{C}^2,$$

we can embed $B(\mathfrak{h}) \otimes \mathfrak{h}_{\mathbb{C}} \otimes \mathbb{C}^2$ in the Hilbert module $\mathfrak{M} = B(\mathfrak{h}, \mathfrak{h} \otimes \mathfrak{h}_{\mathbb{C}} \otimes \mathbb{C}^2)$. We will regard $\mathfrak{h}_{\mathbb{C}} \otimes \mathbb{C}^2$ as a subspace of $\mathfrak{M}$ via the embedding

$$\mathfrak{h}_{\mathbb{C}} \ni k \longmapsto \mathrm{id}_{\mathfrak{h}} \otimes k \in \mathfrak{M}.$$

Note that we have $O \cdot k = k \cdot O = O \otimes k$ and $\langle A, k \rangle = \langle \overline{k}, \overline{A} \rangle$ for all $k \in \mathfrak{h}_{\mathbb{C}} \otimes \mathbb{C}^2$, $O \in B(\mathfrak{h})$, $A \in \mathfrak{M}$, where the conjugation in $\mathfrak{M}$ is defined by $\overline{O \otimes k} = O^* \otimes \overline{k}$.

Proposition 11.3.6   *Let $O \in \mathcal{S}$ and $k \in \mathfrak{h}_{\mathbb{C}} \otimes \mathbb{C}^2$. Then $O \in \mathrm{Dom}\, D_k$ and*

$$D_k O = \langle \overline{k}, DO \rangle = \langle \overline{DO}, k \rangle.$$

Proof: For $h \in \mathfrak{h} \otimes \mathbb{R}^2$ and $\varphi \in \mathrm{Dom}\, O_h$ such that also $\frac{\partial \varphi}{\partial x}, \frac{\partial \varphi}{\partial y} \in \mathrm{Dom}\, O_h$, we get

$$\langle \overline{k}, DO_h(\varphi) \rangle = \left\langle \begin{pmatrix} \overline{k}_1 \\ \overline{k}_2 \end{pmatrix}, \begin{pmatrix} O_h\left(\frac{\partial \varphi}{\partial x}\right) \otimes h_1 \\ O_h\left(\frac{\partial \varphi}{\partial y}\right) \otimes h_2 \end{pmatrix} \right\rangle$$

$$= O_h\left(\langle \overline{k}_1, h_1 \rangle \frac{\partial \varphi}{\partial x} + \langle \overline{k}_2, h_2 \rangle \frac{\partial \varphi}{\partial y} \right)$$

$$= \frac{i}{2}[Q(k_1) - P(k_2), O_h(\varphi)] = D_k O,$$

where we used Proposition 11.3.4. The first equality of the proposition now follows, since both

$$O \longmapsto D_k O = \frac{i}{2}[Q(k_1) - P(k_2), O] \quad \text{and} \quad O \longmapsto \langle \overline{k}, DO \rangle$$

are derivation operators. The second equality follows immediately. $\square$

The next result is the noncommutative analogue of Equation (9.9).

Theorem 11.3.7   *For any $k \in \mathfrak{h}_{\mathbb{C}} \otimes \mathbb{C}^2$ and $O \in \mathcal{S}$ we have*

$$\mathbf{E}\big[\langle \overline{k}, DO \rangle\big] = \frac{1}{2}\,\mathbf{E}\big[\{P(k_1) + Q(k_2), O\}\big]$$

where $\{\cdot, \cdot\}$ denotes the anti commutator $\{X, Y\} = XY + YX$.

Proof: This formula is a consequence of the fact that

$$Q(h)\mathbf{\Omega} = h = iP(h)\mathbf{\Omega}, \qquad h \in \mathfrak{h}_{\mathbb{C}},$$

which implies

$$\mathbf{E}\left[\langle \bar{k}, DO \rangle\right] = \frac{i}{2}\Big(\langle (Q(\bar{k}_1) - P(\bar{k}_2))\Omega, O\Omega \rangle - \langle \Omega, O\big(Q(k_1) - P(k_2)\big)\Omega \rangle\Big)$$

$$= \frac{i}{2}\Big(\langle \bar{k}_1 + i\bar{k}_2, O\Omega \rangle - \langle \Omega, O(k_1 + ik_2)\rangle\Big)$$

$$= \frac{1}{2}\Big(\langle (P(\bar{k}_1) + Q(\bar{k}_2))\Omega, O\Omega \rangle + \langle \Omega, O\big(P(k_1) + Q(k_2)\big)\Omega \rangle\Big)$$

$$= \frac{1}{2}\,\mathbf{E}\left[\{P(k_1) + Q(k_2), O\}\right].$$

$\square$

We can also derive an analogue of the commutative integration by parts formula (9.10).

Corollary 11.3.8 *Let $k \in \mathfrak{h}_{\mathbb{C}} \otimes \mathbb{C}^2$, and $O_1, \dots, O_n \in \mathcal{S}$, then*

$$\frac{1}{2}\,\mathbf{E}\left[\left\{P(k_1) + Q(k_2), \prod_{m=1}^{n} O_m\right\}\right] = \mathbf{E}\left[\sum_{m=1}^{n} \prod_{j=1}^{m-1} O_j \langle \bar{k}, DO_m \rangle \prod_{j=m+1}^{n} O_j\right],$$

where the products are ordered such that the indices increase from the left to the right.

Proof: This is obvious, since $O \longmapsto \langle \bar{k}, DO \rangle$ is a derivation. $\square$

11.3.3 Closability

Corollary 11.3.8 can be used for $n = 3$ to show the closability of D from $B(\mathfrak{h})$ to $\mathfrak{M}$. This also implies that D is also closable in stronger topologies, such as, *e.g.*, the norm topology and the strong topology. We will denote the closure of D again by the same symbol.

Corollary 11.3.9 *The derivation operator D is a closable operator from $B(\mathfrak{h})$ to the $B(\mathfrak{h})$-Hilbert module $\mathfrak{M} = B(\mathfrak{h}, \mathfrak{h} \otimes \mathfrak{h}_{\mathbb{C}} \otimes \mathbb{C}^2)$ with respect to the weak topologies.*

Proof: We have to show that for any net $(A_\iota)_{\iota \in I}$ in $\mathcal{S}$ with $A_\iota \xrightarrow{\iota \in I} 0$ and $DA_\iota \xrightarrow{\iota \in I} \alpha \in \mathfrak{M}$, we get $\alpha = 0$. Let $f, g \in \mathfrak{h}_{\mathbb{C}}$. Set

$$f_1 = \frac{f + \bar{f}}{2}, \quad f_2 = \frac{f - \bar{f}}{2i}, \quad g_1 = \frac{g + \bar{g}}{2}, \quad \text{and} \quad g_2 = \frac{g - \bar{g}}{2i}.$$

Then we have

$$U(f_1, f_2)\Omega = e^{-\|f\|^2/2}\mathcal{E}(f) \quad \text{and} \quad U(g_1, g_2)\Omega = e^{-\|g\|^2/2}\mathcal{E}(g).$$

Thus we get

$$e^{(\|f\|^2 + \|g\|^2)/2} \langle \mathcal{E}(f) \otimes \bar{h}, \alpha \mathcal{E}(g) \rangle$$

$$= e^{(\|f\|^2 + \|g\|^2)/2} \langle \mathcal{E}(f), \langle \bar{h}, \alpha \rangle \mathcal{E}(g) \rangle$$

$$= \lim_{\iota \in I} \mathbf{E}\left[U(-f_1, -f_2) \langle \bar{h}, DA_\iota \rangle U(g_1, g_2) \right]$$

$$= \lim_{\iota \in I} \mathbf{E}\left[\frac{1}{2}\{P(h_1) + Q(h_2), U(-f_1, -f_2)A_\iota U(g_1, g_2)\} \right.$$

$$\left. - \langle \bar{h}, DU(-f_1, -f_2) \rangle A_\iota U(g_1, g_2) - U(-f_1, -f_2)A_\iota \langle \bar{h}, DU(g_1, g_2) \rangle \right]$$

$$= \lim_{\iota \in I} \left(\langle \psi_1, A_\iota \psi_2 \rangle + \langle \psi_3, A_\iota \psi_4 \rangle - \langle \psi_5, A_\iota \psi_6 \rangle - \langle \psi_7, A_\iota \psi_8 \rangle \right)$$

$$= 0$$

for all $h \in \mathfrak{h}_{\mathbb{C}} \otimes \mathbb{C}^2$, where

$$\begin{cases} \psi_1 = \frac{1}{2}U(f_1, f_2)\big(P(\bar{h}_1) + Q(\bar{h}_2)\big)\Omega, & \psi_2 = U(g_1, g_2)\Omega, \\[2mm] \psi_3 = U(f_1, f_2)\Omega, & \psi_4 = \frac{1}{2}U(g_1, g_2)\big(P(h_1) + Q(h_2)\big)\Omega, \\[2mm] \psi_5 = \big(D_h U(-f_1, -f_2)\big)^* \Omega, & \psi_6 = U(g_1, g_2)\Omega, \\[2mm] \psi_7 = U(f_1, f_2)\Omega, & \psi_8 = D_h U(g_1, g_2)\Omega. \end{cases}$$

But this implies $\alpha = 0$, since $\{\mathcal{E}(f) \otimes \bar{h} | f \in \mathfrak{h}_{\mathbb{C}}, h \in \mathfrak{h}_{\mathbb{C}} \otimes \mathbb{C}^2\}$ is dense in $\mathfrak{h} \otimes \mathfrak{h}_{\mathbb{C}} \otimes \mathbb{C}^2$. □

Since D is a derivation, the next proposition implies that $\mathrm{Dom}\, D$ is a $*$-subalgebra of $B(\mathfrak{h})$.

Proposition 11.3.10 *Let $O \in \mathrm{Dom}\, D$. Then $O^* \in \mathrm{Dom}\, D$ and*

$$DO^* = \overline{DO}.$$

Proof: It is not difficult to check this directly on the Weyl operators $U(h_1, h_2)$, $h \in \mathfrak{h} \otimes \mathbb{R}^2$. We get $U(h_1, h_2)^* = U(-h_1, -h_2)$ and

$$D\big(U(h_1, h_2)^*\big) = DU(-h_1, -h_2) = -iU(-h_1, -h_2) \otimes h$$
$$= U(h_1, h_2)^* \otimes \overline{(ih)} = \overline{DU(h_1, h_2)}.$$

By linearity and continuity it therefore extends to all of $\mathrm{Dom}\, D$. □

Finally, we show how the operator D can be iterated. Given $\mathfrak{h}$ a complex Hilbert space we can define the derivation operator

$$D: \mathcal{S} \otimes \mathfrak{h} \longrightarrow B(\mathfrak{h}) \otimes \mathfrak{h}_{\mathbb{C}} \otimes \mathbb{C}^2 \otimes \mathfrak{h}$$

by setting

$$D(O \otimes h) = DO \otimes h, \qquad O \in \mathcal{S}, \quad h \in \mathfrak{h}.$$

By closure of D we get an unbounded derivation from the Hilbert module $B(\mathfrak{h}, \mathfrak{h} \otimes \mathfrak{h})$ to $\mathfrak{M}(\mathfrak{h}) = B(\mathfrak{h} \otimes \mathfrak{h}, \mathfrak{h} \otimes \mathfrak{h}_{\mathbb{C}} \otimes \mathbb{C}^2 \otimes \mathfrak{h})$, which allows us to iterate D. It is easy to see that D maps $\mathcal{S} \otimes \mathfrak{h}$ to $\mathcal{S} \otimes \mathfrak{h}_{\mathbb{C}} \otimes \mathbb{C}^2 \otimes \mathfrak{h}$ and so we have

$$D^n(\mathcal{S} \otimes \mathfrak{h}) \subseteq \mathcal{S} \otimes \left(\mathfrak{h}_{\mathbb{C}} \otimes \mathbb{C}^2\right)^{\otimes n} \otimes \mathfrak{h}.$$

In particular, $\mathcal{S} \subseteq \operatorname{Dom} D^n$ for all $n \in \mathbb{N}$, and we can define Sobolev-type norms $|| \cdot ||_n$ and semi norms $|| \cdot ||_{\psi,n}$, on $\mathcal{S}$ by

$$||O||_n^2 := ||O^*O|| + \sum_{j=1}^{n} ||\langle D^n O, D^n O \rangle||,$$

and

$$||O||_{\psi,n}^2 := ||O\psi||^2 + \sum_{j=1}^{n} ||\langle \psi, \langle D^n O, D^n O \rangle \psi \rangle||,$$

$\psi \in \mathfrak{h}$. In this way we can define Sobolev-type topologies on $\operatorname{Dom} D^n$.

11.3.4 Divergence operator

We now extend the definition of the "Fréchet derivation" D_k to the case where k is replaced by an element of $\mathfrak{M}$. It becomes now important to distinguish between a right and a left "derivation operator". Furthermore, it is no longer a derivation.

Definition 11.3.11 *Let $u \in \mathfrak{M}$ and $O \in \operatorname{Dom} D$. Then we define the right gradient $\overrightarrow{D}_u O$ and the left gradient $O\overleftarrow{D}_u$ of O with respect to u by*

$$\overrightarrow{D}_u O = \langle \bar{u}, DO \rangle, \qquad and \qquad O\overleftarrow{D}_u = \langle \overline{DO}, u \rangle.$$

We list several properties of the gradient.

Proposition 11.3.12 *(i) Let $X \in B(\mathfrak{h})$, $O, O_1, O_2 \in \operatorname{Dom} D$, and $u \in \mathfrak{M}$. We have*

$$
\begin{cases}
\overrightarrow{D}_{Xu}O = X\overrightarrow{D}_u O, \\[2mm]
\overrightarrow{D}_u(O_1 O_2) = \left(\overrightarrow{D}_u O_1\right) O_2 + \overrightarrow{D}_{uO_1} O_2, \\[2mm]
O\overleftarrow{D}_{uX} = (O\overleftarrow{D}_u)X, \\[2mm]
(O_1 O_2)\overleftarrow{D}_u = O_1 \overleftarrow{D}_{O_2 u} + O_1 \left(O_2 \overleftarrow{D}_u\right).
\end{cases}
$$

(ii) For any $k \in \mathfrak{h}_{\mathbb{C}} \otimes \mathbb{C}^2$ and $O \in \operatorname{Dom} D$, we have

$$
D_k O = \overrightarrow{D}_{\operatorname{id}_{\mathfrak{h}} \otimes k} O = O\overleftarrow{D}_{\operatorname{id}_{\mathfrak{h}} \otimes k}.
$$

Proof: These properties can be deduced easily from the definition of the gradient and the properties of the derivation operator D and the inner product $\langle \cdot, \cdot \rangle$. $\qquad\square$

We may also define a two-sided gradient

$$
\overleftrightarrow{D}_u : \operatorname{Dom} D \times \operatorname{Dom} D \longrightarrow B(\mathfrak{h})
$$

$$
(O_1, O_2) \longmapsto O_1 \overleftrightarrow{D}_u O_2 = O_1 \left(\overrightarrow{D}_u O_2\right) + \left(O_1 \overleftarrow{D}_u\right) O_2.
$$

For $k \in \mathfrak{h}_{\mathbb{C}} \otimes \mathbb{C}^2$ we have $O_1 \overleftrightarrow{D}_{\operatorname{id}_{\mathfrak{h}} \otimes k} O_2 = D_k(O_1 O_2)$.

The algebra $B(\mathfrak{h})$ of bounded operators on the symmetric Fock space $\mathfrak{h}$ and the Hilbert module $\mathfrak{M}$ are not Hilbert spaces with respect to the expectation in the vacuum vector Ω. Therefore, we cannot define the divergence operator or Skorohod integral δ as the adjoint of the derivation D. It might be tempting to try to define δX as an operator such that the condition

$$
\mathbf{E}\left[(\delta X)B\right] = \mathbf{E}\left[\overrightarrow{D}_X B\right] \tag{11.1}
$$

is satisfied for all $B \in \operatorname{Dom} \overrightarrow{D}_X$. However this is not sufficient to characterise δX. In addition, the following Proposition 11.3.13 shows that this is not possible without imposing additional commutativity conditions, see also Proposition 11.3.15.

Proposition 11.3.13 *Let $k \in \mathfrak{h}_{\mathbb{C}} \otimes \mathbb{C}^2$ with $k_1 + ik_2 \neq 0$. There exists no (possibly unbounded) operator M whose domain contains the vacuum vector such that*

$$\mathbf{E}[MB] = \mathbf{E}\,[D_k B]$$

holds for all $B \in \mathrm{Dom}\, D_k$.

Proof: We assume that such an operator M exists and show that this leads to a contradiction. Letting $B \in B(\mathfrak{h})$ be the operator defined by

$$\mathfrak{h} \ni \psi \longmapsto B\psi := \langle k_1 + ik_2, \psi \rangle \mathbf{\Omega},$$

it is easy to see that $B \in \mathrm{Dom}\, D_k$ and that $D_k B$ is given by

$$(D_k B)\psi = \frac{i}{2}\langle k_1 + ik_2, \psi \rangle (k_1 + ik_2) - \frac{i}{2}\langle (Q(\bar{k}_1) - P(\bar{k}_2))(k_1 + ik_2), \psi \rangle \mathbf{\Omega},$$

$\psi \in \mathfrak{h}$. Therefore, if M existed, we would have

$$0 = \langle \mathbf{\Omega}, MB\mathbf{\Omega} \rangle = \mathbf{E}[MB] = \mathbf{E}[D_k B]$$

$$= \langle \mathbf{\Omega}, (D_k B)\mathbf{\Omega} \rangle = -\frac{i}{2}\langle k_1 + ik_2, k_1 + ik_2 \rangle,$$

which is clearly impossible.   □

We now introduce the analogue of smooth elementary $\mathfrak{h}$-valued random variables, as

$$\mathcal{S}_{\mathfrak{h}} = \left\{ \sum_{j=1}^{n} F_j \otimes h^{(j)} : F_1, \ldots, F_n \in \mathcal{S}, h^{(1)}, \ldots, h^{(n)} \in \mathfrak{h}_{\mathbb{C}} \otimes \mathbb{C}^2, n \in \mathbb{N} \right\}.$$

Given $A, B \in B(\mathfrak{h})$ and $u \in \mathcal{S}_{\mathfrak{h}}$ of the form $u = \sum_{j=1}^{n} F_j \otimes h^{(j)}$, defining $A \overset{\leftrightarrow}{\delta u} B$ by

$$A \overset{\leftrightarrow}{\delta u} B := \frac{1}{2} \sum_{j=1}^{n} \left\{ P\left(h_1^{(j)}\right) + Q\left(h_2^{(j)}\right), AF_j B \right\} - A \sum_{j=1}^{n} \left(D_{h^{(j)}} F_j \right) B,$$

shows by Corollary 11.3.8 that

$$\mathbf{E}\left[A \overset{\leftrightarrow}{\delta u} B \right] = \mathbf{E}\left[A \overset{\leftrightarrow}{D}_u B \right]. \tag{11.2}$$

However, $A \overset{\leftrightarrow}{\delta u} B$ can written as a product AXB for some operator X only if A and B commute with $P\left(h_1^{(1)}\right) + Q\left(h_2^{(1)}\right), \ldots, P\left(h_1^{(n)}\right) + Q\left(h_2^{(n)}\right)$. In fact, relations such as (11.1) or (11.2) cannot be satisfied for all $A, B \in \mathrm{Dom}\, D$ unless we impose some commutativity conditions on A and B.

For this reason we now define a divergence operator that satisfies a weakened version of (11.2), see Proposition 11.3.15. This definition will be extended to a larger domain in Remark 11.3.18.

Definition 11.3.14   *We set*

$$
\mathcal{S}_{\mathfrak{h},\delta} = \left\{ \sum_{j=1}^{n} F_j \otimes h^{(j)} : \frac{1}{2} \sum_{j=1}^{n} \left\{ P\left(h_1^{(j)}\right) + Q\left(h_2^{(j)}\right), F_j \right\} - \sum_{j=1}^{n} D_{h^{(j)}} F_j \right.
$$

defines a bounded operator on $\mathfrak{h}$ $\Big\} \subset \mathcal{S}_{\mathfrak{h}}$,

and define the divergence operator $\delta \colon \mathcal{S}_{\mathfrak{h},\delta} \longrightarrow B(\mathfrak{h})$ *by*

$$
\delta(u) = \frac{1}{2} \sum_{j=1}^{n} \left\{ P\left(h_1^{(j)}\right) + Q\left(h_2^{(j)}\right), F_j \right\} - \sum_{j=1}^{n} D_{h^{(j)}} F_j,
$$

for $u = \sum_{j=1}^{n} F_j \otimes h^{(j)} \in \mathcal{S}_{\mathfrak{h},\delta}$.

In case $\mathfrak{h} = L^2(\mathbb{R}_+)$, the divergence operator coincides with the Hudson–Parthasarathy quantum stochastic integral for adapted integrable processes and with the noncausal quantum stochastic integrals defined by Lindsay and Belavkin for integrable processes, see Section 11.4.

It is now easily checked that the relation $\delta(\overline{u}) = \big(\delta(u)\big)^*$ holds for all $u \in \mathcal{S}_{\mathfrak{h},\delta}$.

Proposition 11.3.15   *Let* $u = \sum_{j=1}^{n} F_j \otimes h^{(j)} \in \mathcal{S}_{\mathfrak{h},\delta}$ *and*

$$
A, B \in \operatorname{Dom} D \cap \left\{ P\left(h_1^{(1)}\right) + Q\left(h_2^{(1)}\right), \dots, P\left(h_1^{(n)}\right) + Q\left(h_2^{(n)}\right) \right\}'
$$

i.e., A and B are in the commutant of

$$
\left\{ P\left(h_1^{(1)}\right) + Q\left(h_2^{(1)}\right), \dots, P\left(h_1^{(n)}\right) + Q\left(h_2^{(n)}\right) \right\},
$$

then we have

$$
\mathbf{E}\left[A\delta(u)B\right] = \mathbf{E}\left[A \overleftrightarrow{D}_u B\right].
$$

Remark 11.3.16   *Note that* $\delta \colon \mathcal{S}_{\mathfrak{h},\delta} \longrightarrow B(\mathfrak{h})$ *is the only linear map with this property, since for one single element* $h \in \mathfrak{h}_{\mathbb{C}} \otimes \mathbb{C}^2$, *the sets*

$$
\left\{ A^* \boldsymbol{\Omega} : A \in \operatorname{Dom} D \cap \{P(h_1) + Q(h_2)\}' \right\}
$$

and

$$
\left\{ B\boldsymbol{\Omega} : B \in \operatorname{Dom} D \cap \{P(h_1) + Q(h_2)\}' \right\}
$$

are still total in $\mathfrak{h}$.

Proof: From Corollary 11.3.8 we get

$$\mathbf{E}\left[A \overleftrightarrow{D}_u B\right] = \mathbf{E}\left[A\langle \bar{u}, DB\rangle + \langle \overline{DA}, u\rangle B\right]$$

$$= \mathbf{E}\left[\sum_{j=1}^{n} AF_j\left(D_{h^{(j)}}B\right) + \sum_{j=1}^{n}\left(D_{h^{(j)}}A\right)F_jB\right]$$

$$= \mathbf{E}\left[\frac{1}{2}\sum_{j=1}^{n}\left\{P\left(h_1^{(j)}\right) + Q\left(h_2^{(j)}\right), AF_jB\right\} - \sum_{j=1}^{n}A\left(D_{h^{(j)}}F_j\right)B\right].$$

But since A and B commute with

$$P\left(h_1^{(1)}\right) + Q\left(h_2^{(1)}\right), \ldots, P\left(h_1^{(n)}\right) + Q\left(h_2^{(n)}\right),$$

we can pull them out of the anticommutator, and we get

$$\mathbf{E}\left[A \overleftrightarrow{D}_u B\right]$$

$$= \mathbf{E}\left[\frac{1}{2}\sum_{j=1}^{n}A\left\{P\left(h_1^{(j)}\right) + Q\left(h_2^{(j)}\right), F_j\right\}B - \sum_{j=1}^{n}A\left(D_{h^{(j)}}F_j\right)B\right]$$

$$= \mathbf{E}[A\delta(u)B].$$

$\square$

We now give an explicit formula for the matrix elements between two exponential vectors of the divergence of a smooth elementary element $u \in \mathcal{S}_{\mathfrak{h},\delta}$. This is the analogue of the first fundamental lemma in the Hudson–Parthasarathy calculus, see Theorem 5.3.2 or [87, Proposition 25.1].

Theorem 11.3.17 *Let $u \in \mathcal{S}_{\mathfrak{h},\delta}$. Then we have the following formula*

$$\langle \mathcal{E}(k_1), \delta(u)\mathcal{E}(k_2)\rangle = \left\langle \mathcal{E}(k_1) \otimes \begin{pmatrix} ik_1 - i\bar{k}_2 \\ k_1 + \bar{k}_2 \end{pmatrix}, u\mathcal{E}(k_2)\right\rangle$$

for the evaluation of the divergence $\delta(u)$ of u between two exponential vectors $\mathcal{E}(k_1), \mathcal{E}(k_2)$, for $k_1, k_2 \in \mathfrak{h}_{\mathbb{C}}$.

Remark 11.3.18 *This suggests to extend the definition of δ in the following way: set*

$$\mathrm{Dom}\,\delta = \left\{u \in \mathfrak{M} : \exists M \in \mathfrak{B}(\mathfrak{h}) \text{ such that } \forall k_1, k_2 \in \mathfrak{h}_{\mathbb{C}},\right. \tag{11.3}$$

$$\left. \langle \mathcal{E}(k_1), M\mathcal{E}(k_2)\rangle = \left\langle \mathcal{E}(k_1) \otimes \begin{pmatrix} ik_1 - i\bar{k}_2 \\ k_1 + \bar{k}_2 \end{pmatrix}, u\mathcal{E}(k_2)\right\rangle\right\}$$

and define $\delta(u)$ for $u \in \mathrm{Dom}\,\delta$ to be the unique operator M that satisfies the condition in Equation (11.3).

Proof: Let $u = \sum_{j=1}^{n} F_j \otimes h^{(j)}$. Recalling the definition of D_h we get the following alternative expression for $\delta(u)$,

$$\delta(u) = \frac{1}{2} \sum_{j=1}^{n} \left(P\left(h_1^{(j)}\right) + Q\left(h_2^{(j)}\right) - iQ\left(h_1^{(j)}\right) + iP\left(h_2^{(j)}\right) \right) F_j$$

$$+ \frac{1}{2} \sum_{j=1}^{n} F_j \left(P\left(h_1^{(j)}\right) + Q\left(h_2^{(j)}\right) + iQ\left(h_1^{(j)}\right) - iP\left(h_2^{(j)}\right) \right)$$

$$= \sum_{j=1}^{n} \left(a^+\left(h_2^{(j)} - ih_1^{(j)}\right) F_j + F_j a\left(\overline{h_2^{(j)} - ih_1^{(j)}}\right) \right). \tag{11.4}$$

Evaluating this between two exponential vectors, we obtain

$$\langle \mathcal{E}(k_1), \delta(u)\mathcal{E}(k_2) \rangle = \sum_{j=1}^{n} \langle a(h_2^{(j)} - ih_1^{(j)})\mathcal{E}(k_1), F_j \mathcal{E}(k_2) \rangle$$

$$+ \sum_{j=1}^{n} \langle \mathcal{E}(k_1), F_j a(\overline{h_2^{(j)} + ih_1^{(j)}})\mathcal{E}(k_2) \rangle$$

$$= \sum_{j=1}^{n} \left(\overline{\langle h_2^{(j)} - ih_1^{(j)}, k_1 \rangle} + \overline{\langle h_2^{(j)} + ih_1^{(j)}, k_2 \rangle} \right) \langle \mathcal{E}(k_1), F_j \mathcal{E}(k_2) \rangle$$

$$= \sum_{j=1}^{n} \left(\langle k_1, h_2^{(j)} - ih_1^{(j)} \rangle + \langle \bar{k}_2, h_2^{(j)} + ih_1^{(j)} \rangle \right) \langle \mathcal{E}(k_1), F_j \mathcal{E}(k_2) \rangle$$

$$= \left\langle \mathcal{E}(k_1) \otimes \begin{pmatrix} ik_1 - i\bar{k}_2 \\ k_1 + \bar{k}_2 \end{pmatrix}, u\mathcal{E}(k_2) \right\rangle.$$

$\square$

Corollary 11.3.19 *The divergence operator δ is closable in the weak topology.*

Proof: Let $(u_\iota)_{\iota \in I}$ be a net such that $u_\iota \xrightarrow{\iota \in I} 0$ and $\delta(u_\iota) \xrightarrow{\iota \in I} \beta \in B(\mathfrak{h})$ in the weak topology. Then we get

$$\langle \mathcal{E}(k_1), \beta\mathcal{E}(k_2) \rangle = \lim_{\iota \in I} \langle \mathcal{E}(k_1), \delta(u_\iota)\mathcal{E}(k_2) \rangle$$

$$= \lim_{\iota \in I} \left\langle \mathcal{E}(k_1) \otimes \begin{pmatrix} ik_1 - i\bar{k}_2 \\ k_1 + \bar{k}_2 \end{pmatrix}, u_\iota \mathcal{E}(k_2) \right\rangle$$

$$= 0,$$

for all $k_1, k_2 \in \mathfrak{h}_\mathbb{C}$, and thus $\beta = 0$. $\square$

We have the following analogues of Equations (9.11) and (9.12).

Proposition 11.3.20   *Let* $u, v \in \mathcal{S}_{\mathfrak{h}, \delta}$, $F \in \mathcal{S}$, $h \in \mathfrak{h}_{\mathbb{C}} \otimes \mathbb{C}^2$, *then we have*

$$
\begin{cases}
D_h \circ \delta(u) = \langle \overline{h}, u \rangle + \delta \circ D_h(u), \\[2mm]
\delta(Fu) = F\delta(u) - F\overleftarrow{D}_u + \dfrac{1}{2} \displaystyle\sum_{j=1}^{n} \left[P\left(h_1^{(j)}\right) + Q\left(h_2^{(j)}\right), F \right] F_j, \\[4mm]
\delta(uF) = \delta(u)F - \overrightarrow{D}_u F + \dfrac{1}{2} \displaystyle\sum_{j=1}^{n} \left[F, P\left(h_1^{(j)}\right) + Q\left(h_2^{(j)}\right) \right] F_j.
\end{cases} \qquad (11.5a)
$$

Proof: a) Let $u = \sum_{j=1}^{n} F_j \otimes h^{(j)}$. Setting

$$
\begin{cases}
X_j = \dfrac{1}{2}\left(P\left(h_1^{(j)}\right) + Q\left(h_2^{(j)}\right) \right), \\[3mm]
Y_j = \dfrac{i}{2}\left(Q\left(h_1^{(j)}\right) - P\left(h_2^{(j)}\right) \right), \\[3mm]
Y = \dfrac{i}{2}\left(Q\left(h_1\right) - P\left(h_2\right) \right),
\end{cases}
$$

we have

$$
\delta(u) = \sum_{j=1}^{n} \left((X_j - Y_j)F_j + F_j(X_j + Y_j) \right),
$$

and therefore

$$
D_h\big(\delta(u)\big) = \sum_{j=1}^{n} \left(Y(X_j - Y_j)F_j + YF_j(X_j + Y_j) - (X_j - Y_j)F_jY - F_j(X_j + Y_j)Y \right).
$$

On the other hand, we have $D_h(u) = \sum_{j=1}^{n} (YF_j - F_jY) \otimes h^{(j)}$, and

$$
\delta\big(D_h(u)\big) = \sum_{j=1}^{n} \left((X_j - Y_j)YF_j - (X_j - Y_j)F_jY + YF_j(X_j + Y_j) - F_jY(X_j + Y_j) \right).
$$

Taking the difference of these two expressions, we get

$$
D_h\big(\delta(u)\big) - \delta\big(D_h(u)\big) = \sum_{j=1}^{n} \left([Y, X_j - Y_j]F_j + F_j[Y, X_j + Y_j] \right)
$$

$$
= \sum_{j=1}^{n} \left(\langle \overline{h}_1, h_1^{(j)} \rangle + \langle \overline{h}_2, h_2^{(j)} \rangle \right) F_j = \langle \overline{h}, u \rangle.
$$

b) A straightforward computation gives

$$\delta(Fu) = \sum_{j=1}^{n} \left((X_j - Y_j)FF_j + FF_j(X_j + Y_j) \right)$$

$$= F \sum_{j=1}^{n} \left((X_j - Y_j)F_j + F_j(X_j + Y_j) \right) - \sum_{j=1}^{n} [F, X_j - Y_j]F_j$$

$$= F\delta(u) - \sum_{j=1}^{n} [Y_j, F]F_j + \sum_{j=1}^{n} [X_jF, X]F_j$$

$$= F\delta(u) - \sum_{j=1}^{n} \langle F^*, h^{(j)} \rangle F_j + \sum_{j=1}^{n} [X_j, F]F_j$$

$$= F\delta(u) - F\overleftarrow{D}_u + \sum_{j=1}^{n} [X_j, F]F_j$$

where we used that $[X_j, F] = i \left\langle \left(\dfrac{-\overline{h}_2}{\overline{h}_1} \right), DF \right\rangle$ defines a bounded operator, since $F \in \mathcal{S} \subseteq \mathrm{Dom}\, D$. Equation (11.5a) can be shown similarly. $\quad\square$

If we impose additional commutativity conditions, which are always satisfied in the commutative case, then we get simpler formulas that are more similar to the classical ones.

Corollary 11.3.21 *If* $u = \sum_{j=1}^{n} F_j \otimes h^{(j)} \in \mathcal{S}_{\mathfrak{h}, \delta}$ *and*

$$F \in \mathrm{Dom}\, D \cap \left\{ P\left(h_1^{(1)}\right) + Q\left(h_2^{(1)}\right), \dots, P\left(h_1^{(n)}\right) + Q\left(h_2^{(n)}\right) \right\}'$$

then we have

$$\delta(Fu) = F\delta(u) - F\overleftarrow{D}_u, \quad \text{and} \quad \delta(uF) = \delta(u)F - \overrightarrow{D}_u F.$$

11.3.5 Relation to the commutative case

Here we show that the noncommutative calculus contains the commutative calculus as a particular case, at least in the case of bounded functionals. It is well-known that the symmetric Fock space $\Gamma(\mathfrak{h}_{\mathbb{C}})$ is isomorphic to the complexification $L^2(\Omega; \mathbb{C})$ of the Wiener space $L^2(\Omega)$ over $\mathfrak{h}$, cf. [17, 58, 79]. Such an isomorphism

$$I: L^2(\Omega; \mathbb{C}) \overset{\cong}{\longmapsto} \Gamma(\mathfrak{h}_{\mathbb{C}})$$

can be defined by extending the map

$$I: e^{iW(h)} \longmapsto I\left(e^{iW(h)}\right) = e^{iQ(h)}\Omega = e^{-\|h\|^2/2}\mathcal{E}(ih), \quad h \in \mathfrak{h}.$$

Using this isomorphism, a bounded functional $F \in L^\infty(\Omega; \mathbb{C})$ becomes a bounded operator $M(F)$ on $\Gamma(\mathfrak{h}_\mathbb{C})$, acting simply by multiplication,

$$M(F)\psi = I\left(FI^{-1}(\psi)\right), \quad \psi \in \Gamma(\mathfrak{h}_\mathbb{C}).$$

In particular, we get $M\left(e^{iW(h)}\right) = U(0, h)$ for $h \in \mathfrak{h}$. We can show that the derivation of a bounded differentiable functional coincides with its derivation as a bounded operator.

Proposition 11.3.22 Let $k \in \mathfrak{h}$ and $F \in L^\infty(\Omega; \mathbb{C}) \cap \mathrm{Dom}\, \tilde{D}_k$ such that $\tilde{D}_k F \in L^\infty(\Omega; \mathbb{C})$. Then we have $M(F) \in \mathrm{Dom}\, D_{k_0}$, where $k_0 = \binom{0}{k}$, and

$$M(\tilde{D}_k F) = D_{k_0}\left(M(F)\right).$$

Proof: It is sufficient to check this for functionals of the form $F = e^{iW(h)}$, $h \in \mathfrak{h}$. We get

$$
\begin{aligned}
M(\tilde{D}_k e^{iW(h)}) &= M\left(i\langle k, h \rangle e^{iW(h)}\right) \\
&= i\langle k, h \rangle U(0, h) = i\left\langle \binom{0}{k}, \binom{0}{h} \right\rangle U(0, h) \\
&= D_{k_0} U(0, h) = D_{k_0}\left(M(e^{iW(h)})\right).
\end{aligned}
$$

$\square$

This implies that we also have an analogous result for the divergence.

11.4 The white noise case

Belavkin [13, 14] and Lindsay [68] have defined noncausal quantum stochastic integrals with respect to the creation, annihilation, and conservation processes on the boson Fock space over $L^2(\mathbb{R}_+)$ using the classical derivation and divergence operators. Our divergence operator coincides with their noncausal creation and annihilation integrals for integrable processes up to a coordinate transformation. This immediately implies that for integrable adapted processes our integral coincides with the quantum stochastic creation and annihilation integrals defined by Hudson and Parthasarathy, cf. [54, 87].

Let now $\mathfrak{h} = L^2(T, B, \mu)$, where (T, B, μ) is a measure space such that B is countably generated. In this case we can apply the divergence operator to processes indexed by T, i.e., $B(\mathfrak{h})$-valued measurable functions on T, since they can be interpreted as elements of the Hilbert module, if they are square-integrable. Let $L^2(T, B(\mathfrak{h}))$ denote all $B(\mathfrak{h})$-valued measurable functions $t \longmapsto X_t$ on T with $\int_T \|X_t\|^2 dt < \infty$. Then the definition of the divergence operator becomes

$$\mathrm{Dom}\,\delta = \Big\{ X = (X^1, X^2) \in L^2(T, B(\mathfrak{h})) \oplus L^2(T, B(\mathfrak{h})) :$$

$$\exists M \in B(\mathfrak{h}) \text{ such that } \langle \mathcal{E}(k_1), M\mathcal{E}(k_2) \rangle =$$

$$\int_T \big(i(k_2 - \overline{k}_1) \langle \mathcal{E}(k_1), X^1_t \mathcal{E}(k_2) \rangle + (\overline{k}_1 + k_2) \langle \mathcal{E}(k_1), X^2_t \mathcal{E}(k_2) \rangle \big) d\mu(t),$$

$$k_1, k_2 \in \mathfrak{h}_{\mathbb{C}} \Big\},$$

and $\delta(X)$ is equal to the unique operator satisfying the aforementioned condition.

The definition of noncausal quantum stochastic integrals of [13, 14] and [68] with respect to the creation, annihilation, and conservation processes on the boson Fock space over $L^2(\mathbb{R}_+)$ use the classical derivation and divergence operators D and δ from the Malliavin calculus on the Wiener space $L^2(\Omega)$. Recall that D and δ are defined using the isomorphism between $L^2(\Omega)$ and the Fock space $\Gamma(L^2(\mathbb{R}_+; \mathbb{C}))$ over $L^2(\mathbb{R}_+; \mathbb{C}) = L^2(\mathbb{R}_+)_{\mathbb{C}}$, cf. Chapter 4. Namely, $\tilde{D}$ acts on the exponential vectors as

$$\tilde{D}\mathcal{E}(k) = \mathcal{E}(k) \otimes k, \qquad k \in L^2(\mathbb{R}_+, \mathbb{C}),$$

and $\tilde{\delta}$ is the adjoint of $\tilde{D}$. Note that due to the isomorphism between $\Gamma(L^2(\mathbb{R}_+; \mathbb{C})) \otimes L^2(\mathbb{R}_+; \mathbb{C})$ and $L^2(\mathbb{R}_+; \Gamma(L^2(\mathbb{R}_+; \mathbb{C})))$, the elements of $\Gamma(L^2(\mathbb{R}_+; \mathbb{C})) \otimes L^2(\mathbb{R}_+; \mathbb{C})$ can be interpreted as function on $\mathbb{R}_+$, which allows us to write $(D\mathcal{E}(k))_t = k(t)\mathcal{E}(k)$ almost surely. The action of the annihilation integral $\int F_t dA_t$ on some vector $\psi \in \Gamma(L^2(\mathbb{R}_+; \mathbb{C}))$ is then defined as the Bochner integral

$$\int_{\mathbb{R}_+}^{\mathrm{BL}} F_t da_t^- \psi := \int_{\mathbb{R}_+} F_t (D\psi)_t dt,$$

and that of the creation integral as

$$\int_{\mathbb{R}_+}^{\mathrm{BL}} F_t da_t^+ \psi = \tilde{\delta}(F.\psi).$$

We will also use the notation

$$\delta(X) = \int_T X^1_t dP(t) + \int_T X^2_t dQ(t),$$

and call $\delta(X)$ the Hitsuda–Skorohod integral of X. These definitions satisfy the adjoint relations

$$\left(\int_{\mathbb{R}_+}^{\mathrm{BL}} F_t da_r^- \right)^* \supset \int_{\mathbb{R}_+}^{\mathrm{BL}} F_t^* da_t^+, \text{ and } \left(\int_{\mathbb{R}_+}^{\mathrm{BL}} F_t da_r^+ \right)^* \supset \int_{\mathbb{R}_+}^{\mathrm{BL}} F_t^* da_t^-.$$

It turns out that our Hitsuda–Skorohod integral operator δ coincides, up to a coordinate transformation, with the above creation and annihilation integrals. This immediately implies that for adapted, integrable processes our integral also coincides with the quantum stochastic creation and annihilation integrals defined by Hudson and Parthasarathy, cf. [54, 87].

Proposition 11.4.1 *Let* $(T, B, \mu) = (\mathbb{R}_+, B(\mathbb{R}_+), dx)$, *i.e., the positive half-line with the Lebesgue measure, and let* $X = (X^1, X^2) \in \mathrm{Dom}\,\delta$. *Then we have*

$$\int_{\mathbb{R}_+} X_t^1 dP(t) + \int_{\mathbb{R}_+} X_t^2 dQ(t) = \int_{\mathbb{R}_+}^{\mathrm{BL}} (X_t^2 - iX_t^1) da_t^+ + \int_{\mathbb{R}_+}^{\mathrm{BL}} (X_t^2 + iX_t^1) da_t^-.$$

Proof: To prove this, we show that the Belavkin–Lindsay integrals satisfy the same formula for the matrix elements between exponential vectors. Let $(F_t)_{t \in \mathbb{R}_+} \in L^2(\mathbb{R}_+, B(\mathfrak{h}))$ be such that its creation integral in the sense of Belavkin and Lindsay is defined with a domain containing the exponential vectors. Then we get

$$\left\langle \mathcal{E}(k_1), \int_{\mathbb{R}_+}^{\mathrm{BL}} F_t da_t^+ \mathcal{E}(k_2) \right\rangle = \langle \mathcal{E}(k_1), \tilde{\delta}(F.\mathcal{E}(k_2)) \rangle$$

$$= \langle (\tilde{D}\mathcal{E}(k_1))_., F.\mathcal{E}(k_2) \rangle$$

$$= \int_{\mathbb{R}_+} \overline{k_1(t)} \langle \mathcal{E}(k_1), F_t \mathcal{E}(k_2) \rangle dt.$$

For the annihilation integral we deduce the formula

$$\left\langle \mathcal{E}(k_1), \int_{\mathbb{R}_+}^{\mathrm{BL}} F_t da_t^- \mathcal{E}(k_2) \right\rangle = \left\langle \int_{\mathbb{R}_+}^{\mathrm{BL}} F_t^* da_t^+ \mathcal{E}(k_1), \mathcal{E}(k_2) \right\rangle$$

$$= \int_{\mathbb{R}_+} \overline{k_2(t)} \langle \mathcal{E}(k_2), F_t^* \mathcal{E}(k_1) \rangle dt = \int_{\mathbb{R}_+} k_2(t) \langle \mathcal{E}(k_1), F_t \mathcal{E}(k_2) \rangle dt.$$

$\square$

Let $(T, B, \mu) = (\mathbb{R}_+, B(\mathbb{R}_+), dx)$, i.e., the positive half-line with the Lebesgue measure, and let $X = (X^1, X^2) \in \mathrm{Dom}\,\delta$. Then we have

$$\int_{\mathbb{R}_+} X_t^1 dP(t) + \int_{\mathbb{R}_+} X_t^2 dQ(t) = \int_{\mathbb{R}_+}^{\mathrm{BL}} (X_t^2 - iX_t^1) da_t^+ + \int_{\mathbb{R}_+}^{\mathrm{BL}} (X_t^2 + iX_t^1) da_t^-.$$

The integrals defined by Belavkin and Lindsay are an extension of those defined by Hudson and Parthasarathy.

Corollary 11.4.2 *For adapted processes* $X \in \mathrm{Dom}\,\delta$, *the Hitsuda–Skorohod integral*

$$\delta(X) = \int_T X_t^1 \, dP(t) + \int_T X_t^2 \, dQ(t)$$

coincides with the Hudson–Parthasarathy quantum stochastic integral defined in [54].

11.4.1 Iterated integrals

Here we informally discuss the iterated integrals of deterministic functions, showing a close relation between these iterated integrals and the so-called Wick product or normal-ordered product. Although this involves unbounded operators on which the divergence operator δ has not been formally defined, the construction can be made rigorous by choosing an appropriate common invariant domain for these operators, using *e.g.*, vectors with a finite chaos decomposition. Namely, in order to iterate the operator, we start by defining δ on $B(\mathfrak{h}) \otimes \mathfrak{h}_{\mathbb{C}} \otimes \mathbb{C}^2 \otimes \mathfrak{h}$, where $\mathfrak{h}$ is some Hilbert space, as $\delta \otimes \mathrm{id}_{\mathfrak{h}}$. Next, using Equation (11.4), one can show by induction that

$$\delta^n \begin{pmatrix} h_1^{(1)} \\ h_2^{(1)} \end{pmatrix} \otimes \cdots \otimes \begin{pmatrix} h_1^{(n)} \\ h_2^{(n)} \end{pmatrix}$$

$$= \sum_{I \subseteq \{1,\dots,n\}} \prod_{j \in I} a^+(h_2^{(j)} - ih_1^{(j)}) \prod_{j \in \{1,\dots,n\} \setminus I} a^-(\overline{h_2^{(j)}} + i\overline{h^{(j)}}).$$

for $h^{(1)} = \begin{pmatrix} h_1^{(1)} \\ h_2^{(1)} \end{pmatrix}, \dots, h^{(n)} = \begin{pmatrix} h_1^{(n)} \\ h_2^{(n)} \end{pmatrix} \in \mathfrak{h}_{\mathbb{C}} \otimes \mathbb{C}^2$. This is just the Wick product of $\big(P(h_1^{(1)}) + Q(h_2^{(1)})\big), \dots, \big(P(h_1^{(n)}) + Q(h_2^{(n)})\big)$, *i.e.*,

$$\delta^n \begin{pmatrix} h_1^{(1)} \\ h_2^{(1)} \end{pmatrix} \otimes \cdots \otimes \begin{pmatrix} h_1^{(n)} \\ h_2^{(n)} \end{pmatrix} = \big(P(h_1^{(1)}) + Q(h_2^{(1)})\big) \diamond \cdots \diamond \big(P(h_1^{(n)}) + Q(h_2^{(n)})\big),$$

where the Wick product $\diamond$ is defined in terms of the momentum and position operators on the algebra generated by $\{P(k), Q(k) : k \in \mathfrak{h}_{\mathbb{C}}\}$, by

$$\begin{cases} P(h) \diamond X = X \diamond P(h) = -ia^+(h)X + iXa^-(\overline{h}) \\[2mm] Q(h) \diamond X = X \diamond Q(h) = a^+(h)X + Xa^-(\overline{h}) \end{cases}$$

$X \in \mathrm{alg}\,\{P(k), Q(k) : k \in \mathfrak{h}_{\mathbb{C}}\}$, $h \in \mathfrak{h}_{\mathbb{C}}$. Equivalently, in terms of creation and annihilation we have

$$\begin{cases} a^+(h) \diamond X = X \diamond a^+(h) = a^+(h)X \\[2mm] a^-(h) \diamond X = X \diamond a^-(h) = Xa^-(h). \end{cases}$$

Notes

Another definition of D and δ on noncommutative operator algebra has been considered by Biane and Speicher in the free case [19], where the operator algebra is isomorphic to the full Fock space. In [74], Mai, Speicher, and Weber study the regularity of distributions in free probability. Due to the lack of commutativity, it seems impossible in their approach to use an integration by parts formula, so that they were compelled to find alternative methods. It would be interesting to apply these methods to quantum stochastic differential equations.

Our approach to quantum white noise calculus is too restrictive so far since we require the derivatives DX to be bounded operators. Dealing with unbounded operators is necessary for applications of quantum Malliavin calculus to more realistic physical models. Ji and Obata [59, 60] have defined a creation-derivative and an annihilation-derivative in the setting of quantum white noise theory. Up to a basis change (they derive with respect to a_t^- and a_t^+, while we derive with respect to P and Q), these are the same as our derivation operator. But working in the setting of white noise theory, they can derive much more general (*i.e.*, unbounded) operators.

Exercises

Exercise 11.1 In the framework of Proposition 12.1.4, assume in addition that $X \in \operatorname{Dom} D_k^n$, $(D_k X)^{-1} \in \operatorname{Dom} D_k^n$, and

$$\omega \in \bigcap_{1 \le \kappa \le n} \operatorname{Dom}\big(Q(k_1) - P(k_2)\big)^\kappa \bigcap_{1 \le \kappa \le n} \operatorname{Dom}\big(Q(\bar{k}_1) - P(\bar{k}_2)\big)^\kappa.$$

Show that the density of the distribution $\mu_{X,\Phi}$ of $X \in B(\mathfrak{h})$ in the state Φ is $n - 1$ times differentiable for all $n \ge 2$.

12

Smoothness of densities on real Lie algebras

How can it be that mathematics, being after all a product of human thought which is independent of experience, is so admirably appropriate to the objects of reality?

(A. Einstein, in Geometry and Experience.*)*

In this chapter the noncommutative Malliavin calculus on the Heisenberg–Weyl algebra is extended to the affine algebra via a differential calculus and a noncommutative integration by parts. As previously we obtain sufficient conditions for the smoothness of Wigner type laws of noncommutative random variables, this time with gamma or continuous binomial marginals. The Malliavin calculus on the Heisenberg–Weyl algebra $\{P, Q, I\}$ of Chapter 11 relies on the composition of a function with a couple of noncommutative random variables introduced via the Weyl calculus and on a covariance identity which plays the role of integration by parts formula.

12.1 Noncommutative Wiener space

In this section we use the operator D in the framework of Chapter 11 to give sufficient conditions for the existence and smoothness of densities for operators on $\mathfrak{h}$. The domain of the operator D is rather small because we require $\delta(u)$ to be a bounded operator and "deterministic" non-zero elements $h \in \mathfrak{h}_{\mathbb{C}} \otimes \mathbb{C}^2$ cannot be integrable. As in the classical Malliavin calculus we will rely on a Girsanov transformation, here given by Proposition 10.4.1, and it will also be used to derive sufficient conditions for the existence of smooth densities. In the sequel we let $H^{p,\kappa}(\mathbb{R}^2)$ denote the classical Sobolev space of orders $\kappa \in \mathbb{N}$ and $p \in [2, \infty]$.

Proposition 12.1.1　*Let $\kappa \in \mathbb{N}$, $h \in \mathfrak{h} \otimes \mathbb{R}^2$ with $\langle h_1, h_2 \rangle \neq 0$, and Φ a vector state, i.e., there exists a unit vector $\omega \in \mathfrak{h}$ such that*

$$\Phi(X) = \langle \omega, X\omega \rangle, \qquad X \in B(\mathfrak{h}).$$

If there exists a $k \in \mathfrak{h}_{\mathbb{C}} \otimes \mathbb{C}^2$ such that

$$\omega \in \bigcap_{\kappa_1 + \kappa_2 \leq \kappa} \mathrm{Dom}\big(Q(k_1)^{\kappa_1} P(k_2)^{\kappa_2}\big) \cap \bigcap_{\kappa_1 + \kappa_2 \leq \kappa} \mathrm{Dom}\big(Q(\bar{k}_1)^{\kappa_1} P(\bar{k}_2)^{\kappa_2}\big)$$

and

$$\langle h_1, k_1 \rangle \neq 0 \qquad and \qquad \langle h_2, k_2 \rangle \neq 0,$$

then we have $w_{h,\Phi} \in \bigcap_{2 \leq p \leq \infty} H^{p,\kappa}(\mathbb{R}^2)$.

Proof:　We will show the result for $\kappa = 1$, the general case can be shown similarly (see also the proof of Theorem 12.1.2). Let $\varphi \in \mathcal{S}(\mathbb{R})$ be a Schwartz function, and let $p \in [1, 2]$. Then we have

$$\left| \int \frac{\partial \varphi}{\partial x} dW_{h,\Phi} \right| = \left| \left\langle \omega, O_h\left(\frac{\partial \varphi}{\partial x}\right) \omega \right\rangle \right|$$

$$= \left| \left\langle \omega, \frac{i}{2|\langle k_1, h_1 \rangle|} [Q(\bar{k}_1), O_h(\varphi)] \omega \right\rangle \right|$$

$$\leq \frac{C_{h,p}\big(\|Q(k_1)\omega\| + \|Q(\bar{k}_1)\omega\|\big)}{2|\langle k_1, h_1 \rangle|} \|\varphi\|_p.$$

Similarly, we get

$$\left| \int \frac{\partial \varphi}{\partial y} dW_{h,\Phi} \right| \leq \frac{C_{h,p}\big(\|P(k_2)\omega\| + \|P(\bar{k}_2)\omega\|\big)}{2|\langle k_2, h_2 \rangle|} \|\varphi\|_p,$$

and together these two inequalities imply $w_{h,\Phi} \in H^{p',1}(\mathbb{R}^2)$ for $p' = p/(p-1)$.　□

We will give a more general result of this type in the next Theorem 12.1.2. Namely we show that the derivation operator can be used to obtain sufficient conditions for the regularity of the joint Wigner densities as of noncommutating random variables as in the next Theorem 12.1.2 which generalises Proposition 12.1.1 to arbitrary states.

Theorem 12.1.2　*Let $\kappa \in \mathbb{N}$, $h \in \mathfrak{h} \otimes \mathbb{R}^2$ with $\langle h_1, h_2 \rangle \neq 0$, and suppose that Φ is of the form*

$$\Phi(X) = \mathrm{tr}(\rho X), \qquad X \in B(\mathfrak{h}),$$

for some density matrix ρ. If there exist $k, \ell \in \mathfrak{h}_\mathbb{C} \otimes \mathbb{C}^2$ such that

$$\det \begin{bmatrix} \langle h_1, k_1 \rangle & \langle h_2, k_2 \rangle \\ \langle h_1, \ell_1 \rangle & \langle h_2, \ell_2 \rangle \end{bmatrix} \neq 0,$$

and $\rho \in \bigcap_{\kappa_1 + \kappa_2 \leq \kappa} \mathrm{Dom}\, D_k^{\kappa_1} D_\ell^{\kappa_2}$, and

$$\mathrm{tr}\left(|D_k^{\kappa_1} D_\ell^{\kappa_2} \rho| \right) < \infty, \qquad \kappa_1 + \kappa_2 \leq \kappa,$$

then we have $w_{h,\Phi} \in \bigcap_{2 \leq p \leq \infty} H^{p,\kappa}(\mathbb{R}^2)$.

The absolute value of a normal operator is well-defined via functional calculus. For a non-normal operator X we set $|X| = (X^*X)^{1/2}$. The square root is well-defined via functional calculus, since X^*X is positive and therefore normal.

Proof: Let

$$A := \begin{bmatrix} \langle h_1, k_1 \rangle & \langle h_2, k_2 \rangle \\ \langle h_1, \ell_1 \rangle & \langle h_2, \ell_2 \rangle \end{bmatrix} \text{ and } \begin{bmatrix} X_1 \\ X_2 \end{bmatrix} := \frac{i}{2} A^{-1} \begin{bmatrix} Q(k_1) - P(k_2) \\ Q(\ell_1) - P(\ell_2) \end{bmatrix},$$

then we have

$$[X_1, O_h(\varphi)] = \frac{1}{\det A} \left(\langle h_2, \ell_2 \rangle D_k O_h(\varphi) - \langle h_2, k_2 \rangle D_\ell O_h(\varphi) \right) = O_h\left(\frac{\partial \varphi}{\partial x} \right)$$

and

$$[X_2, O_h(\varphi)] = \frac{1}{\det A} \left(- \langle h_1, \ell_1 \rangle D_k O_h(\varphi) + \langle h_1, k_2 \rangle D_\ell O_h(\varphi) \right) = O_h\left(\frac{\partial \varphi}{\partial y} \right),$$

for all Schwartz functions $\varphi \in \mathcal{S}(\mathbb{R})$. Therefore, we have

$$\left| \int \frac{\partial^{\kappa_1 + \kappa_2} \varphi}{\partial x^{\kappa_1} \partial y^{\kappa_2}} dW_{h,\Phi} \right| = \left| \mathrm{tr}\left(\rho\, O_h\left(\frac{\partial^{\kappa_1 + \kappa_2} \varphi}{\partial x^{\kappa_1} \partial y^{\kappa_2}} \right) \right) \right|$$

$$= \left| \mathrm{tr}\left(\rho \underbrace{[X_1, \ldots [X_1,}_{\kappa_1 \text{ times}} \underbrace{[X_2, \ldots [X_2,}_{\kappa_2 \text{ times}} O_h(\varphi)]]]] \right) \right|$$

$$= \left| \mathrm{tr}\left([X_2, \ldots [X_2, [X_1, \ldots [X_1, \rho]]]] O_h(\varphi) \right) \right|$$

$$\leq C_{\rho, \kappa_1, \kappa_2} ||O_h(\varphi)|| \leq C_{\rho, \kappa_1, \kappa_2} C_{h,p} ||\varphi||_p,$$

for all $p \in [1, 2]$, since $\rho \in \bigcap_{\kappa_1 + \kappa_2 \leq \kappa} \mathrm{Dom}\, D_k^{\kappa_1} D_\ell^{\kappa_2}$ and $\mathrm{tr}(|D_k^{\kappa_1} D_\ell^{\kappa_2} \rho|) < \infty$ for all $\kappa_1 + \kappa_2 \leq \kappa$, and thus

$$C_{\rho, \kappa_1, \kappa_2} = \mathrm{tr}\left| [X_2, \ldots [X_2, [X_1, \ldots [X_1, \rho]]]] \right| < \infty.$$

But this implies that the density of $dW_{h,\Phi}$ is contained in the Sobolev spaces $H^{p,\kappa}(\mathbb{R}^2)$ for all $2 \leq p \leq \infty$. $\qquad \square$

Example 12.1.3   Let $0 < \lambda_1 \le \lambda_2 \le \cdots$ be an increasing sequence of positive numbers and $\{e_j : j \in \mathbb{N}\}$ a complete orthonormal system for $\mathfrak{h}_\mathbb{C}$. Let

$$T_t : \mathfrak{h}_\mathbb{C} \longrightarrow \mathfrak{h}_\mathbb{C}$$

be the contraction semigroup defined by

$$T_t e_j = e^{-t\lambda_j} e_j, \qquad j \in \mathbb{N}, \quad t \in \mathbb{R}_+,$$

with generator $A = \sum_{j \in \mathbb{N}} \lambda_j \mathbb{P}_j$. If the sequence increases fast enough to ensure

that $\sum_{j=1}^{\infty} e^{-t\lambda_j} < \infty$, *i.e.*, if $\mathrm{tr}\, T_t < \infty$ for $t > 0$, then the second quantisation $\rho_t = \Gamma(T_t) \colon \mathfrak{h} \longrightarrow \mathfrak{h}$ is a trace class operator with trace

$$Z_t = \mathrm{tr}\, \rho_t = \sum_{\mathbf{n} \in \mathbb{N}_f^\infty} \langle e_\mathbf{n}, \rho_t e_\mathbf{n} \rangle,$$

where we use $\mathbb{N}_f^\infty$ to denote the finite sequences of non-negative integers and $\{e_\mathbf{n} : \mathbf{n} \in \mathbb{N}_f^\infty\}$ is the complete orthonormal system of $\mathfrak{h}$ consisting of the vectors

$$e_\mathbf{n} = e_1^{\circ n_1} \circ \cdots \circ e_r^{\circ n_r}, \qquad \mathbf{n} = (n_1, \ldots, n_r) \in \mathbb{N}_f^\infty,$$

i.e., the symmetrisation of the tensor $e_1 \otimes \cdots \otimes e_1 \otimes \cdots \otimes e_r \otimes \cdots \otimes e_r$ where each vector e_j appears n_j times. We get

$$Z_t = \sum_{\mathbf{n} \in \mathbb{N}^\infty} \prod_{k=1}^{\infty} e^{-n_k t \lambda_k} = \prod_{k=1}^{\infty} \frac{1}{1 - e^{-t\lambda_k}}$$

for the trace of ρ_t. We shall be interested in the state defined by

$$\Phi(X) = \frac{1}{Z_t} \mathrm{tr}(\rho_t X), \qquad X \in \mathcal{B}(\mathfrak{h}).$$

We get

$$\sum_{\mathbf{n} \in \mathbb{N}_f^\infty} \left| \langle e_\mathbf{n}, |\rho_{t/2} a^\ell(e_j)|^2 e_\mathbf{n} \rangle \right| = \sum_{\mathbf{n} \in \mathbb{N}_f^\infty} \| \rho_{t/2} a^\ell(e_j) \|^2$$

$$= \sum_{\mathbf{n} \in \mathbb{N}^\infty} n_j (n_j - 1) \cdots (n_j - \ell + 1) e^{-(n_j - \ell) t \lambda_j} \prod_{k \ne j} e^{-n_k t \lambda_k}$$

$$\le \sum_{n=0}^{\infty} (n + \ell)^\ell e^{-nt\lambda_j} \prod_{k \ne j} \frac{1}{1 - e^{-t\lambda_k}} < \infty,$$

and therefore $\rho_t a^\ell(e_j)$ defines a bounded operator with finite trace for all $j, \ell \in \mathbb{N}$ and $t > 0$. Similarly, we get

$$\text{tr}\,\big|a^\ell(e_j)\rho_t\big| < \infty, \quad \text{tr}\,\big|\rho_t\big(a^+(e_j)\big)^\ell\big| < \infty, \quad \text{etc.,}$$

and

$$\text{tr}\,\big|P^{\ell_1}(e_{j_1})Q^{\ell_2}(e_{j_2})\rho_t\big| < \infty, \quad \text{tr}\,\big|P^{\ell_1}(e_{j_1})\rho_t Q^{\ell_2}(e_{j_2})\big| < \infty,$$

$t > 0, j_1, j_2, \ell_1, \ell_2 \in \mathbb{N}$.

For a given $h \in \mathfrak{h} \otimes \mathbb{R}^2$ with $\langle h_1, h_2 \rangle \neq 0$ (and thus in particular $h_1 \neq 0$ and $h_2 \neq 0$), we can always find indices j_1 and j_2 such that $\langle h_1, e_{j_1} \rangle \neq 0$ and $\langle h_2, e_{j_2} \rangle \neq 0$. Therefore, we can check that for all $\kappa \in \mathbb{N}$, all assumptions of Theorem 12.1.2 are satisfied with $k = \binom{e_{j_1}}{0}$ and $\ell = \binom{0}{e_{j_2}}$. Finally, we check that the Wigner density $w_{h,\Phi}$ of $\big(P(h_1), Q(h_2)\big)$ with $\langle h_1, h_2 \rangle \neq 0$ in the state $\Phi(\cdot) = Z_t^{-1}\,\text{tr}(\rho_t \cdot)$ belongs to $\bigcap_{\kappa \in \mathbb{N}} \bigcap_{2 \leq p \leq \infty} H^{p,\kappa}(\mathbb{R}^2)$, in particular, its derivatives of all orders exist, and are bounded and square-integrable.

We now show that the aforementioned approach also applies to derive sufficient conditions for the regularity of a single bounded self-adjoint operator. Recall that the distribution of a bounded self-adjoint operator X, in the state Φ is the unique measure $\mu_{X,\Phi}$ on the real line such that

$$\Phi(X^n) = \int_{\mathbb{R}} x^n d\mu_{X,\Phi}, \qquad n \in \mathbb{N}.$$

Such a measure $\mu_{X,\Phi}$ always exists, is supported on the interval $\big[-||X||, ||X||\big]$, and it is unique by the Carleman moment growth condition [25]. In the next proposition, for simplicity we consider only vector states $\Phi(\cdot) = \langle \omega, \cdot\,\omega \rangle$ associated to a unit vector $\omega \in \mathfrak{h}$.

Proposition 12.1.4 *Let $X \in \mathcal{B}(\mathfrak{h})$ and assume that there exists a $k \in \mathfrak{h}_{\mathbb{C}} \otimes \mathbb{C}^2$ such that*

$$\omega \in \text{Dom}\big(Q(k_1) - P(k_2)\big) \cap \text{Dom}\big(Q(\bar{k}_1) - P(\bar{k}_2)\big),$$

$X \in \text{Dom}\,D_k$, $X \cdot D_k X = D_k X \cdot X$, $D_k X$ *invertible and* $(D_k X)^{-1} \in \text{Dom}\,D_k$. *Then the distribution* $\mu_{X,\Phi}$ *of* $X \in \mathcal{B}(\mathfrak{h})$ *in the state* Φ *has a bounded density.*

Proof: Since $X \cdot D_k X = D_k X \cdot X$, we have $D_k p(X) = (D_k X)p'(X)$ for all polynomials p. We therefore get

$$D_k\left((D_k X)^{-1}p(X)\right) = p(X)D_k\left((D_k X)^{-1}\right) + p'(X).$$

The hypotheses of the proposition ensure that

$$\left| \langle \omega, D_k \left((D_k X)^{-1} p(X) \right) \omega \rangle \right|$$

$$\leq \frac{||(Q(k_1) - P(k_2))\omega|| + ||(Q(\bar{k}_1) - P(\bar{k}_2))\omega||}{2} ||(D_k X)^{-1}|| \, ||p(X)||$$

$$\leq C_1 \sup_{x \in [-||X||, ||X||]} |p(x)|,$$

and

$$\left| \langle \omega, p(X) D_k \left((D_k X)^{-1} \right) \omega \rangle \right| \leq \left| \left| D \left((D_k X)^{-1} \right) \right| \right| \, ||p(X)||$$

$$\leq C_2 \sup_{x \in [-||X||, ||X||]} |p(x)|,$$

and therefore allow us to get the estimate

$$\left| \int_{-||X||}^{||X||} p'(x) d\mu_{X,\Phi}(x) \right| = \left| \langle \omega, p'(X)\omega \rangle \right|$$

$$= \left| \left\langle \omega, \left(D_k \left((D_k X)^{-1} p(X) \right) - p(X) D_k \left((D_k X)^{-1} \right) \right) \omega \right\rangle \right|$$

$$\leq (C_1 + C_2) \sup_{x \in [-||X||, ||X||]} |p(x)|$$

for all polynomials p. But this implies that $\mu_{X,\Phi}$ admits a bounded density. $\qquad\square$

12.2 Affine algebra

Recall from Section 7.5 that the Wigner density

$$\tilde{W}_{|\phi\rangle\langle\psi|}(\xi_1, \xi_2) = \frac{1}{2\pi |\xi_2|} \overline{W}_{|\phi\rangle\langle\psi|}(\xi_1, \xi_2)$$

exists, and writing $\xi = \xi_1 X_1^* + \xi_2 X_2^* \in \mathcal{G}^*$ we have

$$W_\rho(\xi)$$

$$= \frac{|\xi_2|^{1/2}}{\sqrt{2\pi}} \int_{\mathbb{R}^2} e^{-i\xi_1 x_1 - i\xi_2 x_2} \mathrm{Tr}[e^{-x_1 X_1 - x_2 X_2} \rho C^{-1}] \sqrt{e^{-x_1/2} \mathrm{sinch} \frac{x_1}{2}} \, dx_1 dx_2,$$

and for $\rho = |\phi\rangle\langle\psi|$,

$$W_{|\phi\rangle\langle\psi|}(\xi)$$

$$= \frac{|\xi_2|^{1/2}}{\sqrt{2\pi}} \int_{\mathbb{R}^2} e^{-i\xi_1 x_1 - i\xi_2 x_2} \langle \hat{U}(e^{x_1 X_1 + x_2 X_2}) C^{-1} \psi | \phi \rangle_{\mathfrak{h}} \sqrt{e^{-x_1/2} \mathrm{sinch} \frac{x_1}{2}} \, dx_1 dx_2$$

$$= \frac{1}{2\pi} \int_{\mathbb{R}^3} e^{-i\xi_1 x_1 - i\xi_2 x_2} \phi(e^{-x_1}\tau) \overline{\psi}(\tau) e^{-i\tau x_2 e^{-x_1}/2} \operatorname{sinch}(x_1/2)$$

$$\times\, e^{-(e^{-x_1}-1)|\tau|} e^{-\beta x_1/2} |\tau|^{\beta-1/2} \sqrt{e^{-x_1/2} \operatorname{sinch}(x_1/2)}\, \frac{d\tau}{\Gamma(\beta)} dx_1 dx_2$$

$$= \int_{\mathbb{R}} \phi\left(\frac{\xi_2 e^{-x/2}}{\operatorname{sinch}\frac{x}{2}}\right) \frac{|\xi_2| e^{-ix\xi_1}}{\operatorname{sinch}\frac{x}{2}} \overline{\psi}\left(\frac{\xi_2 e^{x/2}}{\operatorname{sinch}\frac{x}{2}}\right) e^{-|\xi_2|\frac{\cosh\frac{x}{2}}{\operatorname{sinch}\frac{x}{2}}} \left(\frac{|\xi_2|}{\operatorname{sinch}\frac{x}{2}}\right)^{\beta-1} \frac{dx}{\Gamma(\beta)}.$$

Note that W_ρ takes real values when ρ is self-adjoint. Next, we turn to proving the smoothness of the Wigner function $W_{|\phi\rangle\langle\psi|}$. Let now $H_{1,2}^\sigma(\mathbb{R} \times (0,\infty))$ denote the Sobolev space with respect to the norm

$$\|f\|^2_{H_{1,2}^\sigma(\mathbb{R}\times(0,\infty))} = \int_0^\infty \frac{1}{\xi_2} \int_{\mathbb{R}} |f(\xi_1,\xi_2)|^2 d\xi_1 d\xi_2 \tag{12.1}$$
$$+ \int_0^\infty \xi_2 \int_{\mathbb{R}} (|\partial_1 f(\xi_1,\xi_2)|^2 + |\partial_2 f(\xi_1,\xi_2)|^2) d\xi_1 d\xi_2.$$

Note that if ϕ, ψ have supports in $\mathbb{R}_+$, then $W_{|\phi\rangle\langle\psi|}$ has support in $\mathbb{R} \times (0,\infty)$, and the conclusion of Theorem 12.2.1 reads $W_{|\phi\rangle\langle\psi|} \in H_{1,2}^\sigma(\mathbb{R} \times (0,\infty))$.

Theorem 12.2.1 *Let $\phi, \psi \in \mathrm{Dom}\, X_1 \cap \mathrm{Dom}\, X_2$. Then*

$$\mathbf{1}_{\mathbb{R}\times(0,\infty)} W_{|\phi\rangle\langle\psi|} \in H_{1,2}^\sigma(\mathbb{R}\times(0,\infty)).$$

Proof: We have, for $f \in \mathcal{C}_c^\infty(\mathbb{R} \times (0,\infty))$:

$$\left| \int_{\mathbb{R}^2} f(\xi_1,\xi_2) \overline{W}_{|\phi\rangle\langle\psi|}(\xi_1,\xi_2) d\xi_1 d\xi_2 \right| = 2\pi \left| \langle \phi | O(\xi_2 f(\xi_1,\xi_2)) \psi \rangle_{\mathfrak{h}} \right|$$

$$\leq 2\pi \|\phi\|_{\mathfrak{h}} \|\psi\|_{\mathfrak{h}} \|O(\xi_2 f(\xi_1,\xi_2))\|_{B_2(\mathfrak{h})}$$

$$\leq \sqrt{2\pi} \|\phi\|_{\mathfrak{h}} \|\psi\|_{\mathfrak{h}} \|\xi_2 f(\xi_1,\xi_2)\|_{L_{\mathbb{C}}^2(\mathcal{G}^*; d\xi_1 d\xi_2/|\xi_2|)}$$

$$\leq \sqrt{2\pi} \|\phi\|_{\mathfrak{h}} \|\psi\|_{\mathfrak{h}} \|f\|_{L_{\mathbb{C}}^2(\mathcal{G}^*; \xi_2 d\xi_1 d\xi_2)},$$

and for $x_1, x_2 \in \mathbb{R}$:

$$\left| \int_{\mathbb{R}^2} (x_1 \partial_1 f(\xi_1,\xi_2) + x_2 \partial_2 f(\xi_1,\xi_2)) \overline{W}_{|\phi\rangle\langle\psi|}(\xi_1,\xi_2) d\xi_1 d\xi_2 \right|$$

$$= 2\pi \left| \langle \phi | O(x_1 \xi_2 \partial_1 f(\xi_1,\xi_2) - x_2 \xi_2^2 \partial_2 f(\xi_1,\xi_2)) \psi \rangle_{\mathfrak{h}} \right|$$

$$= 2\pi \left| \langle \phi | [x_1 U(X_1) + x_2 U(X_2), O(f)] \psi \rangle_{\mathfrak{h}} \right|$$

$$\leq \sqrt{2\pi} \|\phi\|_{\mathfrak{h}} \|(x_1 U(X_1) + x_2 U(X_2))\psi\| \|f\|_{L_{\mathbb{C}}^2(\mathcal{G}^*; d\xi_1 d\xi_2/|\xi_2|)}.$$

$\square$

Under the same hypothesis we can show that $\mathbf{1}_{\mathbb{R}\times(-\infty,0)} W_{|\phi\rangle\langle\psi|}$ belongs to the Sobolev space $H_{1,2}^\sigma(\mathbb{R} \times (-\infty, 0))$ which is defined similarly to (12.1). Note that the above result and the presence of $\sigma(\xi_1, \xi_2) = 2\pi |\xi_2|$ are consistent with the integrability properties of the gamma law, *i.e.*, if

$$f(\xi_1, \xi_2) = g(\xi_1)\gamma_\beta(\xi_2), \qquad \xi_1 \in \mathbb{R}, \quad \xi_2 > 0, \quad g \neq 0,$$

then $f \in H^\sigma_{1,2}(\mathbb{R} \times (0, \infty))$ if and only if $\beta > 0$.

12.3 Towards a Hörmander-type theorem

In order to further develop the Malliavin calculus for quantum stochastic processes we need to apply the derivative operator D_h to solutions of quantum stochastic differential equations with the aim to find sufficient conditions for their regularity. The goal, which remains to be achieved, would be to prove a Hörmander type theorem for quantum stochastic processes. In this section we sketch an approach that could lead towards such a result.

Let $\mathfrak{h}$ be a Hilbert space carrying a representation $\{P, Q\}$ of the canonical commutation relations and let Φ be a state on $B(\mathfrak{h})$. Recall that the Wigner function W_Φ of (P, Q) in the state Φ satisfies

$$\int_{\mathbb{R}^2} f(u, v) W(u, v) du dv = \Phi(O_{P,Q}(f)),$$

see Definition 7.3.3. Consider the Fock space $\Gamma_s(L^2(\mathbb{R}_+, \ell_2))$ and a unitary cocycle $(U_t)_{t \geq 0}$ on $\mathfrak{h} \otimes \Gamma_s(L^2(\mathbb{R}_+, \ell_2))$ given as the solution of the quantum stochastic differential equation

$$dU_t = \left(\sum_{k \in \mathbb{N}} R_k da_t^+(e_k) + \sum_{k \in \mathbb{N}} (S_{k,l} - \delta_{k,l}) da_t^\circ(E_{k,l}) \right.$$

$$\left. - \sum_{k,l \in \mathbb{N}} R_k^* S_{k,l} da_t^-(e_l) + K dt \right) U(t),$$

with initial condition $U_0 = \mathbf{1}$, where $(e_n)_{n \in \mathbb{N}}$ is an orthonormal basis of ℓ^2 and $E_{k,l} \in B(\ell^2)$ denotes the operator given by

$$E_{k,l} e_j = \begin{cases} e_k, & \text{if } l = j, \\ 0, & \text{otherwise.} \end{cases}$$

For $(U_t)_{t \geq 0}$ to be unitary, the coefficients $(S_{k,l})_{k,l \in \mathbb{N}}$, $(R_k)_{k \in \mathbb{N}}$, which are operators on $\mathfrak{h}$, should be such that $(S_{k,l})_{k,l \in \mathbb{N}} \in B(\mathfrak{h} \otimes \ell^2)$ is unitary and K can be written as

$$K = -iH - \frac{1}{2} \sum_{k \in \mathbb{N}} R_k^* R_k,$$

where H is a Hermitian operator, see, *e.g.*, [87, Theorem 26.3]. The operators Q_t and P_t defined by

$$P_t := U_t^* P \otimes 1U_t, \qquad Q_t := U_t^* Q \otimes 1U_t,$$

satisfy a quantum stochastic differential equation of the form

$$dX_t = \mathcal{L}(X)_t dt + \sum_{k \in \mathbb{N}} \mathcal{R}_k(X)_t da_t^+(e_k)$$

$$+ \sum_{k \in \mathbb{N}} \left(\mathcal{R}_k(X^*)^*\right)_t da_t^-(e_k) + \sum_{k,l \in \mathbb{N}} \mathcal{S}_{k,l}(X)_t da_t^\circ(E_{k,l})$$

with initial condition $X_0 = X \otimes 1$, where

$$\begin{cases} \mathcal{L}(X) = i[H, X] - \dfrac{1}{2} \sum_k \left(R_k^* R_k X + X R_k^* R_k - 2 R_k^* X R_k\right), \\[2mm] \mathcal{L}(X)_t = j_t\left(\mathcal{L}(X)\right) = U_t^* \mathcal{L}(X) U_t, \\[2mm] \mathcal{R}_k(X) = \sum_l S_{lk}^*[X, R_l], \\[2mm] \mathcal{R}_k(X)_t = U_t^* \mathcal{R}_k(X) U_t, \\[2mm] \mathcal{S}_{kl}(X) = \sum_j S_{jk}^* X S_{jl} - \delta_{kl} X, \\[2mm] \mathcal{S}_{kl}(X)_t = U_t^* \mathcal{S}_{kl}(X) U_t. \end{cases}$$

The operators $O_{P_t, Q_t}(f)$ obtained from P_t and Q_t by the Weyl calculus satisfy the same type of quantum stochastic differential equation. To begin, we consider the simpler case

$$dU_t = \left(R da_t^+ dA_t - \frac{1}{2} R^* R dt\right) U_t, \tag{12.2}$$

without the conservation part and with only one degree of freedom, *i.e.*, $\mathfrak{h} = \mathbb{C}$. Then we have

$$dX_t = U_t^* \left(R^* XR - \frac{1}{2} R^* RX - \frac{1}{2} XR^* R\right) U_t dt + U_t^*[X, R] U_t da_t^+$$

$$+ U_t^*[R^*, X] U_t da_t^-$$

$$= U_t^* \left(\left(R^* XR - \frac{1}{2} R^* RX - \frac{1}{2} XR^* R\right) dt + [X, R] da_t^+ + [R^*, X] da_t^-\right) U_t$$

$$\tag{12.3}$$

Next we investigate the differentiation of the solution of a QSDE, in the following two steps.

12.3.1 Derivative of a quantum stochastic integral

In Definition 11.3.1, we introduced the derivation operator D_h for $h = (h_1, h_2) \in L^2(\mathbb{R}_+, \mathbb{C}^2)$ by

$$D_h M = \frac{i}{2} \big[a^-(\overline{h_1 - ih_2}) + a^+(h_1 + ih_2), M \big].$$

Let

$$M_t = \int_0^t F_s \, da_s^-, \quad N_t = \int_0^t G_s \, da_s^+,$$

and suppose that X_t, Y_t, and F_s, G_s are in the domain of D_h, F_s, G_s furthermore adapted. By the quantum Itô formula we get

$$\begin{cases}
a^-(h\mathbf{1}_{[0,t]})M_t = \displaystyle\int_0^t \overline{h(s)} X_s \, da_s^- + \int_0^t a^-(h\mathbf{1}_{[0,s]}) F_s \, da_s^-, \\[2mm]
a^+(h\mathbf{1}_{[0,t]})M_t = \displaystyle\int_0^t h(s) X_s \, da_s^+ + \int_0^t a^+(h\mathbf{1}_{[0,s]}) F_s \, da_s^-, \\[2mm]
a^-(h\mathbf{1}_{[0,t]})N_t = \displaystyle\int_0^t \overline{h(s)} Y_s \, da_s^- + \int_0^t a^-(h\mathbf{1}_{[0,s]}) G_s \, da_s^+ + \int_0^t h(s) G_s \, ds, \\[2mm]
a^+(h\mathbf{1}_{[0,t]})N_t = \displaystyle\int_0^t h(s) Y_s \, da_s^+ + \int_0^t a^+(h\mathbf{1}_{[0,s]}) F_s \, da_s^+,
\end{cases}$$

and similar formulas hold for the products $M_t a^-(h\mathbf{1}_{[0,t]})$, $M_t a^+(h\mathbf{1}_{[0,t]})$, $N_t a^-(h\mathbf{1}_{[0,t]})$, and $N_t a^+(h\mathbf{1}_{[0,t]})$. Therefore, we have

$$\begin{cases}
[a^-(h), M_t] = \displaystyle\int_0^t [a^-(h\mathbf{1}_{[0,s]}), F_s] \, da_s^- = \int_0^t [a^-(h), F_s] \, da_s^-, \\[2mm]
[a^+(h), M_t] = \displaystyle\int_0^t [a^+(h), F_s] \, da_s^- - \int_0^t h(s) F_s \, ds, \\[2mm]
[a^-(h), N_t] = \displaystyle\int_0^t [a^-(h), G_s] \, da_s^+ + \int_0^t \overline{h(s)} G_s \, ds, \\[2mm]
[a^+(h), N_t] = \displaystyle\int_0^t [a^-(h), G_s] \, da_s^+.
\end{cases}$$

Combining, these formulas, we get the following expressions for the derivatives of quantum stochastic integrals,

$$D_h M_t = \frac{i}{2} \big[a^-(\overline{h_1 - ih_2}) + a^+(h_1 + ih_2), M_t \big]$$

$$= \int D_h F_s \, da_s^- - \frac{i}{2} \int_0^t \big(h_1(s) + ih_s(s) \big) F_s \, ds,$$

and

$$D_h N_t = \frac{i}{2} \left[a^- (\overline{h_1 - ih_2}) + a^+ (h_1 + ih_2), N_t \right]$$

$$= \int D_h G_s da_s^+ + \frac{i}{2} \int_0^t \left(h_1(s) - ih_2(s) \right) G_s ds.$$

Time integrals commute with the derivation operator, *i.e.*, we have

$$D_h \int_0^t M_s ds = \int_0^t D_h M_s ds.$$

12.3.2 Derivative of the solution

Let $(U_t)_{t \in \mathbb{R}_+}$ be a solution of Equation (12.2), then we get

$$D_h U_t = \int_0^t \left(R da_s^+ - R^* da_s^- - \frac{1}{2} R^* R ds \right) D_h U_s$$

$$+ \frac{i}{2} \int_0^t \left(h_1(s) + ih_2(s) \right) R^* U_s ds + \frac{i}{2} \int_0^t \left(h_1(s) - ih_2(s) \right) R U_s ds$$

$$= \int_0^t \left(R da_s^+ - R^* da_s^- - \frac{1}{2} R^* R ds \right) D_h U_s + \int_0^t \tilde{R}_s U_s ds,$$

where

$$\tilde{R}_s = \frac{i}{2} \left(h_1(s) + ih_2(s) \right) R^* + \frac{i}{2} \left(h_1(s) - ih_2(s) \right) R.$$

Similarly, we have

$$D_h U_t^* = \int_0^t D_h U_s^* \left(R^* da_s^- - R da_s^+ - \frac{1}{2} R^* R ds \right)$$

$$- \frac{i}{2} \int_0^t U_s^* R^* \left(h_1(s) + ih_2(s) \right) ds - \frac{i}{2} \int_0^t U_s^* R \left(h_1(s) - ih_2(s) \right) ds$$

$$= \int_0^t D_h U_s^* \left(R^* da_s^- - R da_s^+ - \frac{1}{2} R^* R ds \right) - \int_0^t U_s^* \tilde{R}_s ds.$$

Finally, using (12.3), we get

$$D_h \left(j_t(X) \right) = D_h (U_t^* X \otimes 1 U_t) = (D_h U_t^*) X \otimes 1 U_t + U_t^* X \otimes 1 (D_h U_t)$$

$$= \int_0^t (D_h U_s^*) \left(\left(R^* X R - \frac{1}{2} R^* R X - \frac{1}{2} X R^* R \right) ds \right.$$

$$\left. + [X, R] da_s^+ + [R^*, X] da_s^- \right) U_s$$

$$+ \int_0^t U_s^* \left(\left(R^* X R - \frac{1}{2} R^* R X - \frac{1}{2} X R^* R \right) dt \right.$$

$$\left. + [X, R] da_t^+ + [R^*, X] da_s^- \right) (D_h U_s)$$

$$+ \frac{i}{2} \int_0^t \left(h_1(s) - i h_2(s) \right) U_s^* [X, R] U_s ds$$

$$- \frac{i}{2} \int_0^t \left(h_1(s) + i h_2(s) \right) U_s^* [R^*, X] U_s ds$$

$$= \int_0^t D_h\big(j_s(\mathcal{L}(X))\big) ds + \int_0^t D_h\big(j_t(\mathcal{R}(X))\big) da_s^+$$

$$+ \int_0^t D_h\big(j_t(\mathcal{R}(X^*)^*)\big) da_s^- - \int_0^t j_t([\tilde{R}_s, X]) ds.$$

i.e., the "flow" $D_j \circ j_t$ satisfies an equation similar to that of j_t, but with an additional (inhomogenous) term $\int_0^t j_t([\tilde{R}_s, X]) ds$. j_t is homomorphic, but $D_j \circ j_t$ will not be homomorphic in general.

12.3.3 The other flow

Let us also define[1]

$$k_t(X) = U_t \mathbf{1} \otimes X U_t^*,$$

which satisfies the quantum stochastic differential equation

$$k_t(X) = X \otimes 1 + \int_0^t [R, k_s(X)] da_s^+ + \int_0^t [k_s(X), R^*] da_s^-$$

$$+ \int_0^t \left(R^* k_s(X) R - \frac{1}{2} R^* R k_s(X) - \frac{1}{2} k_s(X) R^* R \right) ds$$

as can be shown using the quantum Itô formula and the quantum stochastic differential equations satisfied by U_t and U_t^*. Similarly, $D_h(k_t))$ satisfies the quantum stochastic differential equation

$$D_h\big(k_t(X)\big) = (D_h U_t) X \otimes 1 U_t^* + U_t X \otimes 1 (D_h U_t^*)$$

$$= \int_0^t [R, D_h\big(k_s(X)\big)] da_s^+ + \int_0^t [(D_h k_s(X)), R^*] da_s^-$$

$$+ \int_0^t \left(R^* (D_h k_s(X)) R - \frac{1}{2} R^* R(D_h k_s(X)) - \frac{1}{2}(D_h k_s(X)) R^* R \right) ds$$

$$+ \int_0^t [\tilde{R}_s, (D_h k_s(X))] ds.$$

---

[1] $X \mapsto U_t^* X U_t$ defines an automorphism on $B(\mathfrak{h} \otimes \Gamma_s(L^2(\mathbb{R}_+, \ell_2)))$ with inverse $X \mapsto U_t X U_t^*$. j_t is the restriction of this map to $B(\mathfrak{h}) \otimes \mathbf{1}$, k_t is the restriction of the inverse.

We introduce the shorter notation $Y_t = k_t(X)$, then we have $D_h Y_0 = 0$ and

$$D_h Y_t = \int_0^t \left(R^*(D_h Y_s)R - \frac{1}{2}R^*R(D_h Y_s) - \frac{1}{2}(D_h Y_s)R^*R \right) ds$$

$$+ \int_0^t [R, D_h Y_s]da_s^+ + \int_0^t [D_h Y_s, R^*]da_s^-$$

$$+ \frac{i}{2}\int_0^t \left((h_1(s) - ih_s(s))[Y_s, R] - (h_1(s) + ih_2(s))[R^*, Y_s] \right) ds.$$

The last term is $\int_0^t [\tilde{R}_s, Y_s]ds$, where

$$\tilde{R}_s = \frac{i}{2}h_1(s)(R - R^*) + \frac{1}{2}h_2(s)(R + R^*).$$

We see that $D_h Y_t$ satisfies an inhomogeneous quantum stochastic differential equation, where the inhomogeneity is a function of Y_t. The homogeneous part is the same as for Y_t. We try a variation of constants, *i.e.*, we assume that the solution has the form

$$D_h Y_t = U_t Z_t U_t^*,$$

since the solutions of the homogeneous equation are of the form $U_t Z U_t^*$ (at least for initial conditions acting only on the initial space). For Z_t we make the Ansatz

$$Z_t = \int_0^t F_s da_s^+ + \int_0^t G_s ds + \int_0^t H_s ds$$

with some adapted coefficients F_t, G_t, and H_t. Then the Itô formula yields

$$D_h Y_t = \int_0^t \left(R^* U_s Z_s U_s^* R - \frac{1}{2}R^*R U_s Z_s U_s^* - \frac{1}{2}U_s Z_s U_s^* R^*R \right) ds$$

$$+ \int_0^t [U_s Z_s U_s^*, R]da_s^+ + \int_0^t [R^*, U_s Z_s U_s^*]da_s^-$$

$$+ \int_0^t U_s dZ U_s^* - \int_0^t U_s G_s U_s^* R ds - \int_0^t R^* U_s F_s U_s^* ds.$$

Comparing this equation with the previous equation for $D_h Y_t$, we get

$$\int_0^t U_s dZ_s U_s^* - \int_0^t U_s G_s U_s^* R ds - \int_0^t R^* U_s F_s U_s^* ds = \int_0^t [\tilde{R}, Y_s]ds.$$

Uniqueness of the integral representation of Z implies $F_s = G_s = 0$ (since $\int_0^t U_s dZ_s U_s^* = \int_0^t U_s F_s da_s^+ U_s^* + \int_0^t U_s G_s da_s^- U_s^* + \int_0^t U_s H_s ds U_s^*$, but there are no creation or annihilation integrals on the right-hand side) and

$$\int_0^t U_s H_s ds U_s^* = \int_0^t [\tilde{R}, Y_s]ds, \quad i.e., \quad H_s = U_s^*[\tilde{R}, Y_s]U_s,$$

$0 \le s \le t$. Recalling $Y_t = U_t(X \otimes 1)U_t^*$, we can also rewrite the above as

$$H_s = [\mathcal{R}_s, X \otimes 1], \quad \text{with} \quad \mathcal{R}_s = U_s^* \tilde{R} U_s.$$

Thus we have

$$Z_t = \int_0^t [\mathcal{R}_s, X \otimes 1]ds, \quad \text{and} \quad D_h X_t = U_t \int_0^t [\mathcal{R}_s, X \otimes 1]ds U_t^*.$$

The next step is to take $Y_t = U_t^* O_{P,Q}(f)U_t = O_{P_t,Q_t}(f)$ and to find an expression for $D_h X_t$ involving derivatives of f.

Exercises

Exercise 12.1 Relation to the commutative case. Let

$$
\begin{cases}
Q = B^- + B^+ = \dfrac{1}{2}((a^-_x)^2 + (a^+_x)^2) = \dfrac{P^2 - Q^2}{4}, \\[2ex]
P = i(B^- - B^+) = \dfrac{i}{2}((a^-_x)^2 - (a^+_x)^2) = \dfrac{PQ + QP}{4}.
\end{cases}
$$

1. Show that we have

$$[P,Q] = 2iM, \qquad [P,M] = 2iQ, \qquad [Q,M] = -2iP.$$

2. Show that

$$Q + M = B^- + B^+ + M = \frac{P^2}{2}, \quad Q - M = B^- + B^+ - M = -\frac{Q^2}{2},$$

   i.e., $Q + M$ and $M - Q$ have gamma laws.
3. Give the probability law of $Q + M$ and $Q - M$.
4. Give the probability law of $Q + \alpha M$ when $|\alpha| < 1$ and $|\alpha| > 1$.
5. Find the classical analogues of the integration by parts formula (2.1) written as

$$\mathbf{E}[D_{(1,0)}F] = \frac{1}{2}\mathbf{E}\left[\left\{\frac{P^2}{2}, F\right\} - F\right],$$

   for $\alpha = 1$, and

$$\mathbf{E}[D_{(1,0)}F] = \frac{1}{2}\mathbf{E}\left[F - \left\{\frac{Q^2}{2}, F\right\}\right],$$

   for $\alpha = -1$.

Appendix

I was born not knowing and have had only a little time to change that here and there.

(R.P. Feynman)

This appendix gathers some background and complements on orthogonal polynomials, moments and cumulants, the Fourier transform, adjoint action on Lie algebras, nets, closability of linear operators, and tensor products.

A.1 Polynomials

A.1.1 General idea

Consider a family $(P_n)_{n\in\mathbb{N}}$ of polynomials satisfying the orthogonality relation

$$\int_{-\infty}^{\infty} P_n(x)P_k(x)f(x)\mu(dx) = 0, \qquad n \neq k,$$

with respect to a measure μ on $\mathbb{R}$.

A.1.2 Finite support

We first consider the case where μ is supported on a finite set. If the measure μ is supported on n points, then $L^2_{\mathbb{C}}(\mathbb{R}, \mu)$ has dimension n. If μ is supported on n points $x_1, \ldots, x_n$, then the monomials $1, x, \ldots, x^{n-1}$ correspond to the vectors

$$\begin{pmatrix} 1 \\ 1 \\ \vdots \\ 1 \end{pmatrix}, \begin{pmatrix} x_1 \\ x_2 \\ \vdots \\ x_n \end{pmatrix}, \begin{pmatrix} x_1^2 \\ x_2^2 \\ \vdots \\ x_n^2 \end{pmatrix}, \ldots, \begin{pmatrix} x_1^{n-1} \\ x_2^{n-1} \\ \vdots \\ x_n^{n-1} \end{pmatrix},$$

and the Vandermonde determinant formula

$$\det \begin{bmatrix} 1 & x_1 & x_1^2 & \cdots & x_1^{n-1} \\ 1 & x_2 & x_2^2 & \cdots & x_2^{n-1} \\ \vdots & \vdots & \vdots & \vdots & \vdots \\ 1 & x_n & x_n^2 & \cdots & x_n^{n-1} \end{bmatrix} = \prod_{1 \leq j < k \leq n} (x_j - x_k) \neq 0$$

shows that these vectors are linearly independent. In this case the monomials $1, \ldots, x^{n-1}$ are linearly independent and they form a basis of $L_{\mathbb{C}}^2(\mathbb{R}, \mu)$. By Gram-Schmidt orthogonalisation we can transform the basis $1, x, \ldots, x^{n-1}$ into an orthonormal basis $P_0, \ldots, P_{n-1}$ by letting $P_0 = 1$, and then recursively

$$\tilde{P}_n := x^n - \sum_{k=0}^{n-1} \langle P_k(x), x^n \rangle P_k,$$

for $n \geq 1$, with the normalisation

$$P_n = \frac{\tilde{P}_n}{\sqrt{\langle \tilde{P}_n, \tilde{P}_n \rangle}},$$

where $\langle \cdot, \cdot \rangle$ denotes the inner product of $L_{\mathbb{C}}^2(\mathbb{R}, \mu)$, i.e.,

$$\langle f, g \rangle = \int_{\mathbb{R}} \bar{f}(x) g(x) \mu(dx).$$

Since $L_{\mathbb{C}}^2(\mathbb{R}, \mu)$ has dimension n, it follows that the monomials x^m with $m \geq n$ are linear combinations of $P_0, \ldots, P_{n-1}$. Therefore, we get

$$P_k = 0, \qquad k \geq n.$$

Example A.1.1 Consider $\mu = p\delta_{x_1} + q\delta_{x_2}$ with $p, q > 0$ such that $p + q = 1$ and $x_1, x_2 \in \mathbb{R}$. Here we get $P_0 = 1$ and

$$\begin{cases} \tilde{P}_1 = x - \langle 1, x \rangle = x - px_1 - qx_2, \\ \langle \tilde{P}_1, \tilde{P}_1 \rangle = \int_{-\infty}^{\infty} (x - px_1 - qx_2)^2 \mu(dx), \end{cases}$$

so that $P_0 = 1$ and P_1 form a basis of $L_{\mathbb{C}}^2(\mathbb{R}, \mu)$. This recover the example of 2×2 matrices.

A.1.3 Infinite support

Let us now suppose that μ is not supported on a finite set, in which case we build an orthonormal family $(P_n)_{n \in \mathbb{N}}$ of polynomials. Assuming that the nth polynomial P_n has degree n, $n \in \mathbb{N}$, the polynomials $P_0, \ldots, P_n$ form a basis for the space of polynomials of degree not higher than n. It can be shown, cf. Theorems 4.1 and 4.4 in [27], that a family $(P_n)_{n \in \mathbb{N}}$ of polymials such that $\deg(P_n) = n$ is orthogonal with respect to some measure μ on $\mathbb{R}$ if and only if there exist sequences $(\alpha_n)_{n \in \mathbb{N}}$, $(\beta_n)_{n \in \mathbb{N}}$ such that $(P_n)_{n \in \mathbb{N}}$ satisfies a three-term recurrence relation of the form

$$x P_n(x) = P_{n+1}(x) + \alpha_n P_n(x) + \beta_n P_{n-1}(x), \quad n \geq 1.$$

As an important particular case we have

$$x P_n(x) = P_{n+1}(x) + (\alpha + \beta) n P_n(x) + n(t + \alpha \beta (n - 1)) P_{n-1}(x),$$

for the Meixner polynomials (1934). When $\alpha = \beta = t = 1$ we have

$$a P_n(x) = P_{n+1}(x) + 2n P_n(x) + n^2 P_{n-1}(x),$$

and we find the Laguerre polynomials (1879). When $\beta = 0$ and $\alpha = t = 1$ we have

$$x P_n(x) = P_{n+1}(x) + n P_n(x) + n P_{n-1}(x),$$

and we find the Charlier polynomials that can be associated to numbers of partitions without singletons, cf. [11]. When $\alpha = 0$, $\beta = 0$, $t = 1$ we find

$$x P_n(x) = P_{n+1}(x) + n P_{n-1}(x),$$

which yields the Hermite polynomials that correspond to numbers of pairings, cf. [11]. Let now μ be a probability measure on $\mathbb{R}$ whose moments of all orders are finite, *i.e.*,

$$\int_{-\infty}^{\infty} |x|^n \mu(dx) < \infty, \quad n \in \mathbb{N}.$$

The Legendre polynomials are associated with μ the uniform distribution, and this generalises to the family of Gegenbauer polynomials (or ultraspherical polynomials) in case μ is the measure with density $(1 - x^2)^{\alpha - 1/2} \mathbf{1}_{[-1,1]}$ with respect to the Lebesgue measure, $\alpha > -1/2$. Important special cases include the arcsine, uniform, Wigner's semicircle distributions. The Jacobi polynomials *vs.* the beta distribution constitute another generalisation.

Next we review in detail some important particular cases.

A.1.3.1 Hermite polynomials

Definition A.1.2 *The Hermite polynomial $H_n(x; \sigma^2)$ of degree $n \in \mathbb{N}$ and parameter $\sigma^2 > 0$ is defined with*

$$H_0(x; \sigma^2) = 1, \quad H_1(x; \sigma^2) = x, \quad H_2(x; \sigma^2) = x^2 - \sigma^2,$$

and more generally from the recurrence relation

$$H_{n+1}(x; \sigma^2) = x H_n(x; \sigma^2) - n\sigma^2 H_{n-1}(x; \sigma^2), \qquad n \geq 1. \qquad \text{(A.1)}$$

In particular we have

$$H_n(x; 0) = x^n, \qquad n \in \mathbb{N}.$$

The generating function of Hermite polynomials is defined as

$$\psi_\lambda(x, \sigma^2) = \sum_{n=0}^{\infty} \frac{\lambda^n}{n!} H_n(x; \sigma^2), \quad \lambda \in (-1, 1).$$

Proposition A.1.3   *The following statements hold on the Hermite polynomials:*

  i) Generating function:

$$\psi_\lambda(x, \sigma^2) = e^{\lambda x - \lambda^2 \sigma^2/2}, \quad x, \lambda \in \mathbb{R}.$$

ii) Derivation rule:

$$\frac{\partial H_n}{\partial x}(x; \sigma^2) = n H_{n-1}(x; \sigma^2),$$

iii) Creation rule:

$$H_{n+1}(x; \sigma^2) = \left(x - \sigma^2 \frac{\partial}{\partial x} \right) H_n(x; \sigma^2).$$

Proof: The recurrence relation (A.1) shows that the generating function ψ_λ satisfies the differential equation

$$\begin{cases} \dfrac{\partial \psi_\lambda}{\partial \lambda}(x, \sigma^2) = (x - \lambda \sigma^2)\psi_\lambda(x, \sigma^2), \\[2mm] \psi_0(x, \sigma^2) = 1, \end{cases}$$

which proves (*i*). From the expression of the generating function we deduce (*ii*), and by rewriting (A.1) we obtain (*iii*). $\square$

Next is the orthonormality properties of the Hermite polynomials with respect to the Gaussian density:

$$\int_{-\infty}^{\infty} H_n(x; \sigma^2) H_m(x; \sigma^2) e^{-x^2/(2\sigma^2)} \frac{dx}{\sqrt{2\pi\sigma^2}} = \mathbf{1}_{\{n=m\}} n! \sigma^{2n}.$$

We have

$$\frac{\partial H_n}{\partial x}(x; \sigma^2) = n H_{n-1}(x; \sigma^2),$$

and the partial differential equation

$$\frac{\partial H_n}{\partial s}(x;s) = -\frac{1}{2}\frac{\partial^2 H_n}{\partial x^2}(x;s),$$

i.e, the heat equation with initial condition

$$H_n(x;0) = x^n, \qquad x \in \mathbb{R}, \quad n \in \mathbb{N}.$$

A.1.3.2 Poisson–Charlier polynomials

Definition A.1.4 *Let the Charlier polynomial of order $n \in \mathbb{N}$ and parameter $\lambda \geq 0$ be defined by*

$$C_0(k,\lambda) = 1, \qquad C_1(k,\lambda) = k - \lambda, \qquad k \in \mathbb{R}, \quad \lambda \in \mathbb{R}_+,$$

and the recurrence relation

$$C_{n+1}(k,\lambda) = (k - n - \lambda)C_n(k,\lambda) - n\lambda C_{n-1}(k,\lambda), \qquad n \geq 1.$$

Let

$$p_k(\lambda) = \frac{\lambda^k}{k!}e^{-\lambda}, \qquad k \in \mathbb{N}, \quad \lambda \in \mathbb{R}_+,$$

denote the Poisson probability density, which satisfies the finite difference differential equation

$$\frac{\partial p_k}{\partial \lambda}(\lambda) = -\Delta p_k(\lambda), \tag{A.2}$$

where Δ is the difference operator

$$\Delta f(k) := f(k) - f(k-1), \qquad k \in \mathbb{N}.$$

Let also

$$\psi_\lambda(k,t) = \sum_{n=0}^{\infty} \frac{t^n}{n!}C_n(k,\lambda), \qquad \lambda \in (-1,1),$$

denote the generating function of Charlier polynomials.

Proposition A.1.5 *For all $k \in \mathbb{Z}$ and $\lambda \in \mathbb{R}_+$ we have the relations*

$$\begin{cases} C_n(k,\lambda) = \dfrac{\lambda^n}{p_k(\lambda)}\dfrac{\partial^n p_k}{\partial \lambda^n}(\lambda), & \text{(A.3a)} \\[3mm] C_n(k+1,\lambda) - C_n(k,\lambda) = nC_{n-1}(k,\lambda), & \text{(A.3b)} \\[3mm] C_{n+1}(k,\lambda) = kC_n(k-1,\lambda) - \lambda C_n(k,\lambda), & \text{(A.3c)} \end{cases}$$

and the generating function $\psi_\lambda(k,t)$ *satisfies*

$$\psi_\lambda(k,t) = e^{-\lambda k}(1+t)^k, \tag{A.4}$$

$\lambda, t > 0$, $k \in \mathbb{N}$.

Proof: Relation (A.3c) follows from (A.3a) and (A.2) as

$$C_{n+1}(k,\lambda) = \frac{\lambda^{n+1}}{p_k(\lambda)} \frac{\partial^{n+1} p_k}{\partial \lambda^{n+1}}(\lambda)$$

$$= -\frac{\lambda^{n+1}}{p_k(\lambda)} \frac{\partial^n p_k}{\partial \lambda^n}(\lambda) + \frac{\lambda^{n+1}}{p_k(\lambda)} \frac{\partial^n p_{k-1}}{\partial \lambda^n}(\lambda)$$

$$= -\lambda \frac{\lambda^n}{p_k(\lambda)} \frac{\partial^n p_k}{\partial \lambda^n}(\lambda) + k \frac{\lambda^n}{p_{k-1}(\lambda)} \frac{\partial^n p_{k-1}}{\partial \lambda^n}(\lambda)$$

$$= -\lambda C_n(k,\lambda) + k C_n(k-1,\lambda).$$

Finally, using Relation (A.3c) we have

$$\frac{\partial \psi_\lambda}{\partial t}(k,t) = \sum_{n=1}^{\infty} \frac{t^{n-1}}{(n-1)!} C_n(k,\lambda)$$

$$= \sum_{n=0}^{\infty} \frac{t^n}{n!} C_{n+1}(k,\lambda)$$

$$= -\lambda \sum_{n=0}^{\infty} \frac{t^n}{n!} C_n(k-1,\lambda) + k \sum_{n=0}^{\infty} \frac{t^n}{n!} C_n(k,\lambda)$$

$$= -\lambda \sum_{n=0}^{\infty} \frac{t^n}{n!} C_n(k-1,\lambda) + k \sum_{n=0}^{\infty} \frac{t^n}{n!} C_n(k,\lambda)$$

$$= -\lambda \psi_\lambda(k,t) + k \psi_\lambda(k-1,t),$$

$\lambda \in (-1,1)$, hence the generating function $\psi_\lambda(k,t)$ satisfies the differential equation

$$\frac{\partial \psi_\lambda}{\partial t}(k,t) = -\lambda \psi_\lambda(k,t) + k \psi_\lambda(k-1,t), \quad \psi_0(k,t) = 1, \quad k \geq 1,$$

which yields (A.4) by induction on k. $\square$

We also have

$$\frac{\partial^k p_k}{\partial \lambda^k}(\lambda) = (-\Delta)^k p_k(\lambda).$$

We also have the orthogonality properties of the Charlier polynomials with respect to the Poisson distribution with respect to the inner product

$$\langle u, v \rangle := \sum_{k=0}^{\infty} p_k(\lambda) u(k) v(k) = e^{-\lambda} \sum_{k=0}^{\infty} \frac{\lambda^k}{k!} u(k) v(k),$$

with $\lambda = \sigma(A)$, *i.e.*,

$$\langle C_n(\cdot, \lambda), C_m(\cdot, \lambda) \rangle_{\ell^2(\mathbb{N}, p \cdot (\lambda))} = e^{-\lambda} \sum_{k=0}^{\infty} \frac{\lambda^k}{k!} C_n(k, \lambda) C_m(k, \lambda)$$

$$= n! \lambda^n \delta_{n,m}.$$

The exponential vector

$$\sum_{n=0}^{\infty} \frac{\lambda^n}{n!} C_n(\omega(A), \sigma(A)) = e^{-\lambda \sigma(A)} (1 + \lambda)^{\omega(A)}$$

$$= \psi_\lambda(\omega(A), \sigma(A)).$$

A.1.3.3 Meixner polynomials

In this case we have the recurrence relation

$$(n + 1) P_{n+1} + (2\beta n + \beta m_0 - x) P_n + (n + m_0 - 1) P_{n-1} = 0,$$

with initial condition $P_{-1} = 0$, $P_1 = 1$, for the rescaled polynomials

$$P_n = \prod_{k=1}^{n} \sqrt{\frac{k}{k + m_0}} p_n.$$

According to the value of β we have to distinguish three cases.

1. $|\beta| = 1$: In this case we have, up to rescaling, *Laguerre polynomials*, *i.e.*,

$$P_n(x) = (-\beta)^n L_n^{(m_0 - 1)}(\beta x)$$

where the Laguerre polynomials $L_n^{(\alpha)}$ are defined as in [63, Equation (1.11.1)]. The measure μ can be obtained by normalising the measure of orthogonality of the Laguerre polynomials, and it is equal to

$$\mu(dx) = \frac{|x|^{m_0 - 1}}{\Gamma(m_0)} e^{-\beta x} \mathbf{1}_{\beta \mathbb{R}_+} dx.$$

If $\beta = +1$, then this measure is, up to a normalisation parameter, the usual χ^2-distribution (with parameter m_0) of probability theory.

2. $|\beta| < 1$: In this case we find the *Meixner-Pollaczek polynomials* after rescaling,

$$P_n(x) = P_n^{(m_0/2)}\left(\frac{x}{2\sqrt{1-\beta^2}}; \pi - \arccos\beta\right).$$

For the definition of these polynomials see, *e.g.*, [63, Equation (1.7.1)]. For the measure μ we get

$$\mu(dx) = C\exp\left(\frac{(\pi - 2\arccos\beta)x}{2\sqrt{1-\beta^2}}\right)\left|\Gamma\left(\frac{m_0}{2} + \frac{ix}{2\sqrt{1-\beta^2}}\right)\right|^2 dx,$$

where C has to be chosen such that μ is a probability measure.

3. $|\beta| > 1$: In this case we get *Meixner polynomials*

$$P_n(x) = (-c\,\mathrm{sgn}\beta)^n M_n\left(\frac{x\mathrm{sgn}\beta}{1/c - c} - \frac{m_0}{2}; m_0; c^2\right)\prod_{k=1}^{n}\frac{k + m_0 - 1}{k}$$

after rescaling, where $c = |\beta| - \sqrt{\beta^2 - 1}$.

The definition of these polynomials can be found, *e.g.*, in [63, Equation (1.9.1)]. The density μ is again the measure of orthogonality of the polynomials P_n (normalised to a probability measure). We therefore get

$$\mu = C\sum_{n=0}^{\infty}\frac{c^{2n}}{n!}(m_0)_n\delta_{x_n},$$

where

$$x_n = \left(n + \frac{m_0}{2}\right)\left(\frac{1}{c} - c\right)\mathrm{sgn}\beta, \qquad n \in \mathbb{N},$$

and

$$C^{-1} = \sum_{n=0}^{\infty}\frac{c^{2n}}{n!}(m_0)_n = \frac{1}{(1 - c^2)^{m_0}}.$$

Here,

$$(m_0)_n = m_0(m_0 + 1)\cdots(m_0 + n - 1)$$

denotes the Pochhammer symbol.

A.2 Moments and cumulants

In this section we provide some combinatorial background on the relationships between the moments and cumulants of random variables and we refer the reader to [89] and [90] for more information.

The cumulants $(\kappa_n^X)_{n\geq 1}$ of a random variable X have been defined in [110] and were originally called the "semi-invariants" of X due to the property $\kappa_n^{X+Y} = \kappa_n^X + \kappa_n^Y$, $n \geq 1$, when X and Y are independent random variables. Precisely, given the moment generating function

$$\mathbf{E}[e^{tX}] = \sum_{n=0}^{\infty} \frac{t^n}{n!} \, \mathbf{E}[X^n], \qquad t \in \mathbb{R},$$

of a random variable X, the *cumulants* of X are the coefficients $(\kappa_n^X)_{n\geq 1}$ appearing in the series expansion of the logarithmic moment generating function of X, *i.e.*, we have

$$\log(\mathbf{E}[e^{tX}]) = \sum_{n=1}^{\infty} \kappa_n^X \frac{t^n}{n!}, \qquad t \in \mathbb{R}.$$

Given $j_1, \ldots, j_k, n \in \mathbb{N}$ such that $j_1 + \cdots + j_k = n$, recall the definition of the *multinomial coefficient*

$$\binom{n}{j_1, \ldots, j_k} = \frac{n!}{j_1! \cdots j_k!}.$$

In addition to the multinomial identity

$$(x_1 + \cdots + x_k)^n = \sum_{\substack{j_1,\ldots,j_k \geq 0 \\ j_1+\cdots+j_k=n}} \binom{n}{j_1, \ldots, j_k} x_1^{j_1} \cdots x_k^{j_k}, \qquad n \in \mathbb{N}, \quad \text{(A.5)}$$

we note the combinatorial identity

$$\left(\sum_{n=1}^{\infty} x_n \right)^k = \sum_{n=k}^{\infty} \sum_{\substack{d_1+\cdots+d_k=n \\ d_1 \geq 1,\ldots,d_k \geq 1}} x_{d_1} \cdots x_{d_k}.$$

This expression translates into the classical identity

$$E[X^n] = \sum_{a=1}^{n} \sum_{P_1 \cup \cdots \cup P_a = \{1,\ldots,n\}} \kappa_{|P_1|}^X \cdots \kappa_{|P_a|}^X, \qquad \text{(A.6)}$$

based on the Faà di Bruno formula, links the moments $\mathbb{E}[X^n]$ of a random variable X with its cumulants $(\kappa_n^X)_{n\geq 1}$, cf. *e.g.*, Theorem 1 of [71], and also [67] or § 2.4 and Relation (2.4.4) page 27 of [72]. In (A.6), the sum runs over the partitions $P_1, \ldots, P_a$ of $\{1, \ldots, n\}$ with cardinal $|P_i|$. By inversion of the cumulant formula (A.6), the cumulant κ_n^X can also be computed from the moments μ_n^X of X, cf. Theorem 1 of [71], and also [67] or § 2.4 and Relation (2.4.3) page 27 of [72].

The cumulant formula (A.6) can be inverted to compute the cumulant κ_n^X from the moments μ_n^X of X as

$$\kappa_n^X = \sum_{a=1}^{n} (a-1)! (-1)^{a-1} \sum_{P_1^n \cup \cdots \cup P_a^n = \{1, \ldots, n\}} \mu_{|P_1^n|}^X \cdots \mu_{|P_a^n|}^X,$$

$n \geq 1$, where the sum runs over the partitions $P_1^n, \ldots, P_a^n$ of $\{1, \ldots, n\}$ with cardinal $|P_i^n|$ by the Faà di Bruno formula, cf. Theorem 1 of [71], and also [67] or § 2.4 and Relation (2.4.3) page 27 of [72].

Example A.2.1

a) Gaussian cumulants. When X is centered we have $\kappa_1^X = 0$ and $\kappa_2^X = \mathbf{E}[X^2] = \mathrm{Var}[X]$, and X becomes Gaussian if and only if $\kappa_n^X = 0$, $n \geq 3$, i.e. $\kappa_n^X = \mathbf{1}_{\{n=2\}} \sigma^2$, $n \geq 1$, or

$$(\kappa_1^X, \kappa_2^X, \kappa_3^X, \kappa_4^X, \ldots) = (0, \sigma^2, 0, 0, \ldots).$$

In addition when X is centered Gaussian we have $\kappa_n^X = 0$, $n \neq 2$, and (A.6) can be read as Wick's theorem for the computation of Gaussian moments of $X \simeq \mathcal{N}(0, \sigma^2)$ by counting the pair partitions of $\{1, \ldots, n\}$, cf. [57], as

$$\mathbf{E}[X^n] = \sigma^n \sum_{k=1}^{n} \sum_{\substack{P_1^n \cup \cdots \cup P_k^n = \{1, \ldots, n\} \\ |P_1^n| = 2, \ldots, |P_k^n| = 2}} \kappa_{|P_1^n|}^X \cdots \kappa_{|P_a^n|}^X$$

$$= \begin{cases} \sigma^n (n-1)!!, & n \text{ even}, \\ \\ 0, & n \text{ odd}, \end{cases} \tag{A.7}$$

where the double factorial

$$(n-1)!! = \prod_{1 \leq 2k \leq n} (2k-1) = 2^{-n/2} \frac{n!}{(n/2)!}$$

counts the number of pair-partitions of $\{1, \ldots, n\}$ when n is even.

b) Poisson cumulants. In the particular case of a Poisson random variable $Z \simeq \mathcal{P}(\lambda)$ with intensity $\lambda > 0$ we have

$$\mathbf{E}[e^{tZ}] = \sum_{n=0}^{\infty} e^{nt} \mathbb{P}(Z = n) = e^{-\lambda} \sum_{n=0}^{\infty} \frac{(\lambda e^t)^n}{n!} = e^{\lambda(e^t - 1)}, \quad t \in \mathbb{R}_+,$$

hence $\kappa_n^Z = \lambda$, $n \geq 1$, or

$$(\kappa_1^Z, \kappa_2^Z, \kappa_3^Z, \kappa_4^Z, \ldots) = (\lambda, \lambda, \lambda, \lambda, \ldots),$$

and by (A.6) we have

$$\mathbf{E}_\lambda[Z^n] = A_n(\lambda, \ldots, \lambda) = \sum_{k=0}^n B_{n,k}(\lambda, \ldots, \lambda)$$

$$= \sum_{k=1}^n \sum_{P_1^n \cup \cdots \cup P_k^n = \{1, \ldots, n\}} \lambda^k = \sum_{k=0}^n \lambda^k S(n, k)$$

$$= T_n(\lambda), \tag{A.8}$$

i.e., the n-th Poisson moment with intensity parameter $\lambda > 0$ is given by $T_n(\lambda)$ where T_n is the Touchard polynomial of degree n used in Section 3.2. In particular the moment generating function of the Poisson distribution with parameter $\lambda > 0$ and jump size α is given by

$$t \longmapsto e^{\lambda(e^{\alpha t} - 1)} = \sum_{n=0}^\infty \frac{(\alpha t)^n}{n!} \mathbf{E}_\lambda[Z^n] = \sum_{n=0}^\infty \frac{(\alpha t)^n}{n!} T_n(\lambda).$$

In the case of centered Poisson random variables we note that Z and $Z - \mathbf{E}[Z]$ have same cumulants of order $k \geq 2$, hence for $Z - \mathbf{E}[Z]$, a centered Poisson random variable with intensity $\lambda > 0$, we have

$$\mathbf{E}[(Z - \mathbf{E}[Z])^n] = \sum_{a=1}^n \sum_{\substack{P_1^n \cup \cdots \cup P_a^n = \{1, \ldots, n\} \\ |P_1^n| \geq 2, \ldots, |P_a^n| \geq 2}} \lambda^a = \sum_{k=0}^n \lambda^k S_2(n, k),$$

$n \in \mathbb{N}$, where $S_2(n, k)$ is the number of ways to partition a set of n objects into k non-empty subsets of size at least 2, cf. [99].

A.3 Fourier transform

The Fourier transform $\mathcal{F}\varphi$ of an integrable function $f \in L^2(\mathbb{R}^n)$ is defined as

$$(\mathcal{F}\varphi)(x) := \frac{1}{(2\pi)^{n/2}} \int_{\mathbb{R}^n} e^{i\langle \xi, x \rangle} \varphi(\xi) d\xi, \quad x \in \mathbb{R}^n.$$

The inverse Fourier transform $\mathcal{F}^{-1}$ is given by

$$(\mathcal{F}^{-1}\varphi)(\xi) = \frac{1}{(2\pi)^{n/2}} \int_{\mathbb{R}^n} e^{-i\langle \xi, x \rangle} \varphi(x) dx, \quad \xi \in \mathbb{R}^n,$$

with the property $\mathcal{F}^{-1}(\mathcal{F}\varphi) = \varphi$. In particular when $n = 2$ we have

$$\mathcal{F}\varphi(u, v) = \frac{1}{2\pi} \int_{\mathbb{R}^2} \varphi(x, y) e^{iux+ivy} dxdy$$

and the inverse

$$\mathcal{F}^{-1}\varphi(x, y) = \frac{1}{2\pi} \int_{\mathbb{R}^2} \varphi(u, v) e^{-iux-ivy} dudv.$$

We also note the relation

$$\int_{-\infty}^{\infty} e^{i\xi(x-y)} d\xi \, dy = 2\pi \delta_x(dy), \qquad (A.9)$$

i.e.,

$$\int_{-\infty}^{\infty} \int_{-\infty}^{\infty} e^{i\xi(x-y)} \varphi(y) d\xi \, dy = 2\pi \varphi(x),$$

for φ a sufficiently smooth function in $\mathcal{S}(\mathbb{R})$.

When $n = 1$, given a real-valued random variable X with characteristic function

$$\Psi(u) = \mathbf{E}\left[e^{iuX}\right], \qquad u \in \mathbb{R},$$

and probability density function $\varphi_X(x)$, the inverse Fourier transform

$$\mathcal{F}^{-1}\varphi(x) = \frac{1}{\sqrt{2\pi}} \int_{\mathbb{R}^2} \varphi(u) e^{-iux} du,$$

yields the relation

$$\varphi_X(x) = (\mathcal{F}^{-1}\Psi)(x) = \frac{1}{2\pi} \int_{\mathbb{R}} \mathbf{E}[e^{iuX}] e^{-iux} du,$$

for the probability density function φ_X of X, provided the characteristic function $u \longmapsto \mathbf{E}[e^{iuX}]$ is integrable on $\mathbb{R}$.

When $n = 2$, given a couple (X, Y) of classical random variables with characteristic function

$$\Psi(u, v) = \mathbf{E}\left[e^{iuX+ivY}\right], \qquad u, v \in \mathbb{R},$$

such that Ψ is integrable on $\mathbb{R}^2$, the couple (X, Y) admits a joint probability density function of $\varphi_{(X,Y)}$ given by

$$\varphi_{(X,Y)}(x, y) = (\mathcal{F}^{-1}\Psi)(x, y) = \frac{1}{(2\pi)^2} \int_{\mathbb{R}^2} \Psi(u, v) e^{-iux-ivy} dudv. \qquad (A.10)$$

A.4 Cauchy–Stieltjes transform

Let μ be a probability measure on $\mathbb{R}$, then we define a function

$$G_\mu : \mathbb{C}\backslash\mathbb{R} \longrightarrow \mathbb{C}$$

by

$$G_\mu(z) = \int_\mathbb{R} \frac{1}{z-t} \mu(dt).$$

The function G_μ is called the *Cauchy transform* or *Stieltjes transform* of μ. We have

$$\frac{1}{|z-x|} = \frac{1}{\sqrt{\left(\Re(z)-x\right)^2 + \Im(z)^2}} \leq \frac{1}{|\Im(z)|}$$

so the integral is well-defined for all $x \in \mathbb{C}$ with $\Im(z) \neq 0$ and defines a holomorphic function on $\mathbb{C}\backslash\mathbb{R}$. Furthermore, since

$$\frac{1}{z-x} = \frac{\bar{z}-x}{|z-x|^2},$$

we have $\Im\big(G_\mu(z)\big) < 0$ if $\Im(z) > 0$, and $G_\mu(\bar{z}) = \overline{G_\mu(z)}$. Therefore, it is enough to know G_μ on $\mathbb{C}^+ = \{z \in \mathbb{C} : \Im(z) > 0\}$.

Theorem A.4.1 *[5, Section VI, Theorem 3] Let*

$$G \colon \mathbb{C}^+ \longrightarrow \mathbb{C}^- = \{z \in \mathbb{C} : \Im(z) < 0\}$$

be a holomorphic function. Then there exists a probability measure μ on $\mathbb{R}$ such that

$$G(z) = \int \frac{1}{z-x} \mu(dx)$$

for $z \in \mathbb{C}^+$ if and only if

$$\limsup_{y \to \infty} y|G(iy)| = 1.$$

The measure μ is uniquely determined by G, and it can be recovered by the Stieltjes inversion formula

$$\mu(B) = \frac{1}{\pi} \lim_{\varepsilon \searrow 0} \int_B G(x+i\varepsilon)dx$$

for $B \subseteq \mathbb{R}$ a Borel set such that $\mu(\partial B) = 0$.

If the measure μ has compact support, say in the interval $[-M, M]$ for some $M > 0$, then we can express G_μ in terms of the moments of μ,

$$m_n(\mu) = \int x^n \mu(dx),$$

for $n \in \mathbb{N}$, as a power series

$$G_\mu(z) = \int \frac{1}{z - x} \mu(dx) = \int \sum_{n=0}^{\infty} \frac{x^n}{z^{n+1}} \mu(dx) = \sum_{n=0}^{\infty} \frac{m_n(\mu)}{z^{n+1}},$$

which converges for $|z| > M$.

A.5 Adjoint action

Given X and Y two elements of a Lie algebra, the adjoint actions $\mathrm{Ad}e^X$ and $\mathrm{ad}X$ are defined by

$$e^{\mathrm{ad}X} Y := [X, Y] \qquad \text{and} \qquad \mathrm{Ad}e^X Y := e^X Y e^{-X}.$$

In particular we have

$$\mathrm{Ad}e^X Y = e^{\mathrm{ad}X} Y,$$

and

$$\begin{aligned}
\mathrm{Ad}e^X Y &:= e^X Y e^{-X} \\
&= \sum_{n,m=0}^{\infty} \frac{(-1)^m}{n!m!} X^n Y X^m \\
&= \sum_{k=0}^{\infty} \frac{1}{k!} \sum_{m=0}^{k} \binom{k}{m} (-1)^m X^{k-m} Y X^m \\
&= Y + [X, Y] + \frac{1}{2}[X, [X, Y]] + \cdots \\
&= e^{\mathrm{ad}X} Y.
\end{aligned}$$

The identity

$$\sum_{m=0}^{k} \binom{k}{m} (-1)^m X^{k-m} Y X^m = \underbrace{[X, [X, [\cdots [X, [X, Y]] \cdots]]]}_{k \text{ times}}$$

clearly holds for $k = 0, 1$, and can be extended by induction to all $k \geq 2$, as follows:

$$\underbrace{[X, [X, [\cdots [X, [X, Y]] \cdots]]]}_{k+1 \text{ times}} = \left[X, \underbrace{[X, [X, [\cdots [X, [X, Y]] \cdots]]]}_{k \text{ times}} \right]$$

$$= \left[X, \sum_{m=0}^{k} \binom{k}{m} (-1)^m X^{k-m} Y X^m \right]$$

$$= \sum_{m=0}^{k} \binom{k}{m} (-1)^m [X, X^{k-m} Y X^m]$$

$$= \sum_{m=0}^{k} \binom{k}{m} (-1)^m (X^{k+1-m} Y X^m - X^{k-m} Y X^{m+1})$$

$$= \sum_{m=0}^{k} \binom{k}{m} (-1)^m X^{k+1-m} Y X^m$$

$$- \sum_{m=1}^{k+1} \binom{k}{m-1} (-1)^{m-1} X^{k+1-m} Y X^m$$

$$= \sum_{m=0}^{k+1} (-1)^m X^{k+1-m} Y X^m \left(\binom{k}{m} + \binom{k}{m-1} \right)$$

$$= \sum_{m=0}^{k+1} (-1)^m X^{k+1-m} Y X^m \binom{k+1}{m},$$

where on the last step we used the Pascal recurrence relation for the binomial coefficients.

A.6 Nets

In a metric space (X, d) a point $x \in X$ is called an *adherent point* (also called point of closure or contact point) of a set $A \subseteq X$ if and only if there exists a sequence $(x_n)_{n \in \mathbb{N}} \subset A$ that converges to x. This characterisation cannot be formulated in general topological spaces unless we replace sequences by nets, which are a generalisation of sequences in which the index set $\mathbb{N}$ is replaced by more general sets.

A partially ordered set $(I, \leq)$ is called a *directed set*, if for any $j, k \in I$ there exists an element $\ell \in I$ such that $j \leq \ell$ and $k \leq \ell$. A *net* in a set A is a family of elements $(x_i)_{i \in I} \subseteq A$ indexed by a directed set. A net $(x_i)_{i \in I}$ in a topological space X is said to converge to a point $x \in X$ if, for any neighborhood U_x of x in X, there exists an element $i \in I$ such that $x_j \in U_x$ for all $j \in I$ with $i \leq j$.

In a topological space X a point $x \in X$ is said to be an *adherent point* of a set $A \in X$ if and only if there exists a net $(x_i)_{i \in I}$ in A that converges to x. A map $f : X \longrightarrow Y$ between topological spaces is continuous, if and only if for any point $x \in X$ and any net in X converging to x, the composition of f with this net converges to $f(x)$.

A.7 Closability of linear operators

The notion of closability of operators on a normed linear space H consists in minimal hypotheses ensuring that the extension of a densely defined linear operator is consistently defined.

Definition A.7.1 *A linear operator $T : S \longrightarrow H$ from a normed linear space S into H is said to be closable on H if for every sequence $(F_n)_{n \in \mathbb{N}} \subset S$ such that $F_n \longrightarrow 0$ and $TF_n \longrightarrow U$ in H, one has $U = 0$.*

Remark A.7.2 *For linear operators between general topological vector spaces one has to replace sequences by nets.*

For any two sequences $(F_n)_{n \in \mathbb{N}}$ and $(G_n)_{n \in \mathbb{N}}$ both converging to $F \in H$ and such that that $(TF_n)_{n \in \mathbb{N}}$ and $(TG_n)_{n \in \mathbb{N}}$ converge respectively to U and V in H, the closability of T shows that $(T(F_n - G_n))_{n \in \mathbb{N}}$ converges to $U - V$, hence $U = V$.

Letting $\mathrm{Dom}(T)$ denote the space of functionals F for which there exists a sequence $(F_n)_{n \in \mathbb{N}}$ converging to F such that $(TF_n)_{n \in \mathbb{N}}$ converges to $G \in H$, we can extend a closable operator $T : S \longrightarrow H$ to $\mathrm{Dom}(T)$ as in the following definition.

Definition A.7.3 *Given $T : S \longrightarrow H$ a closable operator and $F \in \mathrm{Dom}(T)$, we let*

$$TF = \lim_{n \to \infty} TF_n,$$

where $(F_n)_{n \in \mathbb{N}}$ denotes any sequence converging to F and such that $(TF_n)_{n \in \mathbb{N}}$ converges in H.

A.8 Tensor products

A.8.1 Tensor products of Hilbert spaces

The algebraic tensor product $V \otimes W$ of two vector spaces V and W is the vector space spanned by vectors of the form $v \otimes w$ subject to the linearity relations

$$\begin{cases} (v_1 + v_2) \otimes w = v_1 \otimes w + v_2 \otimes w, \\ v \otimes (w_1 + w_2) = v \otimes w_1 + v \otimes w_2, \\ (\lambda v) \otimes w = \lambda(v \otimes w) = v \otimes (\lambda w), \end{cases}$$

$\lambda \in \mathbb{C}, v, v_1, v_2 \in V, ww, w_1, w_2 \in W$. Given two Hilbert spaces H_1 and H_2, we can consider the sesquilinear map $\langle \cdot, \cdot \rangle_{H_1 \otimes H_2} : (H_1 \otimes H_2) \times (H_1 \otimes H_2) \longrightarrow \mathbb{C}$ defined by

$$\langle h_1 \otimes h_2, k_1 \otimes k_2 \rangle_{H_1 \otimes H_2} := \langle h_1, k_1 \rangle_{H_1} \langle h_2, k_2 \rangle_{H_2}$$

on product vectors and extended to $H_1 \otimes H_2$ by sesquilinearity. It is not difficult to show that this map is Hermitian and positive, *i.e.*, it is an inner product, and therefore it turns $H_1 \otimes H_2$ into a pre-Hilbert space. Completing $H_1 \otimes H_2$ with respect to the norm induced by inner product, we get the Hilbert space

$$H_1 \overline{\otimes} H_2 = \overline{H_1 \otimes H_2},$$

which is the Hilbert space tensor product of H_1 and H_2, with the continuous extension of $\langle \cdot, \cdot \rangle_{H_1 \otimes H_2}$. This construction is associative and can be iterated to define higher-order tensor products. In the sequel we will denote the Hilbert space tensor product simply by $\otimes$, when there is no danger of confusion with the algebraic tensor product.

The tensor product $T_1 \otimes T_2$ of two bounded operators

$$T_1 : H_1 \longrightarrow K_1 \quad \text{and} \quad T_2 : H_2 \longrightarrow K_2,$$

is defined on product vectors $h_1 \otimes h_2 \in H_1 \otimes H_2$ by

$$(T_1 \otimes T_2)(h_1 \otimes h_2) := (T_1 h_1) \otimes (T_2 h_2)$$

and extended by linearity to arbitrary vectors in the algebraic tensor product $H_1 \otimes H_2$. One can show that $T_1 \otimes T_2$ has norm

$$\|T_1 \otimes T_2\| = \|T_1\| \|T_2\|,$$

therefore $T_1 \otimes T_2$ extends to a bounded linear operator between the Hilbert space tensor products $H_1 \overline{\otimes} H_2$ and $K_1 \overline{\otimes} K_2$, which we denote again by $T_1 \otimes T_2$. Tensor products of more than two bounded operators can be defined in the same way.

A.8.2 Tensor products of L^2 spaces

Let (X, μ) and (Y, ν) denote measure spaces. Given $f \in L^2(X, \mu)$ and $g \in L^2(Y, \nu)$, the tensor product $f \otimes g$ of f by g is the function in $L^2(X \times Y, \mu \otimes \nu)$ defined by

$$(f \otimes g)(x, y) = f(x)g(y).$$

In particular, the tensor product $f_n \otimes g_m$ of two functions $f_n \in L^2(X, \sigma)^{\otimes n}$, $g_m \in L^2(X, \sigma)^{\otimes m}$, satisfies

$$f_n \otimes g_m(x_1, \ldots, x_n, y_1, \ldots, y_m) = f_n(x_1, \ldots, x_n)g_m(y_1, \ldots, y_m),$$

$(x_1, \ldots, x_n, y_1, \ldots, y_m) \in X^{n+m}$. Given $f_1, \ldots, f_n \in L^2(X, \mu)$, the symmetric tensor product $f_1 \circ \cdots \circ f_n$ is defined as the symmetrisation of $f_1 \otimes \cdots \otimes f_n$, i.e.,

$$(f_1 \circ \cdots \circ f_n)(t_1, \ldots, t_n) = \frac{1}{n!} \sum_{\sigma \in \Sigma_n} f_1(t_{\sigma(1)}) \cdots f_n(t_{\sigma(n)}), \tag{A.11}$$

$t_1, \ldots, t_n \in X$, where Σ_n denotes the set of permutations of $\{1, \ldots, n\}$. Let now $L^2(X)^{\circ n}$ denote the subspace of $L^2(X)^{\otimes n} = L^2(X^n)$ made of symmetric functions f_n in n variables. As a convention, $L^2(X)^{\circ 0}$ is identified to $\mathbb{R}$. From (A.11), the symmetric tensor product can be extended as an associative operation on $L^2(X)^{\circ n}$.

The tensor power of order n of $L^2([0, T], \mathbb{R}^d)$, $n \in \mathbb{N}$, $d \in \mathbb{N}^*$, is

$$L^2([0, T], \mathbb{R}^d)^{\otimes n} \simeq L^2([0, T]^n, (\mathbb{R}^d)^{\otimes n}).$$

For $n = 2$ we have $(\mathbb{R}^d)^{\otimes 2} = \mathbb{R}^d \otimes \mathbb{R}^d \simeq \mathcal{M}_{d,d}(\mathbb{R})$ (the linear space of square $d \times d$ matrices), hence

$$L^2([0, T], \mathbb{R}^d)^{\otimes 2} \simeq L^2([0, T]^2, \mathcal{M}_{d,d}(\mathbb{R})).$$

More generally, the tensor product $(\mathbb{R}^d)^{\otimes n}$ is isomorphic to $\mathbb{R}^{d^n}$. The generic element of $L^2([0, T], \mathbb{R}^d)^{\otimes n}$ is denoted by

$$f = (f^{(i_1, \ldots, i_n)})_{1 \leq i_1, \ldots, i_n \leq d},$$

with $f^{(i_1, \ldots, i_n)} \in L^2([0, T]^n)$.

Exercise solutions

Weary of Seeking had I grown,
So taught myself the way to Find.
(F. Nietzsche, in Die fröhliche Wissenschaft.*)*

Chapter 1 - Boson Fock space

Exercise 1.1   Moments of the normal distribution.

a) First moment. We note that

$$\langle Q\mathbf{1}, \mathbf{1}\rangle = \langle a^+\mathbf{1}, \mathbf{1}\rangle = \langle \mathbf{1}, a^-\mathbf{1}\rangle = 0.$$

b) Second moment. Next we have

$$\langle Q^2\mathbf{1}, \mathbf{1}\rangle = \langle (a^+ + a^-)^2\mathbf{1}, \mathbf{1}\rangle$$
$$= \langle (a^{+2} + a^{-2} + a^+a^- + a^-a^+)\mathbf{1}, \mathbf{1}\rangle$$
$$= \langle a^-a^+\mathbf{1}, \mathbf{1}\rangle$$
$$= \langle (a^+a^- + \sigma^2)\mathbf{1}, \mathbf{1}\rangle$$
$$= \sigma^2\langle \mathbf{1}, \mathbf{1}\rangle$$
$$= \sigma^2.$$

c) Third moment. We have

$$\langle Q^3\mathbf{1}, \mathbf{1}\rangle = \langle (a^+ + a^-)^3\mathbf{1}, \mathbf{1}\rangle$$
$$= \langle (a^+ + a^-)^2a^+\mathbf{1}, \mathbf{1}\rangle$$
$$= \langle (a^{+2} + a^{-2} + a^+a^- + a^-a^+)a^+\mathbf{1}, \mathbf{1}\rangle$$

$$= \langle (a^{-2}a^+ + a^- a^{+2})\mathbf{1}, \mathbf{1} \rangle$$
$$= \langle (a^-(a^+a^- + \sigma^2) + (a^+a^- + \sigma^2)a^+)\mathbf{1}, \mathbf{1} \rangle$$
$$= 0.$$

d) Fourth moment. Finally we have

$$\langle Q^4\mathbf{1}, \mathbf{1} \rangle = \langle (a^+ + a^-)^4\mathbf{1}, \mathbf{1} \rangle$$
$$= \langle (a^{+2} + a^{-2} + a^+a^- + a^-a^+)(a^{+2} + a^{-2} + a^+a^- + a^-a^+)\mathbf{1}, \mathbf{1} \rangle$$
$$= \langle (a^{-2} + a^-a^+)(a^{+2} + a^-a^+)\mathbf{1}, \mathbf{1} \rangle$$
$$= \langle (a^{-2}a^{+2} + a^{-3}a^+ + a^-a^{+3} + a^-a^+a^-a^+)\mathbf{1}, \mathbf{1} \rangle$$
$$= \langle (a^{-2}a^{+2} + a^{-3}a^+ + a^-a^{+3} + a^-a^+a^-a^+)\mathbf{1}, \mathbf{1} \rangle$$
$$= \langle (a^-(a^+a^- + \sigma^2)a^+ + a^{-2}(a^+a^- + \sigma^2))\mathbf{1}, \mathbf{1} \rangle$$
$$\quad + \langle ((a^+a^- + \sigma^2)a^{+2} + a^-a^+(a^+a^- + \sigma^2))\mathbf{1}, \mathbf{1} \rangle$$
$$= \langle (a^-(a^+a^- + \sigma^2)a^+ + (a^+a^- + \sigma^2)a^{+2} + \sigma^2 a^-a^+)\mathbf{1}, \mathbf{1} \rangle$$
$$= \langle (a^-a^+a^-a^+ + \sigma^2 a^-a^+ + (a^+a^- + \sigma^2)a^{+2} + \sigma^2 a^-a^+)\mathbf{1}, \mathbf{1} \rangle$$
$$= \langle (a^-a^+(a^+a^- + \sigma^2) + \sigma^2 a^-a^+ + \sigma^2 a^-a^+)\mathbf{1}, \mathbf{1} \rangle$$
$$= \langle (\sigma^2 a^-a^+ + \sigma^2 a^-a^+ + \sigma^2 a^-a^+)\mathbf{1}, \mathbf{1} \rangle$$
$$= 3\sigma^2 \langle (a^+a^- + \sigma^2 I)\mathbf{1}, \mathbf{1} \rangle$$
$$= 3\sigma^4 \langle \mathbf{1}, \mathbf{1} \rangle$$
$$= 3\sigma^4,$$

which is the fourth moment of the centered normal distribution $\mathcal{N}(0, \sigma^2)$ with variance σ^2.

We could continue and show that more generally, $\langle Q^n\mathbf{1}, \mathbf{1} \rangle$ coincides with the n-th moment of the centered Gaussian distribution $\mathcal{N}(0, \sigma^2)$ as given in (A.7).

Chapter 2 - Real Lie algebras

Exercise 2.1

1. The operator W_λ acts on the square roots of exponential vectors as

$$W_\lambda \sqrt{\xi(2f)} = \sqrt{\xi((2\kappa + 2f - 4\kappa f))} \exp(iI_1(\zeta)), \qquad |f| < 1/2.$$

Given that $W_0 = I_d$, by independence we only need to prove that

$$\int_0^\infty W_{u+iv} f(\tau) \overline{W_{u+iv} g}(\tau) e^{-\tau} d\tau$$

$$= \int_0^\infty \frac{1}{\sqrt{1-2u}} f\left(\frac{\tau}{1-2u}\right) \exp\left(-\frac{u\tau}{1-2u} + is(1-\tau)\right)$$

$$\times \frac{1}{\sqrt{1-2u}} \overline{g}\left(\frac{\tau}{1-2u}\right) \exp\left(-\frac{u\tau}{1-2u} - is(1-\tau)\right) e^{-\tau} d\tau$$

$$= \int_0^\infty \frac{1}{1-2u} f\left(\frac{\tau}{1-2u}\right) \overline{g}\left(\frac{\tau}{1-2u}\right) \exp\left(-\frac{\tau}{1-2u}\right) d\tau$$

$$= \int_0^\infty f(\tau) \overline{g}(\tau) e^{-\tau} d\tau.$$

Now, for $u + iv, u' + iv' \in \mathbb{C}$ with $|u|, |u'| < 1/2$,

$$W_{u'+iv'} W_{u+iv} f(\tau)$$

$$= W_{z'} \frac{1}{\sqrt{1-2u}} f\left(\frac{\tau}{1-2u}\right) \exp\left(-\frac{u\tau}{1-2u} + iv(1-\tau)\right)$$

$$= \frac{1}{\sqrt{(1-2u)(1-2u')}} f\left(\frac{\tau}{(1-2u)(1-2u')}\right)$$

$$\times \exp\left(-\frac{u\tau}{(1-2u)(1-2u')} - \frac{u'\tau}{1-2u'}\right.$$

$$\left. - iv\left(\frac{\tau}{1-2u'} - 1\right) + iv'(1-\tau)\right)$$

$$= \frac{1}{\sqrt{1-2(u+u'-2uu')}} f\left(\frac{\tau}{1-2(u+u'-2uu')}\right)$$

$$\times \exp\left(-\frac{(u+u'-2uu')\tau}{1-2(u+u'-2uu')u} + i\left(\frac{v}{1-2u'} + v'\right)(1-\tau)\right).$$

Let $z = u + iv \in \mathbb{C}$. For $t \in \mathbb{R}$ close enough to 0 we have

$$\frac{\partial}{\partial t} W_{tz} f(\tau) = \frac{\partial}{\partial t}\left(\frac{1}{\sqrt{1-2tu}} f\left(\frac{\tau}{1-2tu}\right) \exp\left(-\frac{ut\tau}{1-2ut} + ivt(1-\tau)\right)\right)$$

$$= \frac{1}{\sqrt{1-2tu}} \frac{2u\tau}{(1-2ut)^2} f'\left(\frac{\tau}{1-2tu}\right) \exp\left(-\frac{ut\tau}{1-tu} + ivt(1-\tau)\right)$$

$$+ \frac{1}{1-2tu}\left(-\frac{u\tau}{1-2ut} - \frac{u^2 t\tau}{(1-2ut)^2} + iv(1-\tau) + \frac{u}{\sqrt{1-2tu}}\right)$$

$$\times f\left(\frac{\tau}{1-2tu}\right) \exp\left(-\frac{ut\tau}{1-2ut} + ivt(1-\tau)\right).$$

Evaluating this expression at $t = 0$ yields

$$\frac{\partial W_{tz}f}{\partial t}(\tau)|_{t=0} = 2u\tau f'(\tau) + uf(\tau) + f(\tau)(-u\tau + iv(1-\tau))$$
$$= (u - iv)\tau f'(\tau) + f(\tau)(u + iv - (u + iv)\tau) + (u + iv)\tau f'(\tau)$$
$$= (za^+ - \bar{z}a^-)f(\tau).$$

For the second part, we compute

$$W_u W_{is}f(\tau) = \frac{1}{\sqrt{1 - 2u}}f\left(\frac{\tau}{1 - 2u}\right)\exp\left(iv\left(1 - \frac{\tau}{1 - 2u}\right) - \frac{u\tau}{1 - 2u}\right)$$
$$= \frac{1}{\sqrt{1 - 2u}}f\left(\frac{\tau}{1 - 2u}\right)\exp\left(-\frac{u\tau}{1 - 2u} + iv(1 - \tau) - \frac{2ivu\tau}{1 - 2u}\right)$$
$$= \exp\left(-\frac{2ivu\tau}{1 - 2u}\right)W_{is}W_u f(\tau).$$

For the exponential vector $\frac{1}{1-2\alpha}\exp\left(-\frac{2\alpha\tau}{1-2\alpha}\right)$ we have

$$W_z \frac{1}{\sqrt{1 - 2\alpha}}\exp\left(-\frac{\alpha\tau}{1 - 2\alpha}\right)$$
$$= \frac{1}{\sqrt{1 - 2\alpha}}\frac{1}{\sqrt{1 - 2u}}\exp\left(-\frac{2\alpha\tau}{2(1 - 2\alpha)(1 - 2u)} - \frac{u\tau}{1 - 2u} + is(1 - \tau)\right)$$
$$= \frac{1}{\sqrt{(1 - 2(u + \alpha - 2u\alpha))}}\exp\left(-\tau\frac{2\alpha + 2u - 4\alpha u}{2(1 - 2\alpha)(1 - 2u)} + is(1 - \tau)\right),$$

$\alpha \in (-1/2, 1/2)$. The semi-group property holds for

$$W_{is\zeta} = \left(\exp\left(-s\zeta\tilde{Q}\right)\right)_{s\in\mathbb{R}_+} = (\exp(isI_1(\zeta)))_{s\in\mathbb{R}_+},$$

but not for $(W_{tK})_{t\in\mathbb{R}_+}$, which is different from $(\exp(it\tilde{P}))_{t\in\mathbb{R}_+}$.

2. We have

$$W_t W_{is}f(\tau) = \frac{1}{\sqrt{1 - 2t}}f\left(\frac{\tau}{1 - 2t}\right)\exp\left(-is\left(\frac{\tau}{1 - 2t} - 1\right) - \frac{t\tau}{1 - 2t}\right),$$

hence

$$\frac{\partial}{\partial s}W_t W_{is}f(\tau) = -i\left(\frac{\tau}{1 - 2t} - 1\right)W_t W_{is}f(\tau).$$

Now we have

$$\frac{\partial}{\partial t}\frac{\partial}{\partial s}W_t W_{is}f(\tau) = -\frac{2i\tau}{(1 - 2t)^2}W_t W_{is}f(\tau) - i\left(\frac{\tau}{1 - 2t} - 1\right)\frac{\partial}{\partial t}W_t W_{is}f(\tau),$$

and

$$\frac{\partial}{\partial t}\frac{\partial}{\partial s}W_t W_{is}f(\tau)_{|t=s=0} = -2i\tau f(\tau) + i(1-\tau)\frac{\partial}{\partial t}W_t f(\tau)_{|t=0}$$

$$= -2i\tau f(\tau) + i(1-\tau)(2\tau\partial_\tau + (1-\tau))f(\tau)$$

$$= i((2\tau\partial_\tau + (1-\tau))((1-\tau)f)(\tau) = -\tilde{P}\tilde{Q}f(\tau).$$

On the other hand, we have

$$\frac{\partial}{\partial t}\frac{\partial}{\partial s}W_{is}W_t f(\tau)\mid_{t=s=0} = \frac{\partial}{\partial s}W_{is}1_{|s=0}\frac{\partial}{\partial t}W_t f(\tau)_{|t=0} = -\tilde{Q}\tilde{P}f(\tau).$$

Remarks.

Relation (3.13) can be proved using the operator W_z, as a consequence of the aforementioned proposition. We have from (1)

$$-\tilde{Q}\tilde{P} = \frac{\partial}{\partial s}\frac{\partial}{\partial t}W_{is}W_t f(\tau)\mid_{t=s=0}$$

$$= \frac{\partial}{\partial s}\frac{\partial}{\partial t}\left(\exp\left(\frac{2ist\tau}{1-2t}\right)W_t W_{is}f(\tau)\right)\mid_{t=s=0}$$

$$= \left(\frac{4it\tau}{(1-2t)^2} + \frac{2i\tau}{1-2t}\right)\mid_{t=s=0} W_t W_{is}f(\tau)\mid_{t=s=0}$$

$$+ \left(\exp\left(\frac{2is\tau}{1-2t}\right)\frac{\partial}{\partial s}\frac{\partial}{\partial t}W_t W_{is}f(\tau)\right)\mid_{t=s=0}$$

$$= 2i\tau f(\tau) + \frac{\partial}{\partial t}\frac{\partial}{\partial s}W_t W_{is}f(\tau)\mid_{t=s=0}$$

$$= 2i\tau f(\tau) - \tilde{P}\tilde{Q}f(\tau).$$

Chapter 3 - Basic probability distributions on Lie algebras

Exercise 3.1 Define the operators b^- and b^+ by

$$b^- = -ia^-, \qquad b^+ = ia^+.$$

1. The commutation relations and we clearly have

$$b^- e_0 = -ia^- e_0 = 0.$$

2. We have

$$
\begin{aligned}
\langle b^- u, v \rangle_H &= \langle -ia^- u, v \rangle_H \\
&= i \langle a^- u, v \rangle_H \\
&= i \langle u, a^+ v \rangle_H \\
&= \langle u, ia^+ v \rangle_H \\
&= \langle u, b^+ v \rangle_H.
\end{aligned}
$$

3. It suffices to rewrite P as $P = b^- + b^+$ and to note that $\{b, b^+\}$ satisfy the same properties as $\{a^-, a^+\}$. In other words we check that the transformation

$$
a^- \mapsto -ia^+, \qquad a^+ \mapsto ia^-
$$

maps P to Q and satisfies the commutation relation

$$
[ia^-, -ia^+] = (ia^-)(-ia^+) - (-ia^+)(ia^-) = a^- a^+ - a^+ a^- = \sigma^2 I,
$$

and the duality relation

$$
\langle (-ia^+)u, v \rangle = i \langle a^+ u, v \rangle = i \langle u, a^- v \rangle = \langle u, ia^- v \rangle,
$$

hence $P = i(a^- - a^+)$ also has a Gaussian law in the state e_0.

Exercise 3.2 *Moments of the Poisson distribution.*

a) First moment. We note that

$$
\langle Xe_0, e_0 \rangle = \langle (N + a^+ + a^- + E)e_0, e_0 \rangle = \lambda \langle Ee_0, e_0 \rangle = \lambda \langle e_0, e_0 \rangle = \lambda.
$$

b) Similarly we have

$$
\langle X^2 e_0, e_0 \rangle = \lambda \langle e_0, e_0 \rangle + \lambda \langle Xe_0, e_0 \rangle = \lambda + \lambda^2.
$$

c) We have

$$
\begin{aligned}
\langle X^3 e_0, e_0 \rangle &= \langle Xa^- Xe_0, e_0 \rangle + \langle a^- Xe_0, e_0 \rangle + \lambda \langle Xe_0, e_0 \rangle + \lambda \langle X^2 e_0, e_0 \rangle \\
&= \langle X^2 a^- e_0, e_0 \rangle + \langle Xa^- e_0, e_0 \rangle + \lambda \langle Xe_0, e_0 \rangle \\
&\quad + \langle Xa^- e_0, e_0 \rangle + \langle a^- e_0, e_0 \rangle + \lambda \langle e_0, e_0 \rangle \\
&\quad + \lambda \langle Xe_0, e_0 \rangle + \lambda \langle X^2 e_0, e_0 \rangle \\
&= \lambda \langle Xe_0, e_0 \rangle + \lambda \langle e_0, e_0 \rangle + \lambda \langle Xe_0, e_0 \rangle + \lambda \langle X^2 e_0, e_0 \rangle \\
&= \lambda + 3\lambda^2 + \lambda^3.
\end{aligned}
$$

Exercise 3.3 We have

$$\frac{\partial^n}{\partial t^n} \mathbf{E}[e^{tX}] = \frac{\partial^n}{\partial t^n} (1 - t)^{-\alpha}$$
$$= \alpha(\alpha + 1) \cdots (\alpha + n)(1 - t)^{-\alpha+n}, \qquad t < 1,$$

hence

$$\mathbf{E}[X^n] = \frac{\partial^n}{\partial t^n} \mathbf{E}[e^{tX}]_{|t=0} = \alpha(\alpha + 1) \cdots (\alpha + n)(1 - t)^{-\alpha-n}.$$

Exercise 3.4

1. For $n = 1$ we have

$$\langle e_0, (B^+ + B^- + M)e_0 \rangle = \langle B^- e_0, e_0 \rangle + \langle e_0, M e_0 \rangle = \alpha \langle e_0, e_0 \rangle = \alpha,$$

since $\langle e_0, e_0 \rangle = 1$.

2. For $n = 2$ we have

$$\langle e_0, (B^+ + B^- + M)^2 e_0 \rangle = \langle e_0, (B^+ + B^- + M)(B^+ + B^- + M)e_0 \rangle$$
$$= \langle e_0, (B^- + M)(B^+ + M)e_0 \rangle$$
$$= \langle e_0, M^2 e_0 \rangle + \langle e_0, B^- M e_0 \rangle + \langle e_0, B^- B^+ e_0 \rangle + \langle e_0, M B^+ e_0 \rangle$$
$$= \alpha^2 \langle e_0, e_0 \rangle + \alpha \langle e_0, B^- e_0 \rangle + \langle e_0, B^- B^+ e_0 \rangle + \langle e_0, M B^+ e_0 \rangle$$
$$= \alpha^2 \langle e_0, e_0 \rangle + \langle e_0, [B^-, B^+] e_0 \rangle + \langle e_0, B^+ B^- e_0 \rangle$$
$$\quad + \langle e_0, [M, B^+] e_0 \rangle + 2 \langle e_0, B^- e_0 \rangle$$
$$= \alpha^2 \langle e_0, e_0 \rangle + \langle e_0, M e_0 \rangle + 2 \langle e_0, B^+ e_0 \rangle = \alpha(\alpha + 1).$$

3. For $n = 3$ we have

$$\langle e_0, (B^+ + B^- + M)^3 e_0 \rangle = \langle e_0, (B^- + M)(B^+ + B^- + M)(B^+ + M)e_0 \rangle$$
$$= \langle e_0, B^- B^+ (B^+ + M)e_0 \rangle + \langle e_0, B^- B^- B^+ e_0 \rangle + \langle e_0, B^- M B^+ e_0 \rangle$$
$$\quad + \langle e_0, M B^+ (B^+ + M)e_0 \rangle + \langle e_0, M B^- B^+ e_0 \rangle + \langle e_0, M M (B^+ + M)e_0 \rangle$$
$$= \langle e_0, M(B^+ + M)e_0 \rangle + \langle e_0, B^- M e_0 \rangle + 2 \langle e_0, B^- B^+ e_0 \rangle + \langle e_0, B^- B^+ M e_0 \rangle$$
$$\quad + \langle e_0, M B^+ B^+ e_0 \rangle + \langle e_0, M B^+ M e_0 \rangle + \langle e_0, M M e_0 \rangle + \langle e_0, M M M e_0 \rangle$$
$$= \langle e_0, M^2 e_0 \rangle + 2 \langle e_0, M e_0 \rangle + \langle e_0, M^2 e_0 \rangle + \langle e_0, M M e_0 \rangle + \langle e_0, M M M e_0 \rangle$$
$$= \alpha^3 + 3\alpha^2 + 2\alpha$$
$$= \alpha(\alpha + 1)(\alpha + 2).$$

Chapter 4 - Noncommutative random variables

Exercise 4.1

1. We will reproduce here Jacobi's proof of the diagonalisability of Hermitian matrices.

2. Let $A = (a_{ij})_{1 \leq ij \leq n} \in M_n(\mathbb{C})$ be a Hermitian $n \times n$ matrix. We defined $\Sigma : M_n(\mathbb{C}) \to \mathbb{R}$,

$$\Sigma(B) = \sum_{i=1}^{n} \sum_{j=i+1}^{n} |b_{ij}|^2$$

for $B = (b_{ij})_{1 \leq ij \leq n} \in M_n(\mathbb{C})$, and $f_A : U(n) \to \mathbb{R}$,

$$f_A(U) = \Sigma(U^*AU).$$

Since f_A is continuous and $U(n)$ is compact, the extreme value theorem implies that f_A has a minimum value, *i.e.*, there exists a unitary matrix W such that

$$f_A(W) = \Sigma(W^*AW) \leq f_A(U) \quad \text{for all } U \in U(n).$$

We will show by contradiction that the matrix $B := W^*AW$ is diagonal.

We will show that if $B = W^*AW$ is not diagonal, then there exists a matrix U such that

$$f_A(WU) = \Sigma(U^*W^*AWU) = \Sigma(U^*BU) < \Sigma(B) = f_A(W),$$

which contradicts the minimality of $f_A(W)$.

Suppose that $B = (b_{ij})_{1 \leq ij \leq n}$ has an off-diagonal non-zero coefficient $b_{ij} \neq 0$ with $i \neq j$. Then, there exists a unitary 2×2 matrix

$$V = \begin{bmatrix} u & v \\ -\bar{v} & \bar{u} \end{bmatrix}$$

which diagonalises the Hermitian 2×2 matrix $\begin{bmatrix} b_{ii} & b_{ij} \\ b_{ji} & b_{jj} \end{bmatrix}$.

Note that we have $|u|^2 + |v|^2 = 1$, since V is unitary, and the inverse of V is given by

$$V^* = \begin{bmatrix} \bar{u} & -v \\ \bar{v} & u \end{bmatrix}.$$

Define now

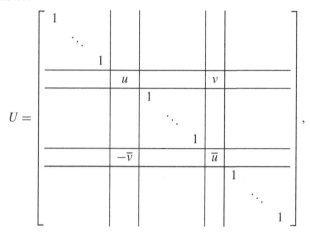

$$U = \begin{bmatrix} 1 & & & & & & \\ & \ddots & & & & & \\ & & 1 & & & & \\ & & & u & & v & \\ & & & & 1 & & \\ & & & & & \ddots & \\ & & & & & & 1 \\ & & & -\bar{v} & & \bar{u} & \\ & & & & & & 1 \\ & & & & & & & \ddots \\ & & & & & & & & 1 \end{bmatrix},$$

where the coefficients that are not marked all vanish. In other words, we embed V into the unitary group $U(n)$ such that it acts non-trivially only on the ith and the jth component. Then conjugation of the matrix B with U will change only the coefficients of the ith and the jth row and column of B, more precisely we get

$U^*BU =$

$$\begin{bmatrix} & ub_{1i} - \bar{v}b_{1j} & & vb_{1i} + \bar{u}b_{1j} & \\ & ub_{i-1,i} - \bar{v}b_{i-1,j} & & vb_{i-1,i} + \bar{u}b_{i-1,j} & \\ \bar{u}b_{i1} - vb_{j1} \;\cdot\;\cdot & * & \cdot\;\cdot\;\cdot & 0 & \cdot\;\cdot\; \bar{u}b_{in} - vb_{jn} \\ & ub_{i+1,i} - \bar{v}b_{i+1,j} & & vb_{i+1,i} + \bar{u}b_{i+1,j} & \\ & ub_{j-1,i} - \bar{v}b_{j-1,j} & & vb_{j-1,i} + \bar{u}b_{j-1,j} & \\ \bar{v}b_{i1} + ub_{j1} \;\cdot\;\cdot & 0 & \cdot\;\cdot\;\cdot & * & \cdot\;\cdot\; \bar{v}b_{in} + ub_{jn} \\ & ub_{j+1,i} - \bar{v}b_{j+1,j} & & vb_{j+1,i} + \bar{u}b_{j+1,j} & \\ & ub_{ni} - \bar{v}b_{nj} & & vb_{ni} + \bar{u}b_{nj} & \end{bmatrix},$$

where the coefficients in the empty blocks are unchanged, and the values marked by $*$ (*i.e.*, $(U^*BU)_{ii}$ and $(U^*BU)_{jj}$) do not matter for our calculations, since they do not occur in the sum defining $\Sigma(U^*BU)$.

We will now prove that $\Sigma(U^*BU) = \Sigma(B) - |b_{ij}|^2 < \Sigma(B)$.

We have

$$\Sigma(U^*BU) = \sum_{1 \leq k < \ell \leq n} |(U^*BU)_{k\ell}|^2 = \frac{1}{2} \sum_{\substack{k,\ell=1,\ldots,n \\ k \neq \ell}} |(U^*BU)_{k\ell}|^2$$

since U^*BU is Hermitian.

If we take the sum over a row different from the ith or jth row, say the kth row, then we have

$$
\begin{aligned}
|ub_{ki} &- \bar{v}b_{kj}|^2 + |vb_{ki} + \bar{u}b_{kj}|^2 \\
&= |u|^2|b_{ki}|^2 - \overline{uv}\overline{b_{ki}}b_{kj} - uvb_{ki}\overline{b_{kj}} \\
&\quad + |v|^2|b_{kj}|^2 + |v|^2|b_{ki}|^2 + \overline{vu}\overline{b_{ki}}b_{kj} + vub_{ki}\overline{b_{kj}} + |u|^2|b_{kj}|^2 \\
&= |b_{ki}|^2 + |b_{kj}|^2
\end{aligned}
$$

for the coefficients in the ith and jth column, and, since the other coefficients are not changed by the conjugation with U, we have

$$
\sum_{\substack{\ell=1,\dots,n \\ \ell \neq k}} |(U^*BU)_{k\ell}|^2 = \sum_{\substack{\ell=1,\dots,n \\ \ell \neq k}} |b_{k\ell}|^2.
$$

For the sum over the ith and jth row, we observe

$$
|\bar{u}b_{i\ell} - vb_{j\ell}|^2 + |\bar{v}b_{i\ell} + ub_{j\ell}|^2 = |b_{i\ell}|^2 + |b_{j\ell}|^2
$$

and

$$
\sum_{\substack{\ell=1,\dots,n \\ \ell \neq ij}} |(U^*BU)_{i\ell}|^2 + \sum_{\substack{\ell=1,\dots,n \\ \ell \neq ij}} |(U^*BU)_{j\ell}|^2 = \sum_{\substack{\ell=1,\dots,n \\ \ell \neq ij}} |b_{i\ell}|^2 + \sum_{\substack{\ell=1,\dots,n \\ \ell \neq ij}} |b_{j\ell}|^2.
$$

Since we chose U such that $(U^*BU)_{ij} = \overline{(U^*BU)_{ji}} = 0$, we get

$$
\begin{aligned}
\sum_{\substack{\ell=1,\dots,n \\ \ell \neq i}} |(U^*BU)_{i\ell}|^2 &+ \sum_{\substack{\ell=1,\dots,n \\ \ell \neq j}} |(U^*BU)_{j\ell}|^2 = \sum_{\substack{\ell=1,\dots,n \\ \ell \neq ij}} |b_{i\ell}|^2 + \sum_{\substack{\ell=1,\dots,n \\ \ell \neq ij}} |b_{j\ell}|^2 \\
&= \sum_{\substack{\ell=1,\dots,n \\ \ell \neq i}} |b_{i\ell}|^2 + \sum_{\substack{\ell=1,\dots,n \\ \ell \neq j}} |b_{j\ell}|^2 - |b_{ij}|^2 - |b_{ji}|^2
\end{aligned}
$$

and finally,

$$
\Sigma(U^*BU) = \Sigma(B) - |b_{ij}|^2 < \Sigma(B),
$$

as desired. This completes the proof, see also [23].

Exercise 4.2

1. This follows by direct computation.
2. We find

$$\Xi_0^n \mathbf{1}_x = \big(\#\{j, x_j = +1\} - \#\{j, x_j = -1\}\big)\mathbf{1}_x$$

for $x \in \Omega_n$, and Ξ_0^n has a binomial distribution on the set
$\{-n, -n+2, \ldots, n-2, n\}$, with density

$$n\mathcal{L}_1(\Xi_0^n) = \sum_{k=0}^{n} \binom{n}{k} p^k q^{n-k} \delta_{2k-n}$$

with respect to the constant function.

3. The law of

$$\Xi_\theta^n = \Xi_0^n + \theta(\Xi_+^n + \Xi_+^n)$$

can be computed from $\exp(\theta \, \Xi_2^n)\mathbf{1}$.

Chapter 5 - Noncommutative stochastic integration

Exercise 5.1   For the first equality expanding the right-hand side gives

$$\frac{1}{n!2^n} \sum_{\epsilon \in \{\pm 1\}^n} \left(\prod_{k=1}^{n} \epsilon_k\right) (\epsilon_1 v_1 + \cdots + \epsilon_n v_n)^{\otimes n}$$

$$= \frac{1}{n!2^n} \sum_{\epsilon \in \{\pm 1\}^n} \left(\prod_{k=1}^{n} \epsilon_k\right) \sum_{j_1,\ldots,j_m=1}^{n} \epsilon_{j_1} v_{j_1} \otimes \cdots \otimes \epsilon_{j_n} v_{j_n}.$$

Next, one checks that the terms with repeated indices vanish. Since an n-tuple of distinct indices $(j_1, \ldots, j_n)$ defines a permutation by $\sigma(k) = j_k$ for $k \in \{1, \ldots, n\}$, the sum becomes

$$\frac{1}{n!2^n} \sum_{\epsilon \in \{\pm 1\}^n} \sum_{\sigma \in \Sigma_n} v_{\sigma(1)} \otimes \cdots v_{\sigma(n)}.$$

The terms in the sum do no longer depend on ϵ and we get the desired result. Note that we can write this polarisation formula equivalently as an expectation

$$\sum_{\sigma \in \Sigma_n} v_{\sigma(1)} \otimes \cdots \otimes v_{\sigma(n)} = \mathbf{E}\left[Z_1 \cdots Z_n (Z_1 v_1 + \cdots + Z_n v_n)^{\otimes n} \right],$$

where $Z_1, \ldots, Z_n$ are independent Bernoulli random variables with

$$\mathbb{P}(Z_k = \pm 1) = \frac{1}{2}, \qquad k = 1, \ldots, n.$$

We refer to Relation (22) of [78] for the second equality.

Chapter 6 - Random variables on real Lie algebras

Exercise 6.1

1. This question is easy.
2. We have

$$
\begin{cases}
\exp(\theta\tilde{R}) = \begin{bmatrix} \cos\theta & -\sin\theta & 0 \\ \sin\theta & \cos\theta & 0 \\ 0 & 0 & 1 \end{bmatrix}, \\[20pt]
\exp(v\tilde{T}_x) = \begin{bmatrix} 1 & 0 & v \\ 0 & 1 & 0 \\ 0 & 0 & 1 \end{bmatrix}, \\[20pt]
\exp(w\tilde{T}_y) = \begin{bmatrix} 1 & 0 & 0 \\ 0 & 1 & w \\ 0 & 0 & 1 \end{bmatrix}.
\end{cases}
$$

$\exp(\theta\tilde{R})$ acts as a rotation on K_2,

$$
\exp(\theta\tilde{R})\begin{bmatrix} x \\ y \\ 1 \end{bmatrix} = \begin{bmatrix} \cos(\theta)x - \sin(\theta)y \\ \sin(\theta)x + \cos(\theta)y \\ 1 \end{bmatrix},
$$

and $\exp(v\tilde{T}_x)$ and $\exp(w\tilde{T}_y)$ act as translations, *i.e.*,

$$
\exp(v\tilde{T}_x)\begin{bmatrix} x \\ y \\ 1 \end{bmatrix} = \begin{bmatrix} x+v \\ y \\ 1 \end{bmatrix}, \quad \exp(w\tilde{T}_y)\begin{bmatrix} x \\ y \\ 1 \end{bmatrix} = \begin{bmatrix} x \\ y+w \\ 1 \end{bmatrix}.
$$

3. We have $M = 2iR$, $E^+ = T_x + iT_y$, $E^- = T_x - iT_y$.
4. The answer follows from the equalities

$$
\begin{cases}
e^{uE^+}Me^{-uE^+} & = \exp\left(u\,\mathrm{ad}(E^+)\right)M = M - 2E^+, \\[6pt]
e^{uM}E^-e^{-uM} & = \exp\left(u\,\mathrm{ad}(M)\right)E^- \\[4pt]
& = E^- - 2uE^- + \frac{4u^2}{2}E^- + \frac{8u^3}{3!}E^- + \cdots \\[4pt]
& = e^{-2u}E^-, \\[6pt]
e^{uE^+}E^-e^{-uE^+} & = \exp\left(u\,\mathrm{ad}(E^+)\right)E^- = E^-.
\end{cases}
$$

5. We have

$$
\frac{d\omega_1}{dt}(t) = (xM + yE^+ + zE^-)\exp(xM + yE^+ + zE^-)
$$

$$
= (xM + yE^+ + zE^-)\omega_1(t).
$$

We compute the derivatives of $\tilde{x}(t)$, $\tilde{y}(t)$, and $\tilde{z}(t)$,

$$\frac{d\tilde{x}}{dt}(t) = x, \qquad \frac{d\tilde{y}}{dt}(t) = ye^{2xt}, \qquad \frac{d\tilde{z}}{dt}(t) = ze^{2xt}.$$

Using (*iv*) we can show that $\omega_2(t)$ satisfies the same differential equation as $\omega_1(t)$,

$$\begin{aligned}
\frac{d\omega_2}{dt}(t) &= ye^{2xt}E^+ \exp\left(\tilde{y}(t)E^+\right) \exp\left(\tilde{x}(t)M\right) \exp\left(\tilde{z}(t)E^-\right) \\
&\quad + \exp\left(\tilde{y}(t)E^+\right)xM \exp\left(\tilde{x}(t)M\right) \exp\left(\tilde{z}(t)E^-\right) \\
&\quad + \exp\left(\tilde{y}(t)E^+\right) \exp\left(\tilde{x}(t)M\right)ze^{2xt}E^- \exp\left(\tilde{z}(t)E^-\right) \\
&= ye^{2xt}E^+ \exp\left(\tilde{y}(t)E^+\right) \exp\left(\tilde{x}(t)M\right) \exp\left(\tilde{z}(t)E^-\right) \\
&\quad + x(M - 2\tilde{y}(t)E^+) \exp\left(\tilde{y}(t)E^+\right) \exp\left(\tilde{x}(t)M\right) \exp\left(\tilde{z}(t)E^-\right) \\
&\quad + ze^{2xt}e^{-2xt}E^- \exp\left(\tilde{y}(t)E^+\right) \exp\left(\tilde{x}(t)M\right) \exp\left(\tilde{z}(t)E^-\right) \\
&= (xM + yE^+ + zE^-)\omega_2(t).
\end{aligned}$$

6. The functions $\omega_1(t)$ and $\omega_2(t)$ have the same initial value for $t = 0$ and satisfy the same differential equation, therefore they agree for all values of t. Taking $t = 1$ we get the desired formula.

Exercise 6.2

1. a) We have

$$\begin{aligned}
e^{za^+}&a^- e^{-za^+} \\
&= a^- + z[a^+, a^-] + \frac{z^2}{2}[a^+, [a^+, a^-]] + \frac{z^3}{3!}[a^+[a^+, [a^+, a^-]]] + \cdots \\
&= a^- - zE + \frac{z^2}{2}[a^+, E] + \frac{z^3}{3!}[a^+[a^+, E]] + \cdots \\
&= a^- - zE.
\end{aligned}$$

   b) Since $[E, a^+] = 0$ we clearly have the commutation relation $e^{za^+}Ee^{-za^+} = Ee^{za^+}e^{-za^+} = E$ by (6.2).

   c) Similarly, since $[E, a^-] = 0$, Relation (6.2) yields the commutation relation $e^{za^-}Ee^{-za^-} = Ee^{za^-}e^{-za^-} = E$.

2. We have

$$\omega_1'(t) = \frac{d}{dt}e^{t(ua^+ + va^- + wE)} = (ua^+ + va^- + wE)e^{t(ua^+ + va^- + wE)}, \qquad t \in \mathbb{R}_+.$$

3. We have

$$
\omega_2'(t) = \frac{d}{dt}\left(e^{uta^+}e^{vta^-}e^{(tw+t^2uv/2)E}\right)
$$

$$
= \left(\frac{d}{dt}e^{uta^+}\right)e^{vta^-}e^{(tw+t^2uv/2)E} + e^{uta^+}\left(\frac{d}{dt}e^{vta^-}\right)e^{(tw+t^2uv/2)E}
$$

$$
+ e^{uta^+}e^{vta^-}\frac{d}{dt}e^{(tw+t^2uv/2)E}
$$

$$
= ua^+e^{uta^+}e^{vta^-}e^{(tw+t^2uv/2)E} + ve^{uta^+}a^-e^{vta^-}e^{(tw+t^2uv/2)E}
$$

$$
+ e^{uta^+}e^{vta^-}(w+tuv)Ee^{(tw+t^2uv/2)E}, \qquad t \in \mathbb{R}_+.
$$

4. We have

$$
\omega_2'(t) = ua^+e^{uta^+}e^{vta^-}e^{(tw+t^2uv/2)E} + ve^{uta^+}a^-e^{vta^-}e^{(tw+t^2uv/2)E}
$$

$$
+ e^{uta^+}e^{vta^-}(w+tuv)Ee^{(tw+t^2uv/2)E}
$$

$$
= ua^+e^{uta^+}e^{vta^-}e^{(tw+t^2uv/2)E} + v(a^- - utE)e^{uta^+}e^{vta^-}e^{(tw+t^2uv/2)E}
$$

$$
+ (w+tuv)Ee^{uta^+}e^{vta^-}e^{(tw+t^2uv/2)E}
$$

$$
= ua^+e^{uta^+}e^{vta^-}e^{(tw+t^2uv/2)E} + va^-e^{uta^+}e^{vta^-}e^{(tw+t^2uv/2)E}
$$

$$
+ we^{uta^+}e^{vta^-}Ee^{(tw+t^2uv/2)E}
$$

$$
= (ua^+ + va^- + w)e^{uta^+}e^{vta^-}Ee^{(tw+t^2uv/2)E}
$$

$$
= (ua^+ + va^- + w)\omega_2(t), \qquad t \in \mathbb{R}_+.
$$

Consequently, $\omega_1(t)$ and $\omega_2(t)$ satisfy the same differential equations

$$
\omega_1'(t) = (ua^+ + va^- + w)\omega_1(t), \qquad t \in \mathbb{R}_+,
$$

and

$$
\omega_2'(t) = (ua^+ + va^- + w)\omega_2(t), \qquad t \in \mathbb{R}_+,
$$

with same initial condition $\omega_1(0) = \omega_2(0) = I$, and this yields
$\omega_1(t) = \omega_2(t), t \in \mathbb{R}_+$, which shows (6.5) for $t = 1$.

5. We have

$$
\langle e_0, e^{ua^+ + va^- + wE}e_0\rangle = \langle e_0, e^{ua^+}e^{va^-}e^{(w+uv/2)E}e_0\rangle
$$

$$
= e^{w+uv/2}\langle e_0, e^{ua^+}e^{va^-}e_0\rangle
$$

$$
= e^{w+uv/2}\langle e_0, e^{ua^+}e_0\rangle
$$

$$= e^{w+uv/2} \langle e^{ua^-} e_0, e_0 \rangle$$
$$= e^{w+uv/2} \langle e_0, e_0 \rangle$$
$$= e^{w+uv/2}.$$

When $u = v = t \in \mathbb{R}$ and $w = 0$ this yields

$$\langle e_0, e^{t(a^- + a^+) + wE} e_0 \rangle = e^{t^2 \sigma^2 / 2},$$

which is the moment generating function of a centered Gaussian random variable with variance σ^2. Similarly when $u = -it$, $v = it$ and $w = 0$ we find

$$\langle e_0, e^{it(a^- - a^+)} e_0 \rangle = \langle e_0, e^{ua^- + va^+} e_0 \rangle = e^{(-it)(it)/2} = e^{t^2 \sigma^2 / 2}. \qquad (6.1)$$

6. It suffices to note that (6.1) is the moment generating function of the centered Gaussian distribution with variance $\sigma^2 > 0$. More generally, one can check that $za^- + \bar{z} a^+$ has a centered Gaussian distribution with variance $\sigma^2 |z|^2$ for all $z \in \mathbb{C}$.

Exercise 6.3

1. A direct calculation can be used for the first part, namely for f polynomial we have

$$\tilde{a}^\circ \exp{(is(1 - \tau))} f(\tau) = \left(-(1 - \tau) \partial_\tau - \tau \partial_\tau^2 \right) \left(f(\tau) e^{is(1-\tau)} \right)$$
$$= -(1 - \tau)(f'(\tau) e^{is(1-\tau)} - is(1 - \tau) f(\tau) e^{is(1-\tau)})$$
$$- \tau (f''(\tau) - isf'(\tau) - isf'(\tau) - s^2 f(\tau)) e^{is(1-\tau)},$$

hence

$$\exp{(-is(1 - \tau))} \, \tilde{a}^\circ \exp{(is(1 - \tau))}$$
$$= (-(1 - \tau) \partial \tau - \tau \partial_\tau^2) f(\tau) + is(\tau \partial_\tau + (1 - \tau)) f(\tau)$$
$$+ is\tau \partial_\tau f(\tau) + s^2 \tau f(\tau)$$
$$= \tilde{a}^\circ f(\tau) + is\tilde{a}^+ f(\tau) - is\tilde{a}^- f(\tau) + s^2 \tau f(\tau).$$

Denoting by $\alpha_x^- = \partial_x$, $\alpha_y^- = \partial_y$, $\alpha_x^+ = x - \partial_x$, $\alpha_y^+ = y - \partial_y$ the annihilation and creation operators on the two-dimensional boson Fock space

$$\Gamma(\mathbb{C}e_1 \oplus \mathbb{C}e_2) \simeq L^2 \left(\mathbb{R}^2, \frac{1}{2\pi} e^{-(x^2+y^2)/2} \right),$$

show that[1]

$$2e^{-t((\alpha_x^+)^2-(\alpha_x^-)^2)/2}\left(\alpha_x^+\alpha_x^- + \alpha_x^-\alpha_x^+\right)e^{it((\alpha_x^+)^2-(\alpha_x^-)^2)/2}$$

$$= e^{2t}((\alpha_x^+)^2 + (\alpha_x^-)^2 + \alpha_x^+\alpha_x^- + \alpha_x^-\alpha_x^+)$$

$$- e^{-2t}((\alpha_x^+)^2 + (\alpha_x^-)^2 - \alpha_x^+\alpha_x^- - \alpha_x^-\alpha_x^+)$$

$$= 2\cosh(2t)(\alpha_x^+\alpha_x^- + \alpha_x^-\alpha_x^+) + 2\sinh(2t)((\alpha_x^+)^2 + (\alpha_x^-)^2),$$

hence (6.6) follows, since $\alpha_u^-\alpha_u^+ = \alpha_u^+\alpha_u^- + 1$, $u = x, y$, and $\alpha_x^+\alpha_x^- + \alpha_x^-\alpha_x^+ + \alpha_y^+\alpha_y^- + \alpha_y^-\alpha_y^+ = 4\tilde{a}^\circ + 2$ by Lemma 2.4.2.

2. From Question 1 the distribution of $\exp\left(-is\tilde{Q}\right)\tilde{a}^\circ\exp\left(is\tilde{Q}\right)$ in the vacuum state $\Omega = \mathbf{1}$ is the same as the distribution of $\tilde{a}^\circ$ in the state $\exp(is\tilde{Q})\Omega$, cf. [17]. In addition, by (3.8) the spectrum of a° is $\mathbb{N}$ and the Laguerre polynomial L_n is its eigenvector of eigenvalue $n \in \mathbb{N}$. In order to determine the distribution of a° in the state $e^{is(1-x)}$, it is necessary and sufficient to decompose $e^{is(1-x)}$ into a series of Laguerre polynomials:

$$\exp(is(1-x)) = \frac{e^{is}}{1+is}\sum_{n=0}^{\infty}\left(\frac{is}{1+is}\right)^n L_n(x),$$

which implies that the distribution of a° in this state is the geometric distribution μ on $\mathbb{N}$ with parameter $s^2/(1+s^2)$, i.e.,

$$\mu(\{n\}) = \left|\frac{e^{is}}{1+is}\left(\frac{is}{1+is}\right)^n\right|^2 = \frac{1}{1+s^2}\left(\frac{s^2}{1+s^2}\right)^n, \quad n \in \mathbb{N}.$$

For the second part, we notice that as shown on page 40 of [17],

$$\int_{-\infty}^{\infty} H_n(x)\exp\left(\frac{t}{2}((\alpha_x^+)^2 - (\alpha_x^-)^2)\right)\Omega\frac{e^{-x^2/2}}{\sqrt{2\pi}}dx = \frac{\sqrt{(2n)!}}{n!2^n}\frac{\tanh(t)^n}{\sqrt{2\cosh(t)}},$$

$n \in \mathbb{N}$, where $(H_n)_{n\in\mathbb{N}}$ is the sequence of Hermite polynomials which are orthonormal with respect to the standard Gaussian density. From [35], we have the relation

$$L_n\left(\frac{x^2 + y^2}{2}\right) = \frac{(-1)^n}{2^n}\sum_{k=0}^{n}\frac{\sqrt{(2k)!}\sqrt{(2n-2k)!}}{k!(n-k)!}H_{2k}(x)H_{2n-2k}(y),$$

[1] cf. page 40 of [17]

hence

$$\langle L_n(\tau) \exp(it\tilde{P})\mathbf{1}, \mathbf{1}\rangle$$

$$= \frac{1}{2\pi} \int_{-\infty}^{\infty} \int_{-\infty}^{\infty} L_n\left(\frac{x^2+y^2}{2}\right) e^{t((\alpha_x^+)^2-(\alpha_x^-)^2)/2} e^{t((\alpha_y^+)^2-(\alpha_y^-)^2)/2} \mathbf{\Omega}$$

$$\times e^{-(x^2+y^2)/2} dx dy$$

$$= \frac{(-1)^n}{2^n\sqrt{2\pi}} \sum_{k=0}^{n} \frac{\sqrt{(2k)!}\sqrt{(2n-2k)!}}{k!(n-k)!} \int_{-\infty}^{\infty} H_k(x) e^{\frac{t}{2}((\alpha_x^+)^2-(\alpha_x^-)^2)} \mathbf{\Omega} e^{-x^2/2} dx$$

$$\times \frac{1}{\sqrt{2\pi}} \int_{-\infty}^{\infty} H_{n-k}(y) e^{t((\alpha_y^+)^2-(\alpha_y^-)^2)/2} \mathbf{\Omega} e^{-y^2/2} dy$$

$$= \frac{(-1)^n}{4^n} \frac{\tanh(t)^n}{\cosh(t)} \sum_{k=0}^{n} \frac{(2k)!(2n-2k)!}{k!^2(n-k)!^2}$$

$$= (-1)^n\sqrt{1-\tanh(t)^2}\,\tanh(t)^n, \tag{6.2}$$

since

$$\sum_{k=0}^{n} \frac{(2k)!(2n-2k)!}{k!^2(n-k)!^2} = 2^{2n}, \qquad n \in \mathbb{N}.$$

Consequently, the distribution of $a°$ in the state $\exp(is\tilde{P})\mathbf{\Omega}$ is the geometric distribution ν on $\mathbb{N}$ with parameter $\tanh^2(s)$, given by

$$\nu(\{n\}) = (1 - \tanh^2(s)) \tanh^{2n}(s), \qquad n \in \mathbb{N}.$$

In other words, this result follows from the fact that the random variables $\alpha_x^+\alpha_x^-$, $\alpha_y^+\alpha_y^-$ are independent and have negative binomial distributions in the states

$$\exp\left(\frac{t}{2}((\alpha_x^+)^2 - (\alpha_x^-)^2)\right) \mathbf{\Omega} \quad \text{and} \quad \exp\left(\frac{t}{2}((\alpha_y^+)^2 - (\alpha_y^-)^2)\right) \mathbf{\Omega},$$

hence their half sum $a°$ has a geometric distribution in the state $\exp(it\tilde{P})$, cf. [93], [97].

3. Applying (6.2) with $n = 0$ we find

$$\mathbf{E}[\exp(it\tilde{P})] = \sqrt{1 - \tanh(t)^2} = \frac{1}{\cosh(t)}, \qquad t \in \mathbb{R}.$$

Chapter 7 - Weyl calculus on real Lie algebras

Exercise 7.1   Quantum optics.

1. The distribution of $N = a^+ a^-$ in the state

$$\epsilon(\alpha) = e^{-|\alpha|^2/2} \sum_{n=0}^{\infty} \frac{\alpha^n}{\sqrt{n!}} e_n$$

is given by its moment generating function

$$\langle \epsilon(\alpha), e^{tN} \epsilon(\alpha) \rangle = e^{-|\alpha|^2} \sum_{n=0}^{\infty} e^{tn} \frac{\alpha^{2n}}{n!} = \exp\left((e^t - 1)|\alpha|^2 \right),$$

i.e., it is a Poisson distribution with parameter $|\alpha|^2$.

2. When $\phi(z) = e^{-z^2/4}/(2\pi)^{1/4}$, the probability density in the pure state $|\phi\rangle\langle\phi|$ is given by

$$
\begin{aligned}
W_{|\phi\rangle\langle\phi|}(x, y) &= \frac{1}{2\pi} \int_{-\infty}^{\infty} \bar{\phi}(x - t)\phi(x + t)e^{iyt} dt \\
&= \frac{1}{(2\pi)^{3/2}} \int_{-\infty}^{\infty} e^{-(x-t)^2/4 - (x+t)^2/4} e^{iyt} dt \\
&= \frac{1}{(2\pi)^{3/2}} \int_{-\infty}^{\infty} e^{-x^2/2 - t^2/2 + ity} dt \\
&= \frac{1}{2\pi} e^{-(x^2 + y^2)/2}, \qquad x, y \in \mathbb{R}.
\end{aligned}
$$

Chapter 8 - Lévy processes on real Lie algebras

Exercise 8.1

1. We have

$$
\begin{aligned}
\int p_n(x) p_m(x) \mu(dx) &= \langle e_0, p_n(X) p_m(X) e_0 \rangle \\
&= \langle p_n(X) e_0, p_m(X) e_0 \rangle \\
&= \delta_{nm}, \qquad n, m \in \mathbb{N}.
\end{aligned}
$$

2. From Section 3.3.2 we get

$$Xe_n = \sqrt{(n+1)(n+m_0)}\, e_{n+1} + \beta(2n + m_0)e_n + \sqrt{n(n + m_0 - 1)}\, e_{n-1},$$

$n \in \mathbb{N}$, and

$$(n+1)P_{n+1} + (2\beta n + \beta m_0 - x)P_n + (n + m_0 - 1)P_{n-1} = 0,$$

with initial condition $P_{-1} = 0$, $P_1 = 1$, for the rescaled polynomials

$$P_n = \prod_{k=1}^{n} \sqrt{\frac{n}{n + m_0}} \, p_n.$$

3. It follows from the results of Section 3.3.2 that when $|\beta| = 1$ we have

$$P_n(x) = (-\beta)^n L_n^{(m_0-1)}(\beta x)$$

where $L_n^{(\alpha)}$ is the Laguerre polynomials, if $|\beta| < 1$ we find the Meixner-Pollaczek polynomials and if $|\beta| > 1$ we get the Meixner polynomials.

4. The three measures μ can be found from the results of Section 3.3.2. In particular, if $\beta = +1$ then μ is, up to a normalisation parameter, the usual χ^2-distribution with parameter m_0.

Chapter 9 - A guide to the Malliavin calculus

Exercise 9.1   First we note that all stochastic integrals with respect to a martingale have expectation equal to zero. Next, if M_t is a normal martingale and u_t is either an adapted processs or an independent process such that $\mathbf{E}[u_t^2] = t$, the Itô isometry shows that

$$\mathrm{Var}\left[\int_0^T u_t dM_t\right] = \mathbf{E}\left[\left(\int_0^T u_t dM_t\right)^2\right] = \mathbf{E}\left[\int_0^T |u_t|^2 dt\right]$$

$$= \int_0^T \mathbf{E}\left[|u_t|^2\right] dt = \int_0^T t dt = \frac{T^2}{2},$$

which is the case in questions $(b) - (c) - (d) - (e)$ since both B_t and $N_t - t$ are normal martingales. For question (a) we note that formally we have

$$\mathrm{Var}\left[\int_0^T B_{e^t} dB_t\right] = \int_0^T \mathbf{E}\left[|B_{e^t}|^2\right] dt = \int_0^T e^t dt = e^T - 1.$$

However, this stochastic integral is not defined as the process B_{e^t} is not adapted since $e^t > t$, $t \in \mathbb{R}_+$.

Exercise 9.2   We have

$$\mathbf{E}\left[\exp\left(\beta \int_0^T B_t dB_t\right)\right] = \mathbf{E}\left[\exp\left(\beta(B_T^2 - T)/2\right)\right]$$

$$= e^{-\beta T/2} \int_{-\infty}^{\infty} \exp\left(\beta \frac{x^2}{2} - \frac{x^2}{2T}\right) \frac{dx}{\sqrt{2\pi T}}$$

$$= e^{-\beta T/2} \int_{-\infty}^{\infty} \exp\left(-(1-\beta T)\frac{y^2}{2}\right) \frac{dy}{\sqrt{2\pi}}$$

$$= \frac{e^{-\beta T/2}}{\sqrt{1-\beta T}}, \qquad \beta < 1/T.$$

Exercise 9.3 We have

$$\mathbf{E}\left[e^{\int_0^T f(s)dB_s} \Big| \mathcal{F}_t\right] = \mathbf{E}\left[e^{\int_0^t f(s)dB_s} e^{\int_t^T f(s)dB_s} \Big| \mathcal{F}_t\right]$$

$$= e^{\int_0^t f(s)dB_s} \mathbf{E}\left[e^{\int_t^T f(s)dB_s} \Big| \mathcal{F}_t\right]$$

$$= e^{\int_0^t f(s)dB_s} \mathbf{E}\left[e^{\int_t^T f(s)dB_s}\right]$$

$$= \exp\left(\int_0^t f(s)dB_s + \frac{1}{2}\int_t^T |f(s)|^2 ds\right),$$

$0 \le t \le T$, since

$$\int_t^T f(s)dB_s \simeq \mathcal{N}\left(0, \int_t^T |f(s)|^2 ds\right).$$

Exercise 9.4 Letting $Y_t = e^{-\alpha t}X_t$ we find $dY_t = e^{-\alpha t}dB_t$, hence

$$Y_t = Y_0 + \int_0^t e^{-\alpha s}dB_s,$$

and

$$X_t = e^{\alpha t}Y_t = e^{\alpha t}Y_0 + e^{\alpha t}\int_0^t e^{-\alpha s}dB_s = e^{\alpha t}X_0 + \int_0^t e^{\alpha(t-s)}dB_s,$$

$0 \le t \le T$.

Exercise 9.5

1. We have $S_t = S_0 e^{rt+\sigma B_t - \sigma^2 t/2}$, $t \in \mathbb{R}_+$.
2. We have

$$f(t, S_t) = \mathbf{E}[(S_T)^2 \mid \mathcal{F}_t]$$

$$= S_t^2 \mathbf{E}[e^{2r(T-t)+2\sigma(B_T-B_t)-\sigma^2(T-t)} \mid \mathcal{F}_t]$$

$$= S_t^2 e^{2r(T-t)-\sigma^2(T-t)} \mathbf{E}[e^{2\sigma(B_T-B_t)} \mid \mathcal{F}_t]$$

$$= S_t^2 e^{2r(T-t)-\sigma^2(T-t)+2\sigma^2(T-t)},$$

$0 \le t \le T$, hence $f(t, x) = x^2 e^{(2r+\sigma^2)(T-t)}$, $0 \le t \le T$.

3. By the tower property of conditional expectations we have

$$\mathbf{E}[f(t, S_t) \mid \mathcal{F}_u] = \mathbf{E}[\mathbf{E}[S_T^2 \mid \mathcal{F}_t] \mid \mathcal{F}_u]$$
$$= \mathbf{E}[S_T^2 \mid \mathcal{F}_u]$$
$$= f(u, S_u), \qquad 0 \le u \le t \le T,$$

hence the process $t \mapsto f(t, S_t)$ is a martingale.

4. By the Itô formula we have

$$f(t, S_t) = f(0, S_0) + \sigma \int_0^t S_u \frac{\partial f}{\partial x}(u, S_u) dB_u$$

$$+ r \int_0^t S_u \frac{\partial f}{\partial x}(u, S_u) du + \frac{\sigma^2}{2} \int_0^t S_u^2 \frac{\partial^2 f}{\partial x^2}(u, S_u) du + \int_0^t \frac{\partial f}{\partial u}(u, S_u) du$$

$$= f(0, S_0) + \sigma \int_0^t \frac{\partial f}{\partial x}(u, S_u) dB_u$$

because the process $f(t, S_t)$ is a martingale. This yields

$$\zeta_t = \sigma S_t \frac{\partial f}{\partial x}(t, S_t) = 2\sigma S_t^2 e^{(2r+\sigma^2)(T-t)}, \qquad t \in [0, T].$$

We also check that $f(t, x)$ satisfies the PDE

$$rx \frac{\partial f}{\partial x}(t, x) + \frac{\sigma^2}{2} x^2 \frac{\partial^2 f}{\partial x^2}(t, x) + \frac{\partial f}{\partial t}(t, x) = 0$$

with terminal condition $f(T, x) = x^2$.

Chapter 10 - Noncommutative Girsanov theorem

Exercise 10.1   We note that

$$\langle e_0, g(e^t Z) e_0 \rangle = \langle e_0, g(Z) e^{Z(1-e^{-t})-ct} e_0 \rangle,$$

which reads as the classical change of variable formula

$$\int_0^\infty g(e^t z) e^{-z} z^{c-1} dz = \int_0^\infty g(z) e^{z(1-e^{-t})-ct} e^{-z} z^{c-1} dz,$$

for a gamma distributed random variable Z with parameter $c > 0$.

Chapter 11 - Noncommutative integration by parts

Exercise 11.1   The proof is similar to that of Proposition 12.1.4. Now using the formula

$$p^{(n)}(X) = D_k^n \left((D_k X)^{-n} p(X) \right) - \sum_{\kappa=0}^{n-1} A_\kappa p^{(\kappa)}(X),$$

where $A_0, \ldots, A_{n-1}$ are bounded operators, to get the necessary estimate

$$\left| \int_{-||X||}^{||X||} p^{(n)}(x) d\mu_{X,\Phi}(x) \right| \leq C \sup_{x \in [-||X||, ||X||]} |p(x)|$$

by induction over $n \geq 2$.

Chapter 12 - Smoothness of densities on real Lie algebras

Exercise 12.1

1. This follows from the computations of Section 2.4.
2. We have

$$Q + M = B^- + B^+ + M = \frac{P^2}{2},$$

and

$$Q - M = B^- + B^+ - M = -\frac{Q^2}{2}.$$

3. The random variables $Q + M$ and $Q - M$ both have gamma laws.
4. The law of $Q + \alpha M$ can be found in Chapter 3, depending on the value of α. For $|\alpha| < 1$, $Q + \alpha M$ has an absolutely continuous law and when $|\alpha| > 1$, $Q + \alpha M$ has a geometric law with parameter c^2 supported by

$$\{-1/2 - \text{sgn}(\alpha)(c - 1/c)k : k \in \mathbb{N}\},$$

with $c = \alpha \, \text{sgn}(\alpha) - \sqrt{\alpha^2 - 1}$.

5. The classical versions of those identities are given by the integration by parts formulas (2.1) for the gamma density.

References

[1] L. Accardi, U. Franz, and M. Skeide. Renormalized squares of white noise and other non-Gaussian noises as Lévy processes on real Lie algebras. *Comm. Math. Phys.*, 228(1):123–150, 2002. (Cited on pages xvii, 17, 38, 40, 96, 132, 147, and 187).

[2] L. Accardi, M. Schürmann, and W.v. Waldenfels. Quantum independent increment processes on superalgebras. *Math. Z.*, 198:451–477, 1988. (Cited on pages xvii and 131).

[3] G.S. Agarwal. *Quantum Optics*. Cambridge University Press, Cambridge, 2013. (Cited on page 130).

[4] N.I. Akhiezer. *The Classical Moment Problem and Some Related Questions in Analysis*. Translated by N. Kemmer. Hafner Publishing Co., New York, 1965. (Cited on page 54).

[5] N.I. Akhiezer and I.M. Glazman. *Theory of Linear Operators in Hilbert Space*. Dover Publications Inc., New York, 1993. (Cited on page 243).

[6] S. Albeverio, Yu. G. Kondratiev, and M. Röckner. Analysis and geometry on configuration spaces. *J. Funct. Anal.*, 154(2):444–500, 1998. (Cited on page 162).

[7] S.T. Ali, N.M. Atakishiyev, S.M. Chumakov, and K.B. Wolf. The Wigner function for general Lie groups and the wavelet transform. *Ann. Henri Poincaré*, 1(4):685–714, 2000. (Cited on pages xvi, xvii, 114, 115, 118, 120, 122, 123, 124, and 191).

[8] S.T. Ali, H. Führ, and A.E. Krasowska. Plancherel inversion as unified approach to wavelet transforms and Wigner functions. *Ann. Henri Poincaré*, 4(6):1015–1050, 2003. (Cited on pages xvi and 124).

[9] G.W. Anderson, A. Guionnet, and O. Zeitouni. *An Introduction to Random Matrices*. Cambridge: Cambridge University Press, 2010. (Cited on page 88).

[10] U. Franz and A. Skalski. Noncommutative Mathematics for Quantum Systems. To appear in the Cambridge IISc Series, 2015. D. Applebaum, Probability on compact Lie groups, volume 70 of Probability Theory and Stochastic Modelling, Springer, 2014" after 45th entry (Cited on pages xvi and 99).

[11] M. Anshelevich. Orthogonal polynomials and counting permutations. www.math.tamu.edu/~manshel/papers/OP-counting-permutations.pdf, 2014. (Cited on page 233).

[12] N.M. Atakishiyev, S.M. Chumakov, and K.B. Wolf. Wigner distribution function for finite systems. *J. Math. Phys.*, 39(12):6247–6261, 1998. (Cited on page 125).

271

[13] V.P. Belavkin. A quantum nonadapted Itô formula and stochastic analysis in Fock scale. *J. Funct. Anal.*, 102:414–447, 1991. (Cited on pages 88, 190, 212, and 213).

[14] V.P. Belavkin. A quantum nonadapted stochastic calculus and nonstationary evolution in Fock scale. In *Quantum Probability and Related Topics VI*, pages 137–179. World Sci. Publishing, River Edge, NJ, 1991. (Cited on pages 88, 190, 212, and 213).

[15] V.P. Belavkin. On quantum Itô algebras. *Math. Phys. Lett.*, 7:1–16, 1998. (Cited on page 88).

[16] C. Berg, J.P.R. Christensen, and P. Ressel. *Harmonic Analysis on Semigroups*, volume 100 of *Graduate Texts in Mathematics*. Springer-Verlag, New York, 1984. Theory of positive definite and related functions. (Cited on page 88).

[17] Ph. Biane. Calcul stochastique non-commutatif. In *Ecole d'Eté de Probabilités de Saint-Flour*, volume 1608 of *Lecture Notes in Mathematics*. Springer-Verlag, Berlin, 1993. (Cited on pages 7, 88, 211, and 264).

[18] Ph. Biane. Quantum Markov processes and group representations. In *Quantum Probability Communications*, QP-PQ, X, pages 53–72. World Sci. Publishing, River Edge, NJ, 1998. (Cited on pages xvii, 132, and 189).

[19] Ph. Biane and R. Speicher. Stochastic calculus with respect to free Brownian motion and analysis on Wigner space. *Probab. Theory Related Fields*, 112(3):373–409, 1998. (Cited on pages 88 and 216).

[20] L.C. Biedenharn and J.D. Louck. *Angular Momentum in Quantum Physics. Theory and Application. With a foreword by P.A. Carruthers.* Cambridge: Cambridge University Press, reprint of the 1981 hardback edition edition, 2009. (Cited on page 72).

[21] L.C. Biedenharn and J.D. Louck. *The Racah-Wigner Algebra in Quantum Theory. With a foreword by P.A. Carruthers. Introduction by G.W. Mackey.* Cambridge: Cambridge University Press, reprint of the 1984 hardback ed. edition, 2009. (Cited on page 72).

[22] J.-M. Bismut. Martingales, the Malliavin calculus and hypoellipticity under general Hörmander's conditions. *Z. Wahrsch. Verw. Gebiete*, 56(4):469–505, 1981. (Cited on pages xvii and 189).

[23] F. Bornemann. Teacher's corner - kurze Beweise mit langer Wirkung. *Mitteilungen der Deutschen Mathematiker-Vereinigung*, 10:55–55, July 2002. (Cited on page 258).

[24] N. Bouleau, editor. *Dialogues autour de la création mathématique*. Association Laplace-Gauss, Paris, 1997.

[25] T. Carleman. *Les Fonctions Quasi Analytiques*. Paris: Gauthier-Villars, Éditeur, Paris, 1926. (Cited on page 221).

[26] M.H. Chang. *Quantum Stochastics*. Cambridge Series in Statistical and Probabilistic Mathematics. Cambridge University Press, Cambridge, 2015. (Cited on pages xviii and 88).

[27] T.S. Chihara. *An Introduction to Orthogonal Polynomials*. Gordon and Breach Science Publishers, New York-London-Paris, 1978. Mathematics and Its Applications, Vol. 13. (Cited on page 233).

[28] S.M. Chumakov, A.B. Klimov, and K.B. Wolf. Connection between two wigner functions for spin systems. *Physical Review A*, 61(3):034101, 2000. (Cited on page 125).

[29] L. Cohen. *Time-Frequency Analysis: Theory and Applications*. Prentice-Hall, New Jersey, 1995. (Cited on pages xvii and 130).

[30] M. Cook. *Mathematicians: An Outer View of the Inner World.* Princeton University Press, USA, 2009. With an introduction by R. C. Gunning.

[31] D. Dacunha-Castelle and M. Duflo. *Probability and Statistics. Vol. I.* Springer-Verlag, New York, 1986. (Cited on page xvii).

[32] D. Dacunha-Castelle and M. Duflo. *Probability and Statistics. Vol. II.* Springer-Verlag, New York, 1986. (Cited on page xvii).

[33] P.A.M. Dirac. *The Principles of Quantum Mechanics.* Oxford, at the Clarendon Press, 1947. 3rd ed.

[34] M. Duflo and C.C. Moore. On the regular representation of a nonunimodular locally compact group.*J. Funct. Anal.*, 21(2):209–243, 1976. (Cited on page 114).

[35] A. Erdélyi, W. Magnus, F. Oberhettinger, and F.G. Tricomi. *Higher Transcendental Functions*, volume 2. McGraw Hill, New York, 1953. (Cited on page 264).

[36] P. Feinsilver and J. Kocik. Krawtchouk matrices from classical and quantum random walks. In *Algebraic methods in statistics and probability. AMS special session on algebraic methods in statistics, Univ. of Notre Dame, IN, USA, April 8–9, 2000*, pages 83–96. Providence, RI: AMS, American Mathematical Society, 2001. (Cited on page 72).

[37] P. Feinsilver and R. Schott. Krawtchouk polynomials and finite probability theory. In *Probability Measures on groups X. Proceedings of the Tenth Oberwolfach conference, held November 4-10, 1990 in Oberwolfach, Germany*, pages 129–135. New York, NY: Plenum Publishing Corporation, 1991. (Cited on page 72).

[38] P. Feinsilver and R. Schott. *Algebraic structures and operator calculus, Vol. I: Representations and Probability Theory*, volume 241 of *Mathematics and Its Applications*. Kluwer Academic Publishers Group, Dordrecht, 1993. (Cited on pages xviii, 92, 94, and 95).

[39] P. Feinsilver and R. Schott. *Algebraic structures and operator calculus. Vol. II*, volume 292 of *Mathematics and its Applications*. Kluwer Academic Publishers Group, Dordrecht, 1994. Special functions and computer science. (Cited on page xv).

[40] P. Feinsilver and R. Schott. *Algebraic structures and operator calculus. Vol. III*, volume 347 of *Mathematics and its Applications*. Kluwer Academic Publishers Group, Dordrecht, 1996. Representations of Lie groups. (Cited on page 99).

[41] R.P. Feynman, R. Leighton, and M. Sands. *The Feynman Lectures on Physics. Vols. 1-3.* Addison-Wesley Publishing Co., Inc., Reading, Mass.-London, 1964-1966. www.feynmanlectures.info (Cited on pages 70 and 72).

[42] U. Franz. Classical Markov processes from quantum Lévy processes. *Inf. Dim. Anal., Quantum Prob., and Rel. Topics*, 2(1):105–129, 1999. (Cited on pages xvii and 132).

[43] U. Franz, R. Léandre, and R. Schott. Malliavin calculus for quantum stochastic processes. *C. R. Acad. Sci. Paris Sér. I Math.*, 328(11):1061–1066, 1999. (Cited on page xviii).

[44] U. Franz, R. Léandre, and R. Schott. Malliavin calculus and Skorohod integration for quantum stochastic processes. *Infin. Dimens. Anal. Quantum Probab. Relat. Top.*, 4(1):11–38, 2001. (Cited on page xviii).

[45] U. Franz and R. Schott. *Stochastic Processes and Operator Calculus on Quantum Groups.* Kluwer Academic Publishers Group, Dordrecht, 1999. (Cited on pages 139 and 147).

[46] U. Franz and N. Privault. Quasi-invariance formulas for components of quantum Lévy processes. Infin. Dimens. Anal. Quantum Probab. Relat. Top., 7(1):131–145, 2004. (Cited on page 16).

[47] C.W. Gardiner and P. Zoller. *Quantum Noise*. Springer Series in Synergetics. Springer-Verlag, Berlin, second edition, 2000. A handbook of Markovian and non-Markovian quantum stochastic methods with applications to quantum optics. (Cited on pages xvii and 131).

[48] C. Gerry and P. Knight. *Introductory Quantum Optics*. Cambridge University Press, Cambridge, 2004. (Cited on page 128).

[49] L. Gross. Abstract Wiener spaces. In *Proceedings of the Fifth Berkeley Symposium on Mathematical Statistics and Probability*, Berkeley, 1967. Univ. of California Press. (Cited on page 173).

[50] A. Guichardet. *Symmetric Hilbert spaces and related topics*, volume 261 of *Lecture Notes in Mathematics*. Springer Verlag, Berlin, Heidelberg, New York, 1972. (Cited on page 88).

[51] A. Guichardet. *Cohomologie des groupes topologiques et des algèbres de Lie.*, volume 2 of *Textes Mathematiques*. CEDIC/Fernand Nathan, Paris, 1980. (Cited on pages 147 and 148).

[52] T. Hida. *Brownian Motion*. Springer Verlag, Berlin 1981. (Cited on page 158).

[53] A.S. Holevo. *Statistical structure of quantum theory*, volume 67 of *Lecture Notes in Physics. Monographs*. Springer-Verlag, Berlin, 2001. (Cited on pages xvii and 131).

[54] R.L. Hudson and K.R. Parthasarathy. Quantum Itô's formula and stochastic evolutions. *Comm. Math. Phys.*, 93(3):301–323, 1984. (Cited on pages 190, 212, 214, and 215).

[55] C.J. Isham. *Lectures on quantum theory*. Imperial College Press, London, 1995. Mathematical and structural foundations. (Cited on page xviii).

[56] Y. Ishikawa. *Stochastic Calculus of Variations for Jump Processes*. de Gruyter, Berlin, 2013. (Cited on page xviii).

[57] L. Isserlis. On a formula for the product-moment coefficient of any order of a normal frequency distribution in any number of variables. *Biometrika*, 12(1-2):134–139, 1918. (Cited on page 240).

[58] S. Janson. *Gaussian Hilbert spaces*, volume 129 of *Cambridge Tracts in Mathematics*. Cambridge University Press, Cambridge, 1997. (Cited on pages 155 and 211).

[59] U.C. Ji and N. Obata. Annihilation-derivative, creation-derivative and representation of quantum martingales. *Commun. Math. Phys.*, 286(2):751–775, 2009. (Cited on page 216).

[60] U.C. Ji and N. Obata. Calculating normal-ordered forms in Fock space by quantum white noise derivatives. *Interdiscip. Inf. Sci.*, 19(2):201–211, 2013. (Cited on page 216).

[61] J.R. Johansson, P.D. Nation, and F. Nori. Qutip: An open-source python framework for the dynamics of open quantum systems. *Computer Physics Communications*, 183(8):1760–1772, 2012. (Cited on page 113).

[62] J.R. Johansson, P.D. Nation, and F. Nori. Qutip 2: A python framework for the dynamics of open quantum systems. *Computer Physics Communications*, 184(4):1234–1240, 2013. (Cited on page 113).

[63] R. Koekoek and R.F. Swarttouw. The Askey-scheme of hypergeometric orthogonal polynomials and its q-analogue. Delft University of Technology, Report 98–17, 1998. (Cited on pages 40, 41, 237, and 238).

[64] A. Korzeniowski and D. Stroock. An example in the theory of hypercontractive semigroups. *Proc. Amer. Math. Soc.*, 94:87–90, 1985. (Cited on pages 18 and 26).

[65] P.S. de Laplace. *Théorie Analytique des Probabilités*. V. Courcier, Imprimeur, 57 Quai des Augustins, Paris, 1814.

[66] M. Ledoux. Concentration of measure and logarithmic Sobolev inequalities. In *Séminaire de Probabilités XXXIII*, volume 1709 of *Lecture Notes in Math.*, pages 120–216. Springer, Berlin, 1999. (Cited on page 18).

[67] V.P. Leonov and A.N. Shiryaev. On a method of calculation of semi-invariants. *Theory Probab. Appl.*, 4:319–329, 1959. (Cited on pages 239 and 240).

[68] J.M. Lindsay. Quantum and non-causal stochastic calculus. *Probab. Theory Related Fields*, 97:65–80, 1993. (Cited on pages 88, 190, 212, and 213).

[69] J.M. Lindsay. Integral-sum kernel operators. In *Quantum Probability Communications (Grenoble, 1998)*, volume XII, page 121. World Scientific, Singapore, 2003. (Cited on page 88).

[70] J.M. Lindsay. Quantum stochastic analysis – an introduction. In *Quantum independent increment processes. I*, volume 1865 of *Lecture Notes in Math.*, pages 181–271. Springer, Berlin, 2005. (Cited on page 88).

[71] E. Lukacs. Applications of Faà di Bruno's formula in mathematical statistics. *Am. Math. Mon.*, 62:340–348, 1955. (Cited on pages 239 and 240).

[72] E. Lukacs. *Characteristic Functions*. Hafner Publishing Co., New York, 1970. Second edition, revised and enlarged. (Cited on pages 239 and 240).

[73] H. Maassen. Quantum markov processes on fock space described by integral kernels. In *Quantum probability and applications II (Heidelberg 1984)*, volume 1136 of *Lecture Notes in Math.*, pages 361–374. Springer, Berlin, 1985. (Cited on page 88).

[74] T. Mai, R. Speicher, and M. Weber. Absence of algebraic relations and of zero divisors under the assumption of full non-microstates free entropy dimension. Preprint arXiv:1502.06357, 2015. (Cited on page 216).

[75] P. Malliavin. Stochastic calculus of variations and hypoelliptic operators. In *Intern. Symp. SDE. Kyoto*, pages 195–253, Tokyo, 1976. Kinokumiya. (Cited on pages xiii, xvii, and 173).

[76] P. Malliavin. *Stochastic analysis*, volume 313 of *Grundlehren der Mathematischen Wissenschaften*. Springer-Verlag, Berlin, 1997. (Cited on page 155).

[77] P. Malliavin. *Stochastic analysis*, volume 313 of *Grundlehren der Mathematischen Wissenschaften*. Springer-Verlag, Berlin, 1997. (Cited on page 169).

[78] S. Mazur and W. Orlicz. Grundlegende Eigenschaften der polynomischen Operationen. I. *Stud. Math.*, 5:50–68, 1934. (Cited on page 259).

[79] P.A. Meyer. *Quantum probability for probabilists*, volume 1538 of *Lecture Notes in Math.* Springer-Verlag, Berlin, 2nd edition, 1995. (Cited on pages 7, 88, 135, 147, 186, and 211).

[80] P.A. Meyer. Quantum probability seen by a classical probabilist. In *Probability towards 2000 (New York, 1995)*, volume 128 of *Lecture Notes in Statist.*, pages 235–248. Springer, New York, 1998. (Cited on page xvi).

[81] A. Nica and R. Speicher. *Lectures on the combinatorics of free probability*, volume 335 of *London Mathematical Society Lecture Note Series*. Cambridge University Press, Cambridge, 2006. (Cited on page 88).

[82] M.A. Nielsen and I.L. Chuang. *Quantum Computation and Quantum Information*. Cambridge University Press, Cambridge, 2000. (Cited on page 70).

[83] D. Nualart. Analysis on Wiener space and anticipating stochastic calculus. In *Ecole d'été de Probabilités de Saint-Flour XXV*, volume 1690 of *Lecture Notes in Mathematics*, pages 123–227. Springer-Verlag, Berlin, 1998. (Cited on page 155).

[84] D. Nualart. *The Malliavin calculus and related topics*. Probability and its Applications. Springer-Verlag, Berlin, second edition, 2006. (Cited on pages xvii, 155, and 173).

[85] H. Oehlmann. Analyse temps-fréquence de signaux vibratoires de boîtes de vitesses. PhD thesis, Université Henri Poincaré Nancy I, 1996. (Cited on page 130).

[86] H. Osswald. *Malliavin calculus for Lévy processes and infinite-dimensional Brownian motion*, volume 191 of *Cambridge Tracts in Mathematics*. Cambridge University Press, Cambridge, 2012. (Cited on pages xviii and 169).

[87] K.R. Parthasarathy. *An Introduction to Quantum Stochastic Calculus*. Birkäuser, 1992. (Cited on pages xv, 7, 47, 48, 82, 86, 88, 135, 186, 208, 212, 214, and 225).

[88] K.R. Parthasarathy. *Lectures on quantum computation, quantum error-correcting codes and information theory*. Tata Institute of Fundamental Research, Mumbai, 2003. Notes by Amitava Bhattacharyya. (Cited on page 72).

[89] G. Peccati and M. Taqqu. *Wiener Chaos: Moments, Cumulants and Diagrams: A survey with Computer Implementation*. Bocconi and Springer Series. Springer, Milan, 2011. (Cited on page 239).

[90] J. Pitman. *Combinatorial stochastic processes*, volume 1875 of *Lecture Notes in Mathematics*. Springer-Verlag, Berlin, 2006. Lectures from the 32nd Summer School on Probability Theory held in Saint-Flour, July 7–24, 2002. (Cited on page 239).

[91] N. Privault. Inégalités de Meyer sur l'espace de Poisson. *C. R. Acad. Sci. Paris Sér. I Math.*, 318:559–562, 1994. (Cited on page 18).

[92] N. Privault. A transfer principle from Wiener to Poisson space and applications. *J. Funct. Anal.*, 132:335–360, 1995. (Cited on page 26).

[93] N. Privault. A different quantum stochastic calculus for the Poisson process. *Probab. Theory Related Fields*, 105:255–278, 1996. (Cited on pages 16 and 265).

[94] N. Privault. Girsanov theorem for anticipative shifts on Poisson space. *Probab. Theory Related Fields*, 104:61–76, 1996. (Cited on page 173).

[95] N. Privault. Une nouvelle représentation non-commutative du mouvement brownien et du processus de Poisson. *C. R. Acad. Sci. Paris Sér. I Math.*, 322:959–964, 1996. (Cited on page 16).

[96] N. Privault. Absolute continuity in infinite dimensions and anticipating stochastic calculus. *Potential Analysis*, 8(4):325–343, 1998. (Cited on page 173).

[97] N. Privault. Splitting of Poisson noise and Lévy processes on real Lie algebras. *Infin. Dimens. Anal. Quantum Probab. Relat. Top.*, 5(1):21–40, 2002. (Cited on pages 16 and 265).

[98] N. Privault. *Stochastic analysis in discrete and continuous settings with normal martingales*, volume 1982 of *Lecture Notes in Mathematics*. Springer-Verlag, Berlin, 2009. (Cited on pages 149, 155, and 174).

[99] N. Privault. Generalized Bell polynomials and the combinatorics of Poisson central moments. *Electron. J. Combin.*, 18(1):Research Paper 54, 10, 2011. (Cited on page 241).

[100] N. Privault and W. Schoutens. Discrete chaotic calculus and covariance identities. *Stochastics Stochastics Rep.*, 72(3-4):289–315, 2002. (Cited on page 72).

[101] L. Pukanszky. *Leçon sur les représentations des groupes*. Dunod, Paris, 1967. (Cited on page 142).

[102] L. Pukanszky. Unitary representations of solvable Lie groups. *Ann. scient. Éc. Norm. Sup.*, 4:457–608, 1971. (Cited on page 142).

[103] R. Ramer. On nonlinear transformations of Gaussian measures. *J. Funct. Anal.*, 15:166–187, 1974. (Cited on page 173).

[104] S. Sakai. *C*-Algebras and W*-Algebras*. Springer-Verlag, New York-Heidelberg, 1971. (Cited on page 179).

[105] M. Schürmann. The Azéma martingales as components of quantum independent increment processes. In J. Azéma, P.A. Meyer, and M. Yor, editors, *Séminaire de Probabilités XXV*, volume 1485 of *Lecture Notes in Math*. Springer-Verlag, Berlin, 1991. (Cited on pages xvii and 132).

[106] M. Schürmann. *White Noise on Bialgebras*. Springer-Verlag, Berlin, 1993. (Cited on pages xvii, 131, 134, 136, 147, and 179).

[107] I. Shigekawa. Derivatives of Wiener functionals and absolute continuity of induced measures. *J. Math. Kyoto Univ.*, 20(2):263–289, 1980. (Cited on page 173).

[108] K.B. Sinha and D. Goswami. *Quantum stochastic processes and noncommutative geometry*, volume 169 of *Cambridge Tracts in Mathematics*. Cambridge University Press, Cambridge, 2007. (Cited on page xviii).

[109] R. F. Streater. Classical and quantum probability. *J. Math. Phys.*, 41(6):3556–3603, 2000. (Cited on pages xvii and 147).

[110] T.N. Thiele. On semi invariants in the theory of observations. *Kjöbenhavn Overs.*, pages 135–141, 1899. (Cited on page 239).

[111] N. Tsilevich, A.M. Vershik, and M. Yor. Distinguished properties of the gamma process and related topics. math.PR/0005287, 2000. (Cited on pages xvii, 180, and 189).

[112] N. Tsilevich, A.M. Vershik, and M. Yor. An infinite-dimensional analogue of the Lebesgue measure and distinguished properties of the gamma process. *J. Funct. Anal.*, 185(1):274–296, 2001. (Cited on pages xvii, 180, and 189).

[113] A.S. Üstünel. *An introduction to analysis on Wiener space*, volume 1610 of *Lecture Notes in Mathematics*. Springer Verlag, Berlin, 1995. (Cited on page 155).

[114] A.M. Vershik, I.M. Gelfand, and M.I. Graev. A commutative model of the group of currents $SL(2, \mathbb{R})^X$ connected with a unipotent subgroup. *Funct. Anal. Appl.*, 17(2):137–139, 1983. (Cited on pages xvii and 189).

[115] N.J. Vilenkin and A.U. Klimyk. *Representation of Lie groups and special functions. Vol. 1*, volume 72 of *Mathematics and its Applications (Soviet Series)*. Kluwer Academic Publishers Group, Dordrecht, 1991. (Cited on page 99).

[116] N.J. Vilenkin and A.U. Klimyk. *Representation of Lie groups and special functions. Vol. 3*, volume 75 of *Mathematics and its Applications (Soviet Series)*. Kluwer Academic Publishers Group, Dordrecht, 1992. (Cited on page 99).

[117] N.J. Vilenkin and A.U. Klimyk. *Representation of Lie groups and special functions. Vol. 2*, volume 74 of *Mathematics and its Applications (Soviet Series)*. Kluwer Academic Publishers Group, Dordrecht, 1993. (Cited on page 99).

[118] N.J. Vilenkin and A.U. Klimyk. *Representation of Lie groups and special functions*, volume 316 of *Mathematics and its Applications*. Kluwer Academic Publishers Group, Dordrecht, 1995. (Cited on page 99).

[119] J. Ville. Théorie et applications de la notion de signal analytique. *Câbles et Transmission*, 2:61–74, 1948. (Cited on page 130).

[120] D. Voiculescu. Lectures on free probability theory. In *Lectures on probability theory and statistics (Saint-Flour, 1998)*, volume 1738 of *Lecture Notes in Math.*, pages 279–349. Berlin: Springer, 2000. (Cited on page 88).

[121] D. Voiculescu, K. Dykema, and A. Nica. *Free random variables. A noncommutative probability approach to free products with applications to random matrices, operator algebras and harmonic analysis on free groups*, volume 1 of *CRM Monograph Series*. American Mathematical Society, Providence, RI, 1992. (Cited on page 88).

[122] W. von Waldenfels. Itô solution of the linear quantum stochastic differential equation describing light emission and absorption. In *Quantum probability and applications to the quantum theory of irreversible processes, Proc. int. Workshop, Villa Mondragone/Italy 1982*, volume 1055 of *Lecture Notes in Math.*, pages 384–411. Springer-Verlag, Berlin, 1984. (Cited on pages xvii and 131).

[123] W. von Waldenfels. *A measure theoretical approach to quantum stochastic processes*, volume 878 of *Lecture Notes in Physics. Monographs*. Springer-Verlag, Berlin, 2014. (Cited on page xviii).

[124] E.P. Wigner. On the quantum correction for thermodynamic equilibrium. *Phys. Rev.*, 40:749–759, 1932. (Cited on page xvi).

[125] M.W. Wong. *Weyl Transforms*. Universitext. Springer-Verlag, Berlin, 1998. (Cited on page 108).

[126] L.M. Wu. L^1 and modified logarithmic Sobolev inequalities and deviation inequalities for Poisson point processes. Preprint, 1998. (Cited on page 167).

Index

Printed in the United States
By Bookmasters